Statistics and Computing

Statistics and Computing

W.N. Venables
B.D. Ripley

Modern Applied Statistics with S

Fourth Edition

With 152 Illustrations

 Springer

W.N. Venables
CSIRO Mathematics and
    Information Science
PO Box 120
Cleveland, Queensland 4163
Australia
bill.venables@csiro.au

B.D. Ripley
Department of Statistics
University of Oxford
1 South Parks Road
Oxford OX1 3TG
England
ripley@stats.ox.ac.uk

Series Editors:
J. Chambers
Bell Labs, Lucent
    Technologies
600 Mountain Avenue
Murray Hill, NJ 07974
USA

W. Eddy
Department of Statistics
Carnegie Mellon University
Pittsburgh, PA 15213
USA

W. Härdle
Institut für Statistik und
    Ökonometrie
Humboldt-Universität zu Berlin
Spandauer Str. 1
D-10178 Berlin
Germany

S. Sheather
Australian Graduate School
    of Management
PO Box 1
Kensington, NSW 2033
Australia

L. Tierney
School of Statistics
University of Minnesota
Vincent Hall
Minneapolis, MN 55455
USA

Library of Congress Cataloging-in-Publication Data
Venables, W.N. (William N.)
    Modern applied statistics with S / W.N. Venables, B.D. Ripley.—4ᵗʰ ed.
        p. cm.—(Statistics and computing)
    Previously published as: Modern applied statistics with S-PLUS, c1999.
    Includes bibliographical references and index.
    ISBN 0-387-95457-0 (alk. paper)
    1. S (Computer system).   2. Statistics—Data processing.   3. Mathematical statistics—Data
processing.   I. Ripley, Brian D., 1952–   II. Venables, W.N. (William N.) Modern applied
statistics with S-PLUS.   III. Title.   IV. Series.
QA276.4.V46   2002
005.369—dc21   2002022925

ISBN 0-387-95457-0   Printed on acid-free paper.

9 8 7 6 5 4

springeronline.com

Preface

S is a language and environment for data analysis originally developed at Bell Laboratories (of AT&T and now Lucent Technologies). It became the statistician's calculator for the 1990s, allowing easy access to the computing power and graphical capabilities of modern workstations and personal computers. Various implementations have been available, currently S-PLUS, a commercial system from the Insightful Corporation[1] in Seattle, and R,[2] an Open Source system written by a team of volunteers. Both can be run on Windows and a range of UNIX / Linux operating systems: R also runs on Macintoshes.

This is the fourth edition of a book which first appeared in 1994, and the S environment has grown rapidly since. This book concentrates on using the current systems to do statistics; there is a companion volume (Venables and Ripley, 2000) which discusses programming in the S language in much greater depth. Some of the more specialized functionality of the S environment is covered in *on-line complements*, additional sections and chapters which are available on the World Wide Web. The datasets and S functions that we use are supplied with most S environments and are also available on-line.

This is not a text in statistical theory, but does cover modern statistical methodology. Each chapter summarizes the methods discussed, in order to set out the notation and the precise method implemented in S. (It will help if the reader has a basic knowledge of the topic of the chapter, but several chapters have been successfully used for specialized courses in statistical methods.) Our aim is rather to show how we analyse datasets using S. In doing so we aim to show both how S can be used and how the availability of a powerful and graphical system has altered the way we approach data analysis and allows penetrating analyses to be performed routinely. Once calculation became easy, the statistician's energies could be devoted to understanding his or her dataset.

The core S language is not very large, but it is quite different from most other statistics systems. We describe the language in some detail in the first three chapters, but these are probably best skimmed at first reading. Once the philosophy of the language is grasped, its consistency and logical design will be appreciated.

The chapters on applying S to statistical problems are largely self-contained, although Chapter 6 describes the language used for linear models that is used in several later chapters. We expect that most readers will want to pick and choose among the later chapters.

This book is intended both for would-be users of S as an introductory guide

---

[1] http://www.insightful.com.
[2] http://www.r-project.org.

and for class use. The level of course for which it is suitable differs from country to country, but would generally range from the upper years of an undergraduate course (especially the early chapters) to Masters' level. (For example, almost all the material is covered in the M.Sc. in Applied Statistics at Oxford.) On-line exercises (and selected answers) are provided, but these should not detract from the best exercise of all, using S to study datasets with which the reader is familiar. Our library provides many datasets, some of which are not used in the text but are there to provide source material for exercises. Nolan and Speed (2000) and Ramsey and Schafer (1997, 2002) are also good sources of exercise material.

The authors may be contacted by electronic mail at

```
MASS@stats.ox.ac.uk
```

and would appreciate being informed of errors and improvements to the contents of this book. Errata and updates are available from our World Wide Web pages (see page 461 for sites).

Acknowledgements:

This book would not be possible without the S environment which has been principally developed by John Chambers, with substantial input from Doug Bates, Rick Becker, Bill Cleveland, Trevor Hastie, Daryl Pregibon and Allan Wilks. The code for survival analysis is the work of Terry Therneau. The S-PLUS and R implementations are the work of much larger teams acknowledged in their manuals.

We are grateful to the many people who have read and commented on draft material and who have helped us test the software, as well as to those whose problems have contributed to our understanding and indirectly to examples and exercises. We cannot name them all, but in particular we would like to thank Doug Bates, Adrian Bowman, Bill Dunlap, Kurt Hornik, Stephen Kaluzny, José Pinheiro, Brett Presnell, Ruth Ripley, Charles Roosen, David Smith, Patty Solomon and Terry Therneau. We thank Insightful Inc. for early access to versions of S-PLUS.

Bill Venables
Brian Ripley
January 2002

Contents

Typographical Conventions

Throughout this book S language constructs and commands to the operating system are set in a monospaced typewriter font `like this`. The character ~ may appear as ˜ on your keyboard, screen or printer.

We often use the prompts `$` for the operating system (it is the standard prompt for the UNIX Bourne shell) and `>` for S. However, we do *not* use prompts for continuation lines, which are indicated by indentation. One reason for this is that the length of line available to use in a book column is less than that of a standard terminal window, so we have had to break lines that were not broken at the terminal.

Paragraphs or comments that apply to only one S environment are signalled by a marginal mark:

- This is specific to S-PLUS (version 6 or later).   S+

- This is specific to S-PLUS under Windows.   S+Win

- This is specific to R.   R

Some of the S output has been edited. Where complete lines are omitted, these are usually indicated by

        . . . .

in listings; however most *blank* lines have been silently removed. Much of the S output was generated with the options settings

```
options(width = 65, digits = 5)
```

in effect, whereas the defaults are around 80 and 7. Not all functions consult these settings, so on occasion we have had to manually reduce the precision to more sensible values.

Chapter 1

Introduction

Statistics is fundamentally concerned with the understanding of structure in data. One of the effects of the information-technology era has been to make it much easier to collect extensive datasets with minimal human intervention. Fortunately, the same technological advances allow the users of statistics access to much more powerful 'calculators' to manipulate and display data. This book is about the modern developments in applied statistics that have been made possible by the widespread availability of workstations with high-resolution graphics and ample computational power. Workstations need software, and the S[1] system developed at Bell Laboratories (Lucent Technologies, formerly AT&T) provides a very flexible and powerful environment in which to implement new statistical ideas. Lucent's current implementation of S is exclusively licensed to the Insightful Corporation[2], which distributes an enhanced system called S-PLUS.

An Open Source system called R[3] has emerged that provides an independent implementation of the S language. It is similar enough that almost all the examples in this book can be run under R.

An S environment is an integrated suite of software facilities for data analysis and graphical display. Among other things it offers

- an extensive and coherent collection of tools for statistics and data analysis,

- a language for expressing statistical models and tools for using linear and non-linear statistical models,

- graphical facilities for data analysis and display either at a workstation or as hardcopy,

- an effective object-oriented programming language that can easily be extended by the user community.

The term *environment* is intended to characterize it as a planned and coherent system built around a language and a collection of low-level facilities, rather than the 'package' model of an incremental accretion of very specific, high-level and

---

[1]The name S arose long ago as a compromise name (Becker, 1994), in the spirit of the programming language C (also from Bell Laboratories).

[2]http://www.insightful.com

[3]http://www.r-project.org

sometimes inflexible tools. Its great strength is that functions implementing new statistical methods can be built on top of the low-level facilities.

Furthermore, most of the environment is open enough that users can explore and, if they wish, change the design decisions made by the original implementors. Suppose you do not like the output given by the regression facility (as we have frequently felt about statistics packages). In S you can write your own summary routine, and the system one can be used as a template from which to start. In many cases sufficiently persistent users can find out the exact algorithm used by listing the S functions invoked. As R is Open Source, *all* the details are open to exploration.

Both S-PLUS and R can be used under Windows, many versions of UNIX and under Linux; R also runs under MacOS (versions 8, 9 and X), FreeBSD and other operating systems.

We have made extensive use of the ability to extend the environment to implement (or re-implement) statistical ideas within S. All the S functions that are used and our datasets are available in machine-readable form and come with all versions of R and Windows versions of S-PLUS; see Appendix C for details of what is available and how to install it if necessary.

System dependencies

We have tried as far as is practicable to make our descriptions independent of the computing environment and the exact version of S-PLUS or R in use. We confine attention to versions 6 and later of S-PLUS, and 1.7.0 or later of R.

Clearly some of the details must depend on the environment; we used S-PLUS 6.0 on Solaris to compute the examples, but have also tested them under S-PLUS for Windows version 6.1 release 1, and using S-PLUS 6.1 on Linux. The output will differ in small respects, for the Windows run-time system uses scientific notation of the form 4.17e-005 rather than 4.17e-05.

Where timings are given they refer to S-PLUS 6.0 running under Linux on one processor of a dual 1 GHz Pentium III PC.

One system dependency is the mouse buttons; we refer to buttons 1 and 2, usually the left and right buttons on Windows but the left and middle buttons on UNIX / Linux (or perhaps both together of two). Macintoshes only have one mouse button.

Reference manuals

The basic S references are Becker, Chambers and Wilks (1988) for the basic environment, Chambers and Hastie (1992) for the statistical modelling and first-generation object-oriented programming and Chambers (1998); these should be supplemented by checking the on-line help pages for changes and corrections as S-PLUS and R have evolved considerably since these books were written. Our aim is not to be comprehensive nor to replace these manuals, but rather to explore much further the use of S to perform statistical analyses. Our companion book, Venables and Ripley (2000), covers many more technical aspects.

Graphical user interfaces (GUIs)

S-PLUS for Windows comes with a GUI shown in Figure B.1 on page 458. This
has menus and dialogs for many simple statistical and graphical operations, and
there is a Standard Edition that only provides the GUI interface. We do not
discuss that interface here as it does not provide enough power for our material.
For a detailed description see the system manuals or Krause and Olson (2000) or
Lam (2001).

The UNIX / Linux versions of S-PLUS 6 have a similar GUI written in Java,
obtained by starting with Splus -g: this too has menus and dialogs for many
simple statistical operations.

The Windows, Classic MacOS and GNOME versions of R have a much simpler console.

Command line editing

All of these environments provide command-line editing using the arrow keys,
including recall of previous commands. However, it is not enabled by default in
S-PLUS on UNIX / Linux: see page 447.

1.1   A Quick Overview of S

Most things done in S are permanent; in particular, data, results and functions are
all stored in operating system files.[4] These are referred to as *objects*.

Variables can be used as scalars, matrices or arrays, and S provides extensive
matrix manipulation facilities. Furthermore, objects can be made up of collections
of such variables, allowing complex objects such as the result of a regression
calculation. This means that the result of a statistical procedure can be saved
for further analysis in a future session. Typically the calculation is separated
from the output of results, so one can perform a regression and then print various
summaries and compute residuals and leverage plots from the saved regression
object.

Technically S is a function language. Elementary commands consist of either
expressions or *assignments*. If an expression is given as a command, it is evaluated, printed and the value is discarded. An assignment evaluates an expression
and passes the value to a variable but the result is not printed automatically. An
expression can be as simple as 2 + 3 or a complex function call. Assignments
are indicated by the *assignment operator* <-. For example,

```
> 2 + 3
[1] 5
> sqrt(3/4)/(1/3 - 2/pi^2)
[1] 6.6265
> library(MASS)
```

---

[4]These should not be manipulated directly, however. Also, R works with an in-memory workspace
containing copies of many of these objects.

```
> mean(chem)
[1] 4.2804
> m <- mean(chem); v <- var(chem)/length(chem)
> m/sqrt(v)
[1] 3.9585
```

Here > is the S prompt, and the [1] states that the answer is starting at the first element of a vector.

More complex objects will have printed a short summary instead of full details. This is achieved by an object-oriented programming mechanism; complex objects have *classes* assigned to them that determine how they are printed, summarized and plotted. In current versions of S *all* objects have classes, for example numeric vectors have class "numeric".

S can be extended by writing new functions, which then can be used in the same way as built-in functions (and can even replace them). This is very easy; for example, to define functions to compute the standard deviation[5] and the two-tailed P value of a t statistic, we can write

```
std.dev <- function(x) sqrt(var(x))
t.test.p <- function(x, mu = 0) {
    n <- length(x)
    t <- sqrt(n) * (mean(x) - mu) / std.dev(x)
    2 * (1 - pt(abs(t), n - 1)) # last value is returned
}
```

It would be useful to give both the t statistic and its P value, and the most common way of doing this is by returning a list; for example, we could use

```
t.stat <- function(x, mu = 0) {
    n <- length(x)
    t <- sqrt(n) * (mean(x) - mu) / std.dev(x)
    list(t = t, p = 2 * (1 - pt(abs(t), n - 1)))
}
z <- rnorm(300, 1, 2)   # generate 300 N(1, 4) variables.
t.stat(z)
$t:
[1] 8.2906
$p:
[1] 3.9968e-15

unlist(t.stat(z, 1))  # test mu=1, compact result
        t       p
-0.56308 0.5738
```

The first call to t.stat prints the result as a list; the second tests the non-default hypothesis $\mu = 1$ and using unlist prints the result as a numeric vector with named components.

Linear statistical models can be specified by a version of the commonly used notation of Wilkinson and Rogers (1973), so that

---

[5]S-PLUS and R have functions stdev and sd, respectively.

```
time ~ dist + climb
time ~ transplant/year + age + prior.surgery
```

refer to a regression of `time` on both `dist` and `climb`, and of `time` on year within each transplant group and on age, with a different intercept for each type of prior surgery. This notation has been extended in many ways, for example to survival and tree models and to allow smooth non-linear terms.

1.2   Using S

How to initialize and start up your S environment is discussed in Appendix A.

Bailing out

One of the first things we like to know with a new program is how to get out of trouble. S environments are generally very tolerant, and can be interrupted by Ctrl-C.[6] (Use Esc on GUI versions under Windows.) This will interrupt the current operation, back out gracefully (so, with rare exceptions, it is as if it had not been started) and return to the prompt.

You can terminate your S session by typing

```
q()
```

at the command line or from Exit on the File menu in a GUI environment.

On-line help

There is a help facility that can be invoked from the command line. For example, to get information on the function `var` the command is

```
> help(var)
```

A faster alternative (to type) is

```
> ?var
```

For a feature specified by special characters and in a few other cases (one is `"function"`), the argument must be enclosed in double or single quotes, making it an entity known in S as a character string. For example, two alternative ways of getting help on the list component extraction function, `[[`, are

```
> help("[[")
> ?"[["
```

Many S commands have additional help for *name*.`object` describing their result: for example, `lm` under S-PLUS has a help page for `lm.object`.

Further help facilities for some versions of S-PLUS and R are discussed in Appendix A. Many versions can have their manuals on-line in PDF format; look under the Help menu in the Windows versions.

---

[6]This means hold down the key marked Control or Ctrl and hit the second key.

1.3   An Introductory Session

The best way to learn S is by using it. We invite readers to work through the following familiarization session and see what happens. First-time users may not yet understand every detail, but the best plan is to type what you see and observe what happens as a result.

Consult Appendix A, and start your S environment.

The whole session takes most first-time users one to two hours at the appropriate leisurely pace. The left column gives commands; the right column gives brief explanations and suggestions.

A few commands differ between environments, and these are prefixed by # R: or # S:. Choose the appropriate one(s) and omit the prefix.

`library(MASS)`	A command to make our datasets available. Your local advisor can tell you the correct form for your system.
`?help`	Read the help page about how to use help.
`# S: trellis.device()`	Start up a suitable device.

`x <- rnorm(1000)` `y <- rnorm(1000)`	Generate 1 000 pairs of normal variates
`truehist(c(x,y+3), nbins=25)`	Histogram of a mixture of normal distributions. Experiment with the number of bins (25) and the shift (3) of the second component.
`?truehist`	Read about the optional arguments.
`contour(dd <- kde2d(x,y))`	2D density plot.
`image(dd)`	Greyscale or pseudo-colour plot.

`x <- seq(1, 20, 0.5)` `x`	Make $x = (1, 1.5, 2, \ldots, 19.5, 20)$ and list it.
`w <- 1 + x/2` `y <- x + w*rnorm(x)`	w will be used as a 'weight' vector and to give the standard deviations of the errors.

`dum <- data.frame(x, y, w)`
`dum`
`rm(x, y, w)`
	Make a *data frame* of three columns named x, y and w, and look at it. Remove the original x, y and w.

`fm <- lm(y ~ x, data = dum)` `summary(fm)`	Fit a simple linear regression of y on x and look at the analysis.

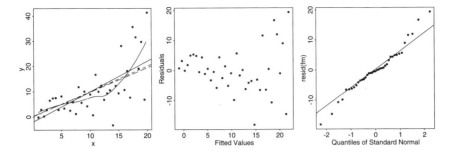

Figure 1.1: Four fits and two residual plots for the artificial heteroscedastic regression data.

```
fm1 <- lm(y ~ x, data = dum,
    weight = 1/w^2)
summary(fm1)
```
Since we know the standard deviations, we can do a weighted regression.

```
lrf <- loess(y ~ x, dum)
```
Fit a smooth regression curve using a modern regression function.

```
attach(dum)
```
Make the columns in the data frame visible as variables.

```
plot(x, y)
```
Make a standard scatterplot. To this plot we will add the three regression lines (or curves) as well as the known true line.

```
lines(spline(x, fitted(lrf)),
    col = 2)
```
First add in the local regression curve using a spline interpolation between the calculated points.

```
abline(0, 1, lty = 3, col = 3)
```
Add in the true regression line (intercept 0, slope 1) with a different line type and colour.

```
abline(fm, col = 4)
```
Add in the unweighted regression line. `abline()` is able to extract the information it needs from the fitted regression object.

```
abline(fm1, lty = 4, col = 5)
```
Finally add in the weighted regression line, in line type 4. This one should be the most accurate estimate, but may not be, of course. One such outcome is shown in Figure 1.1.

You may be able to make a hardcopy of the graphics window by selecting the Print option from a menu.

```plot(fitted(fm), resid(fm),```
```   xlab = "Fitted Values",```
```   ylab = "Residuals")```
	A standard regression diagnostic plot to check for heteroscedasticity, that is, for unequal variances. The data are generated from a heteroscedastic process, so can you see this from this plot?

```qqnorm(resid(fm))``` ```qqline(resid(fm))```	A normal scores plot to check for skewness, kurtosis and outliers. (Note that the heteroscedasticity may show as apparent non-normality.)
```detach()``` ```rm(fm, fm1, lrf, dum)```	Remove the data frame from the search path and clean up again.

We look next at a set of data on record times of Scottish hill races against distance and total height climbed.

```hills```	List the data.
```# S: splom(~ hills)``` ```# R: pairs(hills)```	Show a matrix of pairwise scatterplots (Figure 1.2).
```# S: brush(hills)``` Click on the ```Quit``` button in the graphics window to continue.	Try highlighting points and see how they are linked in the scatterplots (Figure 1.3). Also try rotating the points in 3D.
```attach(hills)```	Make columns available by name.

```plot(dist, time)```
```identify(dist, time,```
```   row.names(hills))```
	Use mouse button 1 to identify outlying points, and button 2 to quit. Their row numbers are returned. On a Macintosh click outside the plot to quit.

```abline(lm(time ~ dist))```	Show least-squares regression line.

```# R: library(lqs)```
```abline(lqs(time ~ dist),```
```   lty = 3, col = 4)```
	Fit a very resistant line. See Figure 1.4.

```detach()```	Clean up again.

We can explore further the effect of outliers on a linear regression by designing our own examples interactively. Try this several times.

```plot(c(0,1), c(0,1), type="n")``` ```xy <- locator(type = "p")```	Make our own dataset by clicking with button 1, then with button 2 (outside the plot on a Macintosh) to finish.

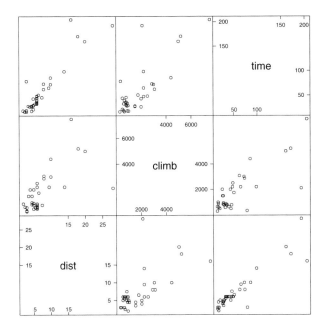

Figure 1.2: Scatterplot matrix for data on Scottish hill races.

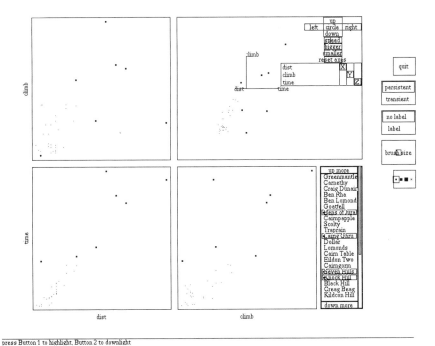

Figure 1.3: Screendump of a `brush` plot of dataset `hills` (UNIX).

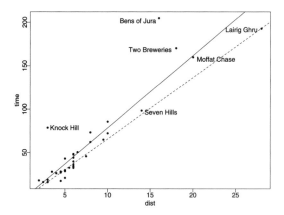

Figure 1.4: Annotated plot of time versus distance for `hills` with regression line and resistant line (dashed).

`abline(lm(y ~ x, xy), col = 4)`	Fit least-squares, a robust regression
`abline(rlm(y ~ x, xy,`	and a resistant regression line. Repeat
`   method = "MM"),`	to try the effect of outliers, both verti-
`   lty = 3, col = 3)`	cally and horizontally.
`abline(lqs(y ~ x, xy),`	
`   lty = 2, col = 2)`	
`rm(xy)`	Clean up again.

We now look at data from the 1879 experiment of Michelson to measure the speed of light. There are five experiments (column `Expt`); each has 20 runs (column `Run`) and `Speed` is the recorded speed of light, in km/sec, less 299 000. (The currently accepted value on this scale is 734.5.)

`attach(michelson)`	Make the columns visible by name.
`search()`	The *search path* is a sequence of places, either directories or data frames, where S-PLUS looks for objects required for calculations.
`plot(Expt, Speed,` `   main="Speed of Light Data",` `   xlab="Experiment No.")`	Compare the five experiments with simple boxplots. The result is shown in Figure 1.5.
`fm <- aov(Speed ~ Run + Expt)` `summary(fm)`	Analyse as a randomized block design, with *runs* and *experiments* as factors.

	Df	Sum of Sq	Mean Sq	F Value	Pr(F)
Run	19	113344	5965	1.1053	0.36321
Expt	4	94514	23629	4.3781	0.00307
Residuals	76	410166	5397		

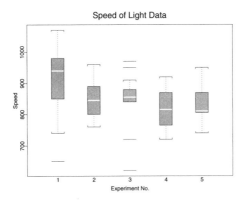

Figure 1.5: Boxplots for the speed of light data.

```
fm0 <- update(fm, .~ .  - Run)   Fit the sub-model omitting the non-
anova(fm0, fm)                   sense factor, runs, and compare using
                                 a formal analysis of variance.
```

Fit the sub-model omitting the nonsense factor, *runs*, and compare using a formal analysis of variance.

```
      Analysis of Variance Table
      Response: Speed

              Terms Resid. Df    RSS Test Df Sum of Sq F Value    Pr(F)
      1        Expt        95 523510
      2 Run + Expt        76 410166 +Run 19    113344  1.1053 0.36321
```

```
detach()                         Clean up before moving on.
rm(fm, fm0)
```

The S environment includes the equivalent of a comprehensive set of statistical tables; one can work out P values or critical values for a wide range of distributions (see Table 5.1 on page 108).

```
1 - pf(4.3781, 4, 76)            P value from the ANOVA table above.
qf(0.95, 4, 76)                  corresponding 5% critical point.
```

P value from the ANOVA table above.

corresponding 5% critical point.

```
q()                              Quit your S environment. R will ask
                                 if you want to save the workspace: for
                                 this session you probably do not.
```

Quit your S environment. R will ask if you want to save the workspace: for this session you probably do not.

1.4   What Next?

We hope that you now have a flavour of S and are inspired to delve more deeply. We suggest that you read Chapter 2, perhaps cursorily at first, and then Sections 3.1–7 and 4.1–3. Thereafter, tackle the statistical topics that are of interest to you. Chapters 5 to 16 are fairly independent, and contain cross-references where they do interact. Chapters 7 and 8 build on Chapter 6, especially its first two sections.

Chapters 3 and 4 come early, because they are about S not about statistics, but are most useful to advanced users who are trying to find out what the system is really doing. On the other hand, those programming in the S language will need the material in our companion volume on S programming, Venables and Ripley (2000).

Note to R users

The S code in the following chapters is written to work with S-PLUS 6. The changes needed to use it with R are small and are given in the scripts available on-line in the `scripts` directory of the MASS package for R (which should be part of every R installation).

Two issues arise frequently:

- R's standard datasets need to be loaded explicitly into R, as in

    ```
    data(ability.cov)
    data(iris3)
    data(swiss)
    ```

    So if dataset `foo` appears to be missing, make sure that you have run `library(MASS)` and then try `data(foo)`. All of our own datasets are made available by `library(MASS)`.

- Not as many of the packages are attached by default, so R needs more use of the `library` function.

Note too that R has a different random number stream and so results depending on random partitions of the data may be quite different from those shown here.

Chapter 2

Data Manipulation

Statistics is fundamentally about understanding data. We start by looking at how data are represented in S, then move on to importing, exporting and manipulating data.

2.1 Objects

Two important observations about the S language are that

> 'Everything in S is an object.'
> 'Every object in S has a class.'

So data, intermediate results and even the result of a regression are stored in S objects, and the class[1] of the object both describes what the object contains and what many standard functions do with it.

Objects are usually accessed by name. Syntactic S names for objects are made up from the letters,[2] the digits 0–9 in any non-initial position and also the period, ' . ', which behaves as a letter except in names such as .37 where it acts as a decimal point. There is a set of reserved names

```
FALSE Inf NA NaN NULL TRUE
break else for function if in next repeat while
```

and in S-PLUS return, F and T. It is a good idea, and sometimes essential, to   S+
avoid the names of system objects like

```
c q s t C D F I T diff mean pi range rank var
```

Note that S is *case sensitive*, so Alfred and alfred are distinct S names, and that the underscore, ' _ ', is *not* allowed as part of a standard name. (Periods are often used to separate words in names: an alternative style is to capitalize each word of a name.)

Normally objects the users create are stored in a workspace. How do we create an object? Here is a simple example, some powers of π. We make use of the sequence operator ' : ' which gives a sequence of integers.

---

[1] In R all objects have classes only from version 1.7.0.

[2] In R and S-PLUS 6.1 the set of letters is determined by the locale, and so may include accented letters.

```
> -2:2
[1] -2 -1  0  1  2
> powers.of.pi <- pi^(-2:2)
> powers.of.pi
[1] 0.10132 0.31831 1.00000 3.14159 9.86960
> class(powers.of.pi)
[1] "numeric"
```

which gives a *vector* of length 5. It contains real numbers, so has class called
`"numeric"`. Notice how we can examine an object by typing its name. This is
the same as calling the function `print` on it, and the function `summary` will give
different information (normally less, but sometimes more).

```
> print(powers.of.pi)
[1] 0.10132 0.31831 1.00000 3.14159 9.86960
> summary(powers.of.pi)
   Min. 1st Qu. Median   Mean 3rd Qu.    Max.
 0.1013 0.3183  1.0000 2.8862 3.1416  9.8696
```

S+   In S-PLUS the object `powers.of.pi` is stored in the file system under the
`.Data` directory in the project directory, and is available in the project until
deleted with

    `rm(powers.of.pi)`

or over-written by assigning something else to that name. (Under some settings,
S+Win   S-PLUS 6 for Windows prompts the user at the end of the session to save to
the main workspace all, none or some of the objects created or changed in that
session.)

R   R stores objects in a workspace kept in memory. A prompt[3] at the end of the
session will ask if the workspace should be saved to disk (in a file `.RData`);
a new session will restore the saved workspace. Should the R session crash
the workspace will be lost, so it can be saved during the session by running
`save.image()` or from a file menu on GUI versions.

    S has no scalars, but the building blocks for storing data are *vectors* of various
types. The most common classes are

- `"character"`, a vector of character strings of varying (and unlimited)
  length. These are normally entered and printed surrounded by double
  quotes, but single quotes can be used.

- `"numeric"`, a vector of real numbers.

- `"integer"`, a vector of (signed) integers.

- `"logical"`, a vector of logical (true or false) values. The values are output
R   as T and F in S-PLUS and as TRUE and FALSE in R, although each system
  accepts both conventions for input.

- `"complex"`, a vector of complex numbers.

---

[3]Prompting for saving and restoring can be changed by command-line options.

- `"list"`, a vector of S objects.

We have not yet revealed the whole story; for the first five classes there is an additional possible value, NA, which means *not available*. See pages 19 and 53 for the details.

The simplest way to access a part of a vector is by number, for example,

```
> powers.of.pi[5]
[1] 9.8696
```

Vectors can also have *names*, and be accessed by name.

```
names(powers.of.pi) <- -2:2
powers.of.pi
      -2      -1 0      1      2
 0.10132 0.31831 1 3.1416 9.8696
powers.of.pi["2"]
      2
 9.8696
> class(powers.of.pi)
[1] "named"     # still "numeric" in R
```

The class in S-PLUS changes to reflect the additional structure of the object. There are several ways to remove the names.

```
> as.vector(powers.of.pi)   # or c(powers.of.pi)
[1] 0.10132 0.31831 1.00000 3.14159 9.86960
> names(powers.of.pi) <- NULL
> powers.of.pi
[1] 0.10132 0.31831 1.00000 3.14159 9.86960
```

This introduces us to another object NULL, which represent nothing, the empty set.

Factors

Another vector-like class is much used in S. Factors are sets of labelled observations with a pre-defined set of labels, not all of which need occur. For example,

```
> citizen <- factor(c("uk", "us", "no", "au", "uk", "us", "us"))
> citizen
[1] uk us no au uk us us
```

Although this is entered as a character vector, it is printed without quotes. Internally the factor is stored as a set of codes, and an attribute giving the *levels*:

```
> unclass(citizen)
[1] 3 4 2 1 3 4 4
attr(, "levels"):
[1] "au" "no" "uk" "us"
```

If only some of the levels occur, all are printed (and they always are in R).   R

```
> citizen[5:7]
[1] uk us us
Levels:
[1] "au" "no" "uk" "us"
```

(An extra argument may be included when subsetting factors to include only those
levels that occur in the subset. For example, `citizen[5:7, drop=T]`.)

Why might we want to use this rather strange form? Using a factor indicates
to many of the statistical functions that this is a categorical variable (rather than
just a list of labels), and so it is treated specially. Also, having a pre-defined set
of levels provides a degree of validation on the entries.

By default the levels are sorted into alphabetical order, and the codes assigned
accordingly. Some of the statistical functions give the first level a special status,
so it may be necessary to specify the levels explicitly:

```
> citizen <- factor(c("uk", "us", "no", "au", "uk", "us", "us"),
                     levels = c("us", "fr", "no", "au", "uk"))
> citizen
[1] uk us no au uk us us
Levels:
[1] "us" "fr" "no" "au" "uk"
```

Function `relevel` can be used to change the ordering of the levels to make a
specified level the first one; see page 383.

Sometimes the levels of a categorical variable are naturally ordered, as in

```
> income <- ordered(c("Mid", "Hi", "Lo", "Mid", "Lo", "Hi", "Lo"))
> income
[1] Mid Hi  Lo  Mid Lo  Hi  Lo

 Hi < Lo < Mid
> as.numeric(income)
[1] 3 1 2 3 2 1 2
```

Again the effect of alphabetic ordering is not what is required, and we need to set
the levels explicitly:

```
> inc <- ordered(c("Mid", "Hi", "Lo", "Mid", "Lo", "Hi", "Lo"),
                 levels = c("Lo", "Mid", "Hi"))
> inc
[1] Mid Hi  Lo  Mid Lo  Hi  Lo

 Lo < Mid < Hi
```

Ordered factors are a special case of factors that some functions (including
`print`) treat in a special way.

The function `cut` can be used to create ordered factors by sectioning continuous variables into discrete class intervals. For example,

```
> erupt <- cut(geyser$duration, breaks = 0:6)
> erupt <- ordered(erupt, labels=levels(erupt))
```

```
> erupt
  [1] 4+ thru 5 2+ thru 3 3+ thru 4 3+ thru 4 3+ thru 4
  [6] 1+ thru 2 4+ thru 5 4+ thru 5 2+ thru 3 4+ thru 5
  ....
  0+ thru 1 < 1+ thru 2 < 2+ thru 3 < 3+ thru 4 < 4+ thru 5 <
    5+ thru 6
```

(R labels these differently.) Note that the intervals are of the form $(n, n + 1]$, so   R
an eruption of 4 minutes is put in category 3+ thru 4. We can reverse this by
the argument left.include = T.[4]

Data frames

A data frame is the type of object normally used in S to store a data matrix. It
should be thought of as a list of variables of the same length, but possibly of
different types (numeric, factor, character, logical, ...). Consider our data frame
painters:

```
> painters
            Composition Drawing Colour Expression School
  Da Udine           10       8     16          3      A
  Da Vinci           15      16      4         14      A
  Del Piombo          8      13     16          7      A
  Del Sarto          12      16      9          8      A
  Fr. Penni           0      15      8          0      A
  ....
```

which has four numerical variables and one factor variable. It is printed in a way
specific to data frames. The components are printed as columns and there is a set
of names, the row.names, common to all variables. The row names should be
unique.[5]

```
> row.names(painters)
  [1] "Da Udine"      "Da Vinci"          "Del Piombo"
  [4] "Del Sarto"     "Fr. Penni"         "Guilio Romano"
  [7] "Michelangelo"  "Perino del Vaga"   "Perugino"
  ....
```

The column names are given by the names function.
    Applying summary gives a summary of each column.

```
> summary(painters) # try it!
```

    Data frames are by far the commonest way to store data in an S environment.
They are normally imported by reading a file or from a spreadsheet or database.
However, vectors of the same length can be collected into a data frame by the
function data.frame.

```
mydat <- data.frame(MPG, Dist, Climb, Day = day)
```

---

[4]In R use right = FALSE.
[5]In S-PLUS there is a dup.names.ok argument which should be avoided.

However, all character columns are converted to factors unless their names are included in I() so, for example,

```
mydat <- data.frame(MPG, Dist, Climb, Day = I(day))
```

preserves day as a character vector, Day.

The row names are taken from the names of the first vector found (if any) which has names without duplicates, otherwise numbers are used.

Sometimes it is convenient to make the columns of the data frame available by name. This is done by attach and undone by detach:

```
> attach(painters)
> School
 [1] A A A A A A A A A A B B B B B B B C C C C C C D D D D D D D
[30] D D D E E E E E E E F F F F G G G G G G G H H H H
> detach("painters")
```

Be wary of masking system objects,[6] and detach as soon as you are done with this.

Matrices and arrays

A data frame may be printed like a matrix, but it is not a matrix. Matrices like vectors[7] have all their elements of the same type. Indeed, a good way to think of a matrix in S is as a vector with some special instructions as to how to lay it out. The matrix function generates a matrix:

```
> mymat <- matrix(1:30, 3, 10)
> mymat
     [,1] [,2] [,3] [,4] [,5] [,6] [,7] [,8] [,9] [,10]
[1,]    1    4    7   10   13   16   19   22   25    28
[2,]    2    5    8   11   14   17   20   23   26    29
[3,]    3    6    9   12   15   18   21   24   27    30
> class(mymat)
[1] "matrix"
> dim(mymat)
[1]  3 10
```

Note that the entries are (by default) laid out down columns, the class is one not seen before, and the dim function gives the dimensions. Use argument byrow = T to fill the matrix along rows.

Matrices have two dimensions: arrays have one, two, three or more dimensions. We can create an array using the array function or by assigning the dimension.

```
> myarr <- mymat
> dim(myarr) <- c(3, 5, 2)
> class(myarr)
[1] "array"
```

---

[6]See page 43.
[7]Except lists, but a list can be matrix in S.

```
> myarr

, , 1
    [,1] [,2] [,3] [,4] [,5]
[1,]   1    4    7   10   13
[2,]   2    5    8   11   14
[3,]   3    6    9   12   15

, , 2
    [,1] [,2] [,3] [,4] [,5]
[1,]  16   19   22   25   28
[2,]  17   20   23   26   29
[3,]  18   21   24   27   30
> dim(myarr)
[1] 3 5 2
```

Matrices and arrays can also have names for the dimensions, known as *dim-names*. The simple way to add them is just to assign them, using NULL where we do not want to specify a set of names.

```
> dimnames(myarr) <- list(letters[1:3], NULL, c("(i)", "(ii)"))
> myarr

, , (i)
  [,1] [,2] [,3] [,4] [,5]
a   1    4    7   10   13
b   2    5    8   11   14
c   3    6    9   12   15

, , (ii)
  [,1] [,2] [,3] [,4] [,5]
a  16   19   22   25   28
b  17   20   23   26   29
c  18   21   24   27   30
```

We can begin to see how arrays may be useful to represent multi-dimensional tables.

Missing and special values

We have already mentioned the special value NA. If we assign NA to a new variable we see that it is *logical*

```
> newvar <- NA
> class(NA)
[1] "logical"
```

It is important to realize that S works in a three-valued logic, and that

```
> newvar > 3
[1] NA
```

is perfectly logical in that system. As we do not know the value of `newvar` (it is 'not available') we cannot know if it bigger or smaller than 3. In all such cases S does not guess, it returns `NA`.

R   There are missing numeric, integer, complex and (R only) character values, and these are also printed as `NA`. Coercion occurs as needed, so

```
> x <- c(pi, 4, 5)
> x[2] <- NA
> x
[1] 3.1416      NA 5.0000
> class(x)
[1] "numeric"
```

shows that this is a numeric `NA`. To test which elements of an object are missing, use the function `is.na`

```
> is.na(x)
[1] F T F
```

There are other special numeric (and complex) values that can cause confusion. As far as we know all current S environments support IEC 60559 arithmetic,[8] which has the special values `NaN`, `Inf` and `-Inf`. The values `Inf` and `-Inf` can be entered as such and can also occur in arithmetic:

```
> 1/0
[1] Inf
```

The value[9] `NaN` means 'not a number' and represent results such as `0/0`. In S-PLUS they are printed as `NA`, in R as `NaN` and in both `is.na` treats them as missing.

```
> x <- c(-1, 0, 1)/0
> x
[1] -Inf   NA  Inf  ## NaN in R
> is.na(x)
[1] F T F
> x > Inf
[1]  F NA  F
```

Notice the logic is not perfect; even though the second element is missing we know it cannot exceed `Inf`, but S does not.

For more on this topic see Venables and Ripley (2000, p. 22).

2.2 Connections

Connections between a statistical environment and the outside world are increasingly important. Here 'connections' is used both in the general sense and a specific sense, a set of S classes to connect to the world rather than just to files.

---

[8]More commonly referred to as IEEE 754.

[9]There are actually many such values.

There is a class `"connection"` for S objects that provide such connections; this specializes to a class `"file"` for files, but there are also (in some of the implementations) connections to terminals, pipes, fifos, sockets, character vectors, ... We will only scratch the surface here.

Another set of connections are to data repositories, either to import/export data or to directly access data in another system. This is an area in its infancy.

For most users the S environment is only one of the tools available to manipulate data, and it is often productive to use a combination of tools, pre-processing the data before reading into the S environment.

Data entry

For all but the smallest datasets the easiest way to get data into an S environment is to import it from a connection such as a file. For small datasets two ways are

```
> x <- c(2.9, 3.1, 3.4, 3.4, 3.7, 3.7, 2.8, 2.5)

> x <- scan()
1: 2.9 3.1 3.4 3.4 3.7 3.7 2.8 2.5
9:
```

where in the second method input is terminated by an empty line or the end-of-file character (probably Ctrl-D). (To enter a character vector this way use `scan(what = "")`.)

Windows versions of S-PLUS and all versions of R have a spreadsheet-like   S+Win
data window that can be used to enter or edit data frames. It is perhaps easiest to start with a dummy data frame:

```
> mydf <- data.frame(dist = 0., climb = 0., time = 0.)
> Edit.data(mydf)  ## S-PLUS for Windows only
```

Function `Edit.data` brings up a spreadsheet-like grid: see Figure 2.1. It works on matrices and vectors too. Alternatively open an Objects Explorer, right click on the object and select Edit..., or use the Select Data... item on the Data menu.

In R create a dummy data frame and then use   R

```
> fix(mydf) ## R
```

to bring up a data grid. See `?edit.data.frame` for further details.

Importing using `read.table`

The function `read.table` is the most convenient way to read in a rectangular grid of data. Because such grids can have many variations, it has many arguments.

The first argument is called `"file"`, but specifies a connection. The simplest use is to give a character string naming the file. One warning for Windows users: specify the directory separator either as `"/"` or as `"\\"` (but *not* `"\"`).

The basic layout is one record per row of input, with blank rows being ignored. There are a number of issues to consider:

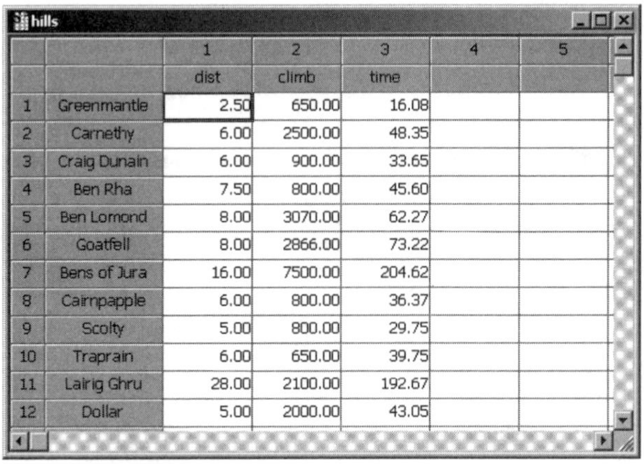

		1	2	3	4	5
		dist	climb	time		
1	Greenmantle	2.50	650.00	16.08		
2	Carnethy	6.00	2500.00	48.35		
3	Craig Dunain	6.00	900.00	33.65		
4	Ben Rha	7.50	800.00	45.60		
5	Ben Lomond	8.00	3070.00	62.27		
6	Goatfell	8.00	2866.00	73.22		
7	Bens of Jura	16.00	7500.00	204.62		
8	Cairnpapple	6.00	800.00	36.37		
9	Scolty	5.00	800.00	29.75		
10	Traprain	6.00	650.00	39.75		
11	Lairig Ghru	28.00	2100.00	192.67		
12	Dollar	5.00	2000.00	43.05		

Figure 2.1: A data-window view (from S-PLUS 6 under Windows) of the first few rows of the `hills` dataset. For details of data windows see page 460.

(a) Separator The argument `sep` specifies how the columns of the file are to be distinguished. Normally looking at the file will reveal the right separator, but it can be important to distinguish between the default `sep = ""` that uses any white space (spaces, tabs or newlines), `sep = " "` (a single space) and `sep = "\t"` (tab).

(b) Row names It is best to have the row names as the first column in the file, or omit them altogether (when the rows are numbered, starting at 1).

The row names can be specified as a character vector argument `row.names`, or as the number or name of a column to be used as the row names. If there is a header one column shorter than the body of the file, the first column in
S+ the file is taken as the row names. Otherwise S-PLUS grabs the first suitable column (non-numeric, no duplicates), or if there is none, numbers the rows. You can force numbered rows by argument `row.names = NULL`.

(c) Header line It is convenient to include a first line of input giving the names for the columns. We recommend that you include the argument `header` explicitly. Conventionally the header line excludes the row names and so has one entry fewer than subsequent rows if the file contains row names. (If `header` is not specified, S sets `header = T` if and only if this is true.) If the names given are not syntatically valid S names they will be converted (by replacing invalid characters by '`.`').

(d) Missing values By default the character string `NA` in the file is assumed to represent missing values, but this can be changed by the argument `na.strings`, a character vector of zero, one or more representations of missing values. To turn this off, use `na.strings = character(0)`.

In otherwise numeric columns, blank fields are treated as missing.

(e) Quoting By default character strings may be quoted by `"` or `'` and in each

case all characters on the line up to the matching quote are regarded as part
of the string.

In R the set of valid quoting characters (which might be none) is specified by   R
the `quote` argument; for `sep = "\n"` this defaults to `quote = ""`, a useful
value if there are singleton quotes in the data file. If no separator is specified,
quotes may be escaped by preceding them with a backslash; however, if a
separator is specified they should be escaped by doubling them, spreadsheet-
style.

(f) **Type conversion**   By default, `read.table` tries to work out the correct
class for each column. If the column contains just numeric (logical) values
and one of the `na.strings` it is converted to `"numeric"` (`"logical"`).
Otherwise it is converted to a factor. The logical argument `as.is` controls
the conversion to factors (only); it can be of length one or give an entry for
each column (excluding row names).

R has more complex type conversion rules, and can produce integer and com-   R
plex columns: it is also possible to specify the desired class for each column.

(g) **White space in character fields**   If a separator is specified, leading and trail-
ing white space in character fields is regarded as part of the field.

Post-processing

There are some adjustments that are often needed after using `read.table`. Character variables will have been read as factors (modulo the use of `as.is`), with levels in alphabetical order. We might want another ordering, or an ordered factor.
Some examples:[10]

```
whiteside$Insul <-
    factor(whiteside$Insul, levels = c("Before", "After"))
Insurance$Group <- ordered(Insurance$Group,
    labels = c("<1l", "1-1.5l", "1.5-2l", ">2l"))
```

Also, numeric variables will have been made into a numeric column, even if they
are really factor levels:

```
Insurance$District <- factor(Insurance$District)
```

Importing from other systems

Often the safest way to import data from another system is to export it as a tab- or
comma-delimited file and use `read.table`. However, more direct methods are
available.

S-PLUS has a function `importData`, and on GUI versions a dialog-box in-   S+
terface via the Import Data... item on its File menu. This can import from a
wide variety of file formats, and also directly from relational databases.[11]   The
file formats include plain text, Excel,[12] Lotus 123 and Quattro spreadsheets, and

---

[10] All from the scripts used to make the MASS library section.
[11] Which databases is system-dependent.
[12] But only up to the long superseded version 4 on UNIX / Linux.

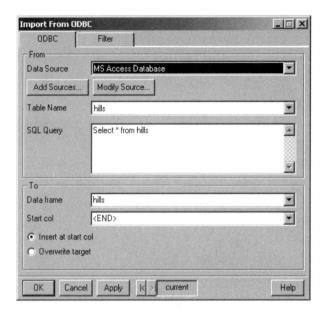

Figure 2.2: S-PLUS 6 GUI interface to importing from ODBC: the Access database is selected from a pop-up dialog box when that type of 'Data Source' is selected.

various SAS, SPSS, Stata, SysStat, Minitab and Matlab formats. Files can be read in sequential blocks of rows via openData and readNextDataRows.

Importing data in the GUI usually brings up a data grid showing the data; it is also saved as an S object. We will illustrate this by importing a copy of our data frame hills from an Access database. The data had been stored in table hills in an Access database, and an ODBC 'Data Source Name' testacc entered via the control panel ODBC applet.[13]

```
hills2 <- importData(type = "ODBC",
    odbcConnection = "DSN=testacc", table = "hills")
```

Users unfamiliar with ODBC will find the GUI interface easier to use; see Figure 2.2.

If you have Microsoft Excel installed, data frames can be linked to ranges of Excel spreadsheets. Open the spreadsheet via the Open item on the File menu (which brings up an embedded Excel window) and select the 'Link Wizard' from the toolbar.

R   R can import from several file formats and relational database systems; see the *R Data Import/Export* manual.

Using scan

Function read.table is an interface to a lower-level function scan. It is rare to use scan directly, but it does allow more flexibility, including the ability to

---

[13] In the Administrative Tools folder in Windows 2000 and XP.

bypass completely the type conversion of `read.table` and to have records span-
ning more than one input line (with argument `multi.line = T`).

Using `readLines`

If all you need is to read a set of character strings from a file, one string per line,
then `readLines` is the fastest and most convenient tool. Often the simplest way
to read complicated formats is to read them line-by-line and then manipulate the
lines inside the S environment.

Data export

Once again there are many possibilities. Function `write.table` will write a data
frame to a connection, most often a file. Its default usage

```
> write.table(painters, file = "painters.dat")
```

writes a data frame, matrix or vector to a file in a comma-separated format with
row and column names, something like (from S-PLUS)

```
row.names,Composition,Drawing,Colour,Expression,School
Da Udine,10, 8,16, 3,A
Da Vinci,15,16, 4,14,A
Del Piombo, 8,13,16, 7,A
Del Sarto,12,16, 9, 8,A
....
```

There are a number of points to consider.

(a) Header line   Note that that is not quite the format of header line that
    `read.table` expects, and R will omit the first field unless argument   R
    `col.names = NA` is supplied.

    The header line can be suppressed in S-PLUS by `dimnames.write = "row"`
    and in R by `col.names = FALSE`.

(b) Row names   These can be suppressed by `dimnames.write = "col"` in
    S-PLUS and `row.names = FALSE` in R. In S-PLUS `dimnames.write = F`
    omits both row and column names.

(c) Separator   The comma is widely used in English-speaking countries as it is
    unlikely to appear in a field, and such files are known as CSV files. In some
    locales the comma is used as a decimal point, and there the semicolon is used
    as a field separator in CSV fields (use `sep = ";"`). A tab (use `sep = "\t"`)
    is often the safest choice.

(d) Missing values   By default missing values are output as `NA`; this can be
    changed by the argument `na`.

(e) Quoting   In S-PLUS character strings are not quoted by default. With ar-   S+
    gument `quote.strings = T` all character strings are double-quoted. Other
    quoting conventions are possible, for example `quote.strings = c("‘",`
    `"’")`. Quotes within strings are not treated specially.

In R character strings *are* quoted by default, this being suppressed by   R
quote = FALSE, or selectively by giving a numeric vector for quote. Embedded quotes are escaped, either as \" or doubled (Excel-style, set by
qmethod = "double").

(f) **Precision**   The precision to which real (and complex) numbers are output is
controlled by the setting of options("digits"). You may need to increase
this.

Using write.table can be very slow for large data frames; if all that is
needed is to write out a numeric or character matrix, function write.matrix in
our library section MASS can be much faster.

S+   S-PLUS has function exportData, and on Windows a dialog-box interface
to it via item Export Data... on the File menu. This can write files in similar formats to importData (but not to databases). The arguments are similar to those of
write.table but with confusingly different names, for example delimiter,
colNames, rowNames and quote (which defaults to true). Files can be exported in blocks of rows via openData and writeNextDataRows.

Saving objects

Sometimes one needs to save an S object for future reference or to take to another
machine. If it is a data frame, the simplest way is often to export to a text file.
However, there are ways to save and restore arbitrary objects.

S+   In S-PLUS the recommended way is to save the object using data.dump and
restore it using data.restore. To save and restore three objects we can use

```
data.dump(c("obj1", "obj2", "obj3"), file = "mydump.sdd")
data.restore(file = "mydump.sdd")
```

Under Windows the .sdd extension is associated with such dumps.

R   In R we can use save and load. A simple usage is

```
save(obj1, obj2, obj3, file = "mydump.rda", ascii = FALSE)
load(file = "mydump.rda")
```

which gives a binary dump (usually in a machine-independent format). To send
the result by email or to ensure a greater chance of machine-independence, use
ascii = TRUE. Compression can be specified *via* compress = TRUE, and is
useful for archival storage of R objects.

Note that none of these methods is guaranteed to work across different architectures (but they usually do) nor across different versions of S-PLUS or R.

More on connections

So far we have made minimal use of connections; by default functions such as
read.table and scan open a connection to a file, read (or write) the file, and
close the connection. However, users can manage the process themselves: suppose we wish to read a file which has a header and some text comments, and then
read and process 1000 records at a time. For example,

```
con <- open("data.dat", "r")    # open the file for reading
header <- scan(con, what=list(some format), n=1, multi.line=T)
## compute the number of comment lines from 'header'
comments <- readLines(con, n = ncomments)
repeat {
    z <- scan(con, what = list(record format), n = 1000)
    if(!length(z[[1]])) break;
    ## process z (which might be less than 1000 records)
}
close(con)
```

This approach is particularly useful with binary files of known format, where `readRaw` (S-PLUS) or `readBin` (R) can be used to read sections of a particular format (say character or float type). It is also helpful for creating formatted output a piece at a time.

Connections can also be used to input from other sources. Suppose the data file contains comment lines starting with `#`. Now R's `read.table` and `scan`   R can handle these directly, but we could also make use of a `pipe` connection by[14]

```
DF <- read.table(pipe("sed -e /^[ \\t]*#/d data.dat"), header = T)
```

A similar approach can be used to edit the data file, for example to change[15] the use of comma as a decimal separator to ' . ' by `sed -e s/,/./g`.

Taking this approach further, a connection can (on suitable systems) read from or write to a fifo or socket and so wait for data to become available, process it and pass it on to a display program.

2.3   Data Manipulation

S-PLUS for Windows has a set of dialog boxes accessed from its **Data** menu for data manipulation. These can be useful for simple operations, but are very limited compared to the S language as used on, say, page 380.

The primary means of data manipulation in S is *indexing*. This is extremely powerful, and most people coming to S take a while to appreciate the possibilities. How indexing works in detail varies by the class of the object, and we only cover the more common possibilities here.

Indexing vectors

We have already seen several examples of indexing vectors. The complete story needs to take into account that indexing can be done on the left-hand side of an assignment (to select parts of a vector to replace) as well on the right-hand side. The general form is `x[ind]` where `ind` is on of the following forms:

---

[14]This may only work on a UNIX-like system.

[15]R has argument `dec` to specify the decimal point character, and S-PLUS 6.1 consults the locale.

1. A vector of positive integers. In this case the values in the index vector normally lie in the set { 1, 2, ..., length(x) }. The corresponding elements of the vector are selected, in that order, to form the result. The index vector can be of any length and the result is of the same length as the index vector. For example, x[6] is the sixth component of x and x[1:10] selects the first 10 elements of x (assuming length(x) ⩾ 10). For another example, we use the dataset letters, a character vector of length 26 containing the lower-case letters:

```
> letters[1:3]
[1] "a" "b" "c"
> letters[c(1:3,3:1)]
[1] "a" "b" "c" "c" "b" "a"
```

Such indices can also include zero. A zero index on the right-hand side passes nothing, and a zero index in a vector to which something is being assigned accepts nothing.

2. A logical vector. The index vector must be of the same length as the vector from which elements are to be selected. Values corresponding to T in the index vector are selected and those corresponding to F or NA are omitted. For example,

```
y <- x[!is.na(x)]
```

creates an object y that will contain the non-missing values of x, in the same order as they originally occurred. Note that if x has any missing values, y will be shorter than x. Also,

```
x[is.na(x)] <- 0
```

replaces any missing values in x by zeros.

3. A vector of negative integers. This specifies the values to be *excluded* rather than included. Thus

```
> y <- x[-(1:5)]
```

drops the first five elements of x. Zero values are handled in the same way as in case **1.**

4. A vector of character strings. This possibility applies only where a vector has names. In that case a subvector of the names vector may be used in the same way as the positive integers in case **1.** For example,

```
> # R: data(state)
> longitude <- state.center$x
> names(longitude) <- state.name
> longitude[c("Hawaii", "Alaska")]
  Hawaii   Alaska
 -126.25  -127.25
```

finds the longitude of the geographic centre of the two most western states of the USA. The names attribute is retained in the result.

5. Empty. This implies all possible values for the index. It is really only useful on the receiving side, where it replaces the contents of the vector but keeps other aspects (the class, the length, the names, ...).

What happens if the absolute value of an index falls outside the range 1, ..., length(x)? In an expression this gives NA if positive and imposes no restriction if negative. In a replacement, a positive index greater than length(x) extends the vector, assigning NAs to any gap, and a negative index less than -length(x) is ignored in **S-PLUS** but an error in **R**.   R

If the sub-vector selected for replacement is longer than the right-hand side, the S *recycling* rule comes into play. That is, the right-hand side is recycled as often as necessary; if this involves partial recycling there will be a warning or error message.

Indexing data frames, matrices and arrays

Matrices and data frames may be indexed by giving two indices in the form mydf[i, j] where i and j can take any of the five forms shown for vectors. If character vectors are used as indices, they refer to column names, row names or dimnames as appropriate.

Matrices are just arrays with two dimensions, and the principle extends to arrays: for a k-dimensional array give k indices from the five forms. Indexing arrays[16] has an unexpected quirk: if one of the dimensions of the result is of length one, it is dropped. Suppress this by adding the argument drop = F. For example,

```
> myarr[1, 2:4, ]
      (i) (ii)
[1,]   4   19
[2,]   7   22
[3,]  10   25
> myarr[1, 2:4, , drop = F]

, , (i)
   [,1] [,2] [,3]
a    4    7   10

, , (ii)
   [,1] [,2] [,3]
a   19   22   25
```

Forgetting drop = F is a common error. Conversely, the function drop will drop all length-one dimensions.

Columns in a data frame are most commonly selected by the $ operator, for example painters$School.

---

[16]But not data frames.

There are several other forms of indexing that you might meet, although we do not recommend them for casual use; they are discussed in Venables and Ripley (2000, pp. 23–27). Columns of a data frame can be selected by using a one-dimensional index, for example `painters[c("Colour", "School")]`. An array is just a vector with dimensions, and so can be indexed as a vector. Arrays and data frames can also be indexed by matrices.

Selecting subsets

A common operation is to select just those rows of a data frame that meet some criteria. This is a job for logical indexing. For example, to select all those rows of the `painters` data frame with `Colour` $\geqslant 17$ we can use

```
> attach(painters)
> painters[Colour >= 17, ]
          Composition Drawing Colour Expression School
   Bassano           6       8     17          0      D
 Giorgione           8       9     18          4      D
 Pordenone           8      14     17          5      D
    Titian          12      15     18          6      D
 Rembrandt          15       6     17         12      G
    Rubens          18      13     17         17      G
  Van Dyck          15      10     17         13      G
```

We often want to select on more than one criterion, and we can combine logical indices by the 'and', 'or' and 'not' operators `&`, `|` and `!`. For example,

```
> painters[Colour >= 15 & Composition > 10, ]
> painters[Colour >= 15 & School != "D", ]
```

Now suppose we wanted to select those from schools A, B and D. We can select a suitable integer index using `match` (see page 53) or a logical index using `is.element`.

```
painters[is.element(School, c("A", "B", "D")), ]
painters[School %in% c("A", "B", "D"), ]   ## R and S+6.1 only
```

One needs to be careful with these checks, and consider what happens if part of the selection criterion is `NA`. Thus `School != "D"` not only omits those known to be in school D, but also any for which the school is unknown, which are kept by `!is.element(School, "D")`.

One thing that does not work as many people expect is `School == c("A", "B", "D")`. That tests the first element against `"A"`, the second against `"B"`, the third against `"C"`, the fourth against `"A"`, and so on.

The `ifelse` function can also be useful in selecting a subset. For example, to select the better of `Colour` and `Expression` we could use a matrix index

```
painters[cbind(1:nrow(painters), ifelse(Colour > Expression, 3, 4))]
```

Partial matching can be useful, and is best done by *regular expressions* (see page 53). For example, to select those painters whose names end in 'io' we can use

```
painters[grep("io$", row.names(painters)), ]
```

We must remember to clean up:

```
> detach("painters")
```

Sub-sampling

Sub-sampling is also done by indexing. For a random sample of m rows of data frame `fgl` we can use

```
fglsub1 <- fgl[sort(sample(1:nrow(fgl), m)), ]
```

Using `sort` keeps the rows in their original order.

Sometimes one wants a, say, 10% sample where this means not a fixed-size random sample of 10% of the original size, but a sample in which each row appears with probability 0.1, independently. For this, use

```
fglsub2 <- fgl[rbinom(nrow(fgl), 1, 0.1) == 1, ]
```

For systematic sampling we can use the `seq` function described on page 50. For example, to sample every 10th row use

```
fglsub3 <- fgl[seq(1, nrow(fgl), by = 10), ]
```

Re-coding missing values

A common problem with imported data is to re-code missing values, which may have been coded as '999' or '.', say. Often this is best avoided by using the `na.strings` argument to `read.table` or by editing the data before input, but this is not possible with direct (e.g., ODBC) connections.

An actual example was an import from SPSS in which 9, 99 and 999 all represented 'missing'. For a vector z this can be recoded by

```
z[is.element(z, c(9, 99, 999))] <- NA
```

If '.' has been used it is likely that the vector has been imported as a character vector, in which case

```
z[z == "."] <- NA
z <- as.numeric(z)
```

may be needed.

Combining data frames or matrices

The functions `cbind` and `rbind` combine data frames, matrices or vectors column-wise and row-wise respectively.

Compatible data frames can be joined by `cbind`, which adds columns of the same length, and `rbind`, which stacks data frames vertically. The result is a data frame with appropriate names and row names; the names can be changed by naming the arguments as on page 191.

The functions can also be applied to matrices and vectors; the result is a matrix. If one just wants to combine vectors to form a data frame, use `data.frame` and *not* cbind; cbind-ing a mixture of numeric and character variables will result in a character matrix.

Repeated use of cbind and (especially) rbind is inefficient; it is better to create a matrix (or data frame) of the desired final size (or an overestimate) and then assign to sections of it using indexing.

Function merge (see page 35) allows more general combinations of data frames.

Sorting

The S function sort at its simplest takes one vector argument and returns a vector of sorted values. The vector to be sorted may be numeric or character, and if there is a names attribute the correspondence of names is preserved. The ordering of tied values is preserved. How character vectors are sorted is locale-specific in R and S-PLUS 6.1; S-PLUS 6.0 uses the ASCII collating sequence.

To reverse the ordering of a vector, use the function rev .

More often we want to sort several values in parallel: for example to sort the painters data frame by the painter's name. We can do that by

```
painters[sort.list(row.names(painters)), ]
```

The function sort.list produces a (positive integer) index vector that will arrange its argument in increasing order. To put a numeric vector x into decreasing order, use sort.list(-x); for any vector x one can use rev(sort.list(x)), but that will also reverse tied entries.

Function order generalizes sort.list to an arbitrary number of arguments. It returns the index vector that would arrange the first in increasing order, with ties broken by the second, and so on. For example, to print employees arranged by age, by salary within age, and by employment number within salary, we might use:

```
attach(Employees)
Employees[order(Age, Salary, No), ]
detach("Employees")
```

All these functions have an argument na.last that determines the handling of missing values. With na.last = NA (the default for sort) missing values are deleted; with na.last = T (the default for sort.list and order) they are put last, and with na.last = F they are put first.

Data transformations

There are of course many possible transformations, and here we only consider a few of the more common ones. Unless otherwise stated, we assume that the data are stored in a data frame.

Individual variables

Individual variables are most easily accessed by the $ operator, which can also be used to add new variables (at the right edge). Here are some examples from later in the book:

```
> hills$ispeed <- hills$time/hills$dist    # ratio of two vars
> Cf$Tetrahydrocortisone <- log(Cf$Tetrahydrocortisone)
> levels(Quine$Eth) <- c("Aboriginal", "Non-aboriginal")
```

One set of transformations that is sometimes needed is to convert a factor with numeric levels to numbers, and *vice versa*. Use

```
a.num <- as.numeric(as.character(a.num.fac))      # or
a.num <- as.numeric(levels(a.mu.fac))[a.num.fac] # more efficient

a.fac <- factor(a.num, levels = sort(unique(a.num)))
```

Not specifying `levels` in the last example would sort the levels as strings, using `"2" > "10"` for example.

To merge levels of a factor, re-assign the levels giving two or more levels the same label.

Sets of columns

Some operations can be applied to whole data frames, for example `log`, so in Chapter 11 we have `log(ir)`. More often we want to take logs of some of the variables, say all numeric variables. We can apply this to the `crabs` dataset we consider in Chapter 11.

```
lcrabs <- crabs  # make a copy
lcrabs[, 4:8] <- log(crabs[, 4:8])
```

One common operation to use in this way is `scale`, which by default centres each variable to have zero mean and then rescales to unit variance. (Either operation can be de-selected via arguments `center` and `scale`.)

Other operations can only be applied to vectors, and so must be applied to each column in turn. This is the purpose of the function `lapply`, so we could scale the columns of `crabs` by

```
scrabs <- crabs  # make a copy
scrabs[, 4:8] <- lapply(scrabs[, 4:8], scale)
## or to just centre the variables
scrabs[, 4:8] <- lapply(scrabs[, 4:8], scale, scale = F)
```

albeit less efficiently. Notice how extra arguments can be given, and are passed on to the function called for each column. For an example where variables are scaled to $[0, 1]$ see page 348.

Suppose we wanted to standardize[17] *all* the numerical variables. We could use

```
scrabs <- crabs  # make a copy
scrabs[ ] <- lapply(scrabs,
    function(x) {if(is.numeric(x)) scale(x) else x})
```

---

[17]Transform to zero mean and unit variance.

using a simple *anonymous function* (and if ... else; see page 58). The right-hand side gives a list without row names, which we use to replace all the columns in the data frame.

We can find out which variables are numeric by

```
> sapply(crabs, is.numeric)
sp sex index FL RW CL CW BD
 F   F       T  T  T  T  T  T
```

Function sapply is very similar to lapply, but attempts to simplify the result to a vector or matrix.

Operations on rows

Operating on each row is much tricker. Whereas each column is a variable of a single class, a row can be rather diverse. However, in the special case that all columns are numeric, or all are character or factor, we can make progress by coercing the data frame to a (numeric or character) matrix.

Function apply operates on arrays,[18] but here we need only the special case of row-wise operations on a matrix. For example, on page 204 we use

```
> house.cpr <- apply(house.pr, 1, cumsum)
```

to form cumulative probabilities row-wise in a matrix giving multinomial probabilities for each case (row).

A data frame used as the first argument of apply will automatically be coerced to a matrix.

Splitting

We can consider groups of rows by splitting a data frame on one or more factors.

The function split takes as arguments a vector, matrix or data frame and a factor defining groups. The value is a list, one component for each group. For example, in Chapter 14 we use

```
boxplot(split(nott, cycle(nott)), names = month.abb)
```

to split a time series by month, and perform a boxplot on each sub-series.

For data frames it is often more convenient to use by. This takes a data frame and splits it by the second argument, INDICES, passing each data frame in turn to its FUN argument. INDICES can be a factor or a list of factors. For example, to summarize the measurements in each sex–species group in crabs we can use

```
> by(crabs[, 4:8], list(crabs$sp, crabs$sex), summary)
crabs$sp:B
crabs$sex:F
        FL              RW              CL              CW
 Min.   : 7.2    Min.   : 6.5    Min.   :14.7    Min.   :17.1
 1st Qu.:11.5    1st Qu.:10.6    1st Qu.:23.9    1st Qu.:27.9
 Median :13.1    Median :12.2    Median :27.9    Median :32.4
 Mean   :13.3    Mean   :12.1    Mean   :28.1    Mean   :32.6
```

---

[18] See page 65 for a fuller description of apply.

```
3rd Qu.:15.3    3rd Qu.:13.9    3rd Qu.:32.8    3rd Qu.:37.8
Max.   :19.2    Max.   :16.9    Max.   :40.9    Max.   :47.9
   ....
```

Function `aggregate` is similar to `by`, but for functions that return a single number so the result can be a data frame. For example,

```
> aggregate(crabs[, 4:8], by = list(sp=crabs$sp, sex=crabs$sex),
            median)
  sp sex    FL     RW     CL     CW     BD
1  B   F 13.15  12.20  27.90  32.35  11.60
2  O   F 18.00  14.65  34.70  39.55  15.65
3  B   M 15.10  11.70  32.45  37.10  13.60
4  O   M 16.70  12.10  33.35  36.30  15.00
```

It is important to ensure that the function used is applicable to each column, which is why we must omit the factor columns here.

Merging

The function `merge` provides a *join* of two data frames as databases. That is, it combines pairs of rows that have common values in specified columns to a row with all the information contained in either data frame, allowing many–many matches. Use `?merge` for full details.

As an example, consider using two data frames as the equivalent of relations in a database, and joining them.

```
> authors
   surname nationality deceased
1    Tukey          US      yes
2 Venables   Australia       no
3  Tierney          US       no
4   Ripley          UK       no
5   McNeil   Australia       no
> books
      name                       title
1    Tukey    Exploratory Data Analysis
2 Venables Modern Applied Statistics ...
3  Tierney                   LISP-STAT
4   Ripley           Spatial Statistics
5   Ripley        Stochastic Simulation
6   McNeil    Interactive Data Analysis

> merge(authors, books, by.x = "surname", by.y = "name")
   surname nationality deceased                     title
1   McNeil   Australia       no  Interactive Data Analysis
2   Ripley          UK       no         Spatial Statistics
3   Ripley          UK       no      Stochastic Simulation
4  Tierney          US       no                  LISP-STAT
5    Tukey          US      yes  Exploratory Data Analysis
6 Venables   Australia       no Modern Applied Statistics ...
```

Figure 2.3: Dialog box to stack `mr` to form `mr2`.

Reshaping

Sometimes spreadsheet data are in a compact format that gives the covariates for each subject followed by all the observations on that subject. Consider the following extract of data frame `mr` from repeated MRI brain measurements

```
Status   Age    V1      V2      V3      V4
     P 23646  45190   50333   55166   56271
    CC 26174  35535   38227   37911   41184
    CC 27723  25691   25712   26144   26398
    CC 27193  30949   29693   29754   30772
    CC 24370  50542   51966   54341   54273
    CC 28359  58591   58803   59435   61292
    CC 25136  45801   45389   47197   47126
    . . . .
```

There are two covariates and up to four measurements on each subject, exported from Excel.

Such data are sometimes said to be in *stacked* form, an unstacking them would give a data frame with variables `Status`, `Age`, `V` and `Replicate`, the latter a factor with levels corresponding to the four columns.

S+Win   The S-PLUS for Windows GUI has operations to perform stacking and un-stacking that call the functions `menuStackColumn` and `menuUnStackColumn`. The dialog box needed for this example is shown in Figure 2.3, which when applied to the seven rows shown above gives

```
> mr2
          V Status   Age Replicate
1 45190       P 23646        V1
2 35535      CC 26174        V1
3 25691      CC 27723        V1
4 30949      CC 27193        V1
5 50542      CC 24370        V1
6 58591      CC 28359        V1
7 45801      CC 25136        V1
8 50333       P 23646        V2
9 38227      CC 26174        V2
    . . . .
```

R has functions `stack` and `unstack` to do the basic work, and `reshape` R
that applies to whole data frames. Here we could use

```
> reshape(mr, idvar = "Person", timevar = "Scan",
    varying = list(c("V1","V2","V3","V4")), direction = "long")
      Status   Age Scan    V1 Person
1.1        P 23646   1 45190      1
2.1       CC 26174   1 35535      2
3.1       CC 27723   1 25691      3
4.1       CC 27193   1 30949      4
5.1       CC 24370   1 50542      5
6.1       CC 28359   1 58591      6
7.1       CC 25136   1 45801      7
   ....
```

2.4 Tables and Cross-Classification

Thus far we have concentrated on data frames, which are the most common
form of data storage in S. Here we look at data manipulations related to cross-
tabulation, using our `quine` data frame for illustration. The study giving rise to
the dataset is described more fully on page 169; the data frame has four factors,
`Sex`, `Eth` (ethnicity—two levels), `Age` (four levels) and `Lrn` (Learner group—
two levels) and a quantitative response, `Days`, the number of days the child in the
sample was away from school in a year.

Cross-tabulation

Sometimes all we need are summary tables. The main function for this purpose is
`table` which returns a cross-tabulation as an array of frequencies (counts). For
example,

```
> attach(quine)
> table(Age)
 F0 F1 F2 F3
 27 46 40 33
> table(Sex, Age)
   F0 F1 F2 F3
F 10 32 19 19
M 17 14 21 14
```

Note that the factor levels become the appropriate `names` or `dimnames` attribute
for the frequency array (and the factor names are printed in R but not in S-PLUS). R
If the arguments given to `table` are not factors they are coerced to factors.

The function `crosstabs` (`xtabs` in R) may also be used. It takes a formula R
and data frame as its first two arguments, so the data frame need not be attached.
A call to `crosstabs` for the same table is

```
> tab <- crosstabs(~ Sex + Age, quine) # xtabs in R
> unclass(tab)
  F0 F1 F2 F3
F 10 32 19 19
M 17 14 21 14
....
```

The print method for objects of class "crosstabs" gives extra information of
no interest to us here, but the object behaves in calculations as a frequency table.

Calculations on cross-classifications

The combination of a vector and a labelling factor or factors is an example of
what is called a *ragged array*, since the group sizes can be irregular. (When the
group sizes are all equal the indexing may be done more efficiently using arrays.)

To calculate the average number of days absent for each age group (used on
page 170) we can use the function tapply, the analogue of lapply and apply
for ragged arrays.

```
> tapply(Days, Age, mean)
    F0     F1     F2     F3
14.852 11.152 21.05 19.606
```

The first argument is the vector for which functions on the groups are required,
the second argument, INDICES, is the factor defining the groups and the third
argument, FUN, is the function to be evaluated on each group. If the function
requires more arguments they may be included as additional arguments to the
function call, as in

```
> tapply(Days, Age, mean, trim = 0.1)
    F0     F1     F2     F3
12.565 9.0789 18.406 18.37
```

for 10% trimmed means.

If the second argument is a *list* of factors the function is applied to each group
of the cross-classification given by the factors. Thus to find the average days
absent for age-by-sex classes we could use

```
> tapply(Days, list(Sex, Age), mean)
      F0     F1     F2     F3
F 18.700 12.969 18.421 14.000
M 12.588  7.000 23.429 27.214
```

To find the standard errors of these we could use an anonymous function as the
third argument, as in

```
> tapply(Days, list(Sex, Age),
         function(x) sqrt(var(x)/length(x)))
      F0     F1     F2     F3
F 4.2086 2.3299 5.3000 2.9409
M 3.7682 1.4181 3.7661 4.5696
```

There is a more complicated example on page 318.

As with table, coercion to factor takes place where necessary.

Frequency tables as data frames

Consider the general problem of taking a set of n factors and constructing the complete n-way frequency table as a data frame, that is, as a frequency vector and a set of n classifying factors. We can illustrate this with the `quine` data frame. First we remove any non-factor components from the data frame.

```
quineFO <- quine[sapply(quine, is.factor)]
```

In S-PLUS the function `table` takes as arguments a series of objects, so we need to construct a suitable call.[19] The function `do.call` takes two arguments: the name of a function (as a character string) and a list. The result is a call to that function with the list supplying the arguments. List names become argument names. Hence we may find the frequency table using

```
tab <- do.call("table", quineFO)
```

The result is a multi-way array of frequencies.

Next we find the classifying factors, which correspond to the indices of this array. A convenient way is to use `expand.grid`.[20] This function takes any number of vectors and generates a data frame consisting of all possible combinations of values, in the correct order to match the elements of a multi-way array with the lengths of the vectors as the index sizes. Argument names become component names in the data frame. Alternatively, `expand.grid` may take a single list whose components are used as individual vector arguments. Hence to find the index vectors for our data frame we may use

```
QuineF <- expand.grid(lapply(quineFO, levels))
```

Finally we put together the frequency vector and classifying factors.

```
> QuineF$Freq <- as.vector(tab)
> QuineF
    Eth Sex Age Lrn Freq
  1   A   F  FO  AL    4
  2   N   F  FO  AL    4
  3   A   M  FO  AL    5
  ....
 30   N   F  F3  SL    0
 31   A   M  F3  SL    0
 32   N   M  F3  SL    0
```

We use `as.vector` to remove all attributes of the table.

---

[19] In R `table(quineFO)` will work.

[20] Which we use in several places in this book; see its index entry.

Chapter 3

The S Language

S is a language for the manipulation of objects. It aims to be both an interactive language (like, for example, a UNIX shell language) and a complete programming language with some convenient object-oriented features. This chapter is intended for reference use by interactive users; Venables and Ripley (2000) covers more aspects of the language for programmers.

3.1 Language Layout

Commands to S are either expressions or assignments. Commands are separated by either a semi-colon, ; , or a newline. A # marks the rest of the line as comments, so comments can be placed at ends of lines.

The S prompt is > unless the command is syntactically incomplete, when the prompt changes to +.[1] The only way to extend a command over more than one line is by ensuring that it *is* syntactically incomplete until the final line.

An expression command is evaluated and (normally) printed. For example,

```
> 1 - pi + exp(1.7)
[1] 3.3324
```

This rule allows any object to be printed by giving its name. Note that pi is the value of π. Giving the name of an object will normally print it or a short summary; this can be done explicitly using the function print, and summary will often give a full description.

An assignment command evaluates an expression and passes the value to a variable but the result is not printed. The recommended assignment symbol[2] is the combination, " <- ", so an assignment in S looks like

```
a <- 6
```

which gives the object a the value 6. To improve readability of your code we strongly recommend that you put at least one space before and after binary operators, especially the assignment symbol. Assignments using the right-pointing

---

[1] These prompts can be altered; see prompt and continue under ?options .

[2] The use of " _ " for assignments is deprecated in both S-PLUS and R, and will be removed as from R 1.8.0. Also, " = " can often be used, but the exceptions are hard to get right in S-PLUS.

combination " -> " are also allowed to make assignments in the opposite direction, but these are never needed and are little used in practice.

An assignment is a special case of an expression with value equal to the value assigned. When this value is itself passed by assignment to another object the result is a multiple assignment, as in

```
b <- a <- 6
```

Multiple assignments are evaluated from right to left, so in this example the value 6 is first passed to a and then to b.

It is useful to remember that the most recently evaluated non-assignment expression in the session is stored as the variable .Last.value[3] and so may be kept by an assignment such as

```
keep <- .Last.value
```

This is also useful if the result of an expression is unexpectedly not printed; just use print(.Last.value).

If you want to both assign a value *and* print it, just enclose the assignment in parentheses, for example,

```
> (z <- 1 - pi + exp(1.7))
[1] 3.3324
```

Executing commands from, or diverting output to, a file

If commands are stored on an external file, say, commands.q in the current directory, they may be executed at any time in an S session with the command

```
source("commands.q")
```

It is often helpful to use the echo argument of source to have the commands echoed. Similarly

```
sink("record.lis")
```

will divert all subsequent output from the session window to an external file record.lis. Function sink can be called repeatedly, with sink() restoring output to the previous diversion, eventually back to the terminal.

Managing S objects

It can be important to understand where S keeps its objects and where it looks for objects on which to operate. The objects that S creates at the interactive level during a session are stored in a workspace (see page 14). On the other hand, objects created at a higher level, such as within a function, are kept in what is known as a *local frame* or *environment* that is only transient, and such objects vanish when the function is exited.

When S looks for an object, it searches through a sequence of places known as the *search path*. Usually the first entry in the search path is the workspace, now called the working *chapter* in S-PLUS. The names of the places currently making up the search path are given by invoking the function search. For example,

---

[3]If the expression consists of a simple name such as x, only, the .Last.value object is not changed.

```
> search()
[1] "/home/ripley/MySwork"   "splus"              "stat"
[4] "data"                   "trellis"            "nlme3"
[7] "main"
```

To get the names of all objects currently held in the first place on the search path, use the command

```
objects()
```

The names of the objects held in any database in the search path can be displayed by giving the `objects` function an argument. For example,

```
objects(2)
```

lists the contents of the database at position 2 of the search path. It is also possible to list selectively by a regular expression (page 53), using the `regexpr.pattern` (S-PLUS) or `pattern` (R) argument of `objects`. Users of S-PLUS under Windows can explore objects graphically; see page 460.

     R
     S+Win

    The databases on the search path can be of two main types. As well as chapters/workspaces, they can also be S lists, usually data frames. The database at position 1 is called the *working database*. A *library section*[4] is a specialized use of a database, discussed in Appendix C.2. Extra chapters, lists or data frames can be added to this list with the `attach` function and removed with the `detach` function. Normally a new database is attached at position 2, and `detach()` removes the entity at position 2, normally the result of the last `attach`. All the higher-numbered databases are moved up or down accordingly. If a list is attached, a copy is used, so any subsequent changes to the original list will not be reflected in the attached copy.

    The function `find(`*object*`)` discovers where an object appears on the search path, perhaps more than once. For example,

```
> find("objects")
[1] "splus"
```

If an object is found more than once in the search path, the first occurrence is used. This can be useful way to override system functions, but more often it occurs by mistake. Use the functions `conflicts` or `masked`[5] to see if this has happened.

    To examine an object which is masked or has a non-standard name, we need to use the function `get`. For example,

```
get("[<-.data.frame", where = 2)   ## pos = 4 in R
```

allows experts to look at indexing for data frames. Function `exists` checks if an object of a given name exists.

    When a command would alter an object that is not on the working database, a copy must be made on the working database first. R does this silently, but S-PLUS does not and will report an error, so a manual copy must be made. Objects are usually altered through assignment with a replacement function, for example (page 154),

---

[4]Or *package* in R usage.
[5]S-PLUS only.

```
> hills <- hills  # only needed in S-PLUS
> hills$ispeed <- hills$time/hills$dist
```

To remove objects permanently from the working database, the function `rm` is used with arguments giving the names of the objects to be discarded, as in

```
rm(x, y, z, ink, junk)
```

If the names of objects to be removed are held in a character vector it may be specified by the named argument `list`. An equivalent form of the preceding command is

```
rm(list = c("x", "y", "z", "ink", "junk"))
```

The function `remove` can be used to remove objects from databases other than the workspace.

3.2   More on S Objects

We saw at the beginning of Chapter 2 that every object has a *class*. It also has a *length* reported by the `length` function,

```
> length(letters)
[1] 26
```

Lists

A list is a vector of other S objects, called *components*. Lists are used to collect together items of different classes. For example, an employee record might be created by

```
Empl <- list(employee = "Anna", spouse = "Fred", children = 3,
             child.ages = c(4, 7, 9))
```

The components of a list are always numbered and may always be referred to as such. If the components were given names (either when created as here or via the `names` function), they may be invoked by name using the $ operator as in

```
> Empl$employee
[1] "Anna"
> Empl$child.ages[2]
[1] 7
```

Names of components may be abbreviated to the minimum number of letters needed to identify them uniquely. Thus `Empl$employee` may be minimally specified as `Empl$e` since it is the only component whose name begins with the letter 'e', but `Empl$children` must be specified as at least `Empl$childr` because of the presence of another component called `Empl$child.ages`. Note that the names of a list are not necessarily unique, when name-matching will give the first occurrence.

Individual components are extracted by the `[[` operator. Here `Empl` is a list of length 4, and the individual components may be referred to as `Empl[[1]]`, `Empl[[2]]`, `Empl[[3]]` and `Empl[[4]]`. We can also use `Empl[["spouse"]]` or even

```
x <- "spouse"; Empl[[x]]
```

It is important to appreciate the difference between [and [[. The [form extracts sub-vectors, so Empl[2] is a list of length one, whereas Empl[[2]] is the component (a character vector of length one).

The function unlist converts a list to an atomic vector:

```
> unlist(Empl)
 employee spouse children child.ages1 child.ages2 child.ages3
  "Anna"   "Fred" "3"        "4"          "7"          "9"
> unlist(Empl, use.names = F)
[1] "Anna" "Fred" "3"     "4"    "7"    "9"
```

which can be useful for a compact printout (as here). (Mixed classes will all be converted to character, giving a character vector.)

Attributes

Most objects[6] can have *attributes*, other objects attached by name to the object. The dimension and dimnames (if any) of an array are attributes:

```
> attributes(myarr)
$dim:
[1] 3 5 2
$dimnames:
$dimnames[[1]]:
[1] "a" "b" "c"
$dimnames[[2]]:
character(0)  ## NULL in R
$dimnames[[3]]:
[1] "(i)"  "(ii)"

> attr(myarr, "dim")
[1] 3 5 2
```

The attributes are a list. Notice the notation for $dimnames[[2]]: this is a zero-length character vector.

Attributes are often used to store ancillary information; we use them to store gradients and Hessians (pages 215 and 442) and probabilities of classification (page 347).

A construct you may see occasionally is

```
z <- structure(x, somename = value)
```

which is a shorthand way of doing

```
z <- x; attr(z, "somename") <- value
```

---

[6] NULL cannot in R.

Concatenation

The concatenate function, c, is used to concatenate vectors, including lists, so

```
Empl <- c(Empl, service = 8)
```

would add a component for years of service.

The function c has a named argument recursive; if this is true the list arguments are unlisted before being joined together. Thus

```
c(list(x = 1:3, a = 3:6), list(y = 8:23, b = c(3, 8, 39)))
```

is a list with four (vector) components, but adding recursive = T gives a vector of length 26. (Try both to see.)

S-PLUS has a function concat that concatenates vectors and omits the names, whereas c keeps all the names (even if this results in duplicates).

Coercion

There is a series of functions named as.xxx that convert to the specified type in the best way possible. For example, as.matrix will convert a numerical data frame to a numerical matrix, and a data frame with any character or factor columns to a character matrix. The function as.character is often useful to generate names and other labels.

Functions is.xxx test if their argument is of the required type. These do not always behave as one might guess; for example, is.vector(powers.of.pi) will be *false* as this tests for a 'pure' vector without any attributes such as names.[7] Similarly, as.vector has the (often useful) side effect of discarding all attributes.

Many of these functions are being superseded by the more general functions[8] as and is, which have as arguments an object and a class.

```
> as(powers.of.pi, "vector")
[1] 0.10132 0.31831 1.00000 3.14159 9.86960
> as(powers.of.pi, "numeric")
[1] 0.10132 0.31831 1.00000 3.14159 9.86960
> is(powers.of.pi, "numeric")
[1] T
> as(powers.of.pi, "character")
[1] "0.101321183642338" "0.318309886183791" "1"
[4] "3.14159265358979"  "9.86960440108936"
> is(powers.of.pi, "vector")
[1] T
> as(powers.of.pi, "integer")
[1] 0 0 1 3 9
> is(mymat, "array")
[1] T
```

Note carefully the last one: mymat does not have class "array", but "matrix" which is a specialized version of "array".

---

[7]R allows names, only.

[8]In package methods in R.

3.3   Arithmetical Expressions

We have seen that a basic unit in S is a vector. Arithmetical operations are performed on numeric (and integer) vectors, element by element. The standard operators + – * / ^ are available, where ^ is the power (or exponentiation) operator (giving x^y).

Vectors may be empty. The expression numeric(0) is both the expression to create an empty numeric vector and the way it is represented when printed. It has length zero. It may be described as "a vector such that if there were any elements in it, they would be numbers!"

Vectors can be complex, and almost all the rules for arithmetical expressions apply equally to complex quantities. A complex number is entered in the form 3.1 + 2.7i, with no space before the i. Functions Re and Im return the real and imaginary parts. Note that complex arithmetic is not used unless explicitly requested, so sqrt(x) for x real and negative produces an error. If the complex square root is desired use sqrt(as.complex(x)) or sqrt(x + 0i).

The recycling rule

The expression y + 2 is a syntactically natural way to add 2 to each element of the vector y, but 2 is a vector of length 1 and y may be a vector of any length. A convention is needed to handle vectors occurring in the same expression but not all of the same length. The value of the expression is a vector with the same length as that of the longest vector occurring in the expression. Shorter vectors are *recycled* as often as need be until they match the length of the longest vector. In particular, a single number is repeated the appropriate number of times. Hence

```
x <- c(10.4, 5.6, 3.1, 6.4, 21.7)
y <- c(x, x)
v <- 2 * x + y + 1
```

generates a new vector v of length 10 constructed by

1. repeating the number 2 five times to match the length of the vector x and multiplying element by element, and
2. adding together, element by element, 2*x repeated twice, y as it stands and 1 repeated ten times.

Fractional recycling is allowed in R, with a warning, but in S-PLUS it is an error.

There is one exception to the recycling rule: normally an operation with a zero-length vector gives a zero-length result.

Some standard S functions

Some examples of standard functions follow.

1. There are several functions to convert to integers; round will normally be preferred, and rounds to the nearest integer. (It can also round to any number of digits in the form round(x, 3). Using a negative number rounds

to a power of 10, so that `round(x, -3)` rounds to thousands.) Each of `trunc`, `floor` and `ceiling` round in a fixed direction, towards zero, down and up, respectively.

2. Other arithmetical operators are `%/%` for integer divide and `%%` for modulo reduction.[9]

3. The common functions are available, including `abs`, `sign`, `log`, `log10`, `sqrt`, `exp`, `sin`, `cos`, `tan`, `acos`, `asin`, `atan`, `cosh`, `sinh` and `tanh` with their usual meanings. Note that the value of each of these is a vector of the same length as its argument. In S-PLUS `logb` is used for 'log to base' whereas in R `log` has a second argument, the base of the logarithms (default e). R also has `log2`.

   Less common functions are `gamma`, `lgamma` ($\log_e \Gamma(x)$) and its derivatives `digamma` and `trigamma`.

4. There are functions `sum` and `prod` to form the sum and product of a whole vector, as well as cumulative versions `cumsum` and `cumprod`.

5. The functions `max(x)` and `min(x)` select the largest and smallest elements of a vector x. The functions `cummax` and `cummin` give cumulative maxima and minima.

6. The functions `pmax( ...)` and `pmin(...)` take an arbitrary number of vector arguments and return the element-by-element maximum or minimum values, respectively. Thus the result is a vector of length that of the longest argument and the recycling rule is used for shorter arguments. For example,

   ```
   xtrunc <- pmax(0, pmin(1, x))
   ```

   is a vector like x but with negative elements replaced by 0 and elements larger than 1 replaced by 1.

7. The function `range(x)` returns `c(min(x), max(x))`. If `range`, `max` or `min` is given several arguments these are first concatenated into a single vector.

8. Two useful statistical functions are `mean(x)` which calculates the sample mean, which is the same as `sum(x)/length(x)`, and `var(x)` which gives the sample variance, `sum((x-mean(x))^2)/(length(x)-1)`.[10]

9. The function `duplicated` produces a logical vector with value T only where a value in its argument has occurred previously and `unique` removes such duplicated values. (These functions also have methods for data frames that operate row-wise.)

---

[9] The result of e1 `%/%` e2 is `floor(e1/e2)` if e2 `!= 0` and 0 if e2 `== 0`. The result of e1 `%%` e2 is e1 - `floor(e1/e2)*e2` if e2`!=0` and e1 otherwise (see Knuth, 1968, §1.2.4). Thus `%/%` and `%%` always satisfy e1 `==` `(e1%/%e2)*e2 + e1%%e2`.

[10] If the argument to `var` is an $n \times p$ matrix the value is a $p \times p$ sample covariance matrix obtained by regarding the rows as sample vectors.

10. Set operations may be done with the functions `union`, `intersect` and `setdiff`, which enact the set operations $A \cup B$, $A \cap B$ and $A \cap \overline{B}$, respectively. Their arguments (and hence values) may be vectors of any mode but, like true sets, they should contain no duplicated values.

Logical expressions

Logical vectors are most often generated by *conditions*. The logical binary operators are `<`, `<=`, `>`, `>=` (which have self-evident meanings), `==` for exact equality and `!=` for exact inequality. If `c1` and `c2` are vector valued logical expressions, `c1 & c2` is their intersection ('and'), `c1 | c2` is their union ('or') and `!c1` is the negation of `c1`. These operations are performed separately on each component with the recycling rule (page 47) applying for short arguments.

The unary operator `!` denotes negation. There is one trap with this in S-PLUS: if used at the beginning of a line it is taken to be a shell escape; extra parentheses can be used to avoid this.   S+

Character vectors may be used in logical comparisons such as `"ann" < "belinda"`, in which case lexicographic ordering is applied using the current collating sequence.

Logical vectors may be used in ordinary arithmetic. They are *coerced* into numeric vectors, false values becoming `0` and true values becoming `1`. For example, assuming the value or values in `sd` are positive

```
N.extreme <- sum(y < ybar - 3*sd | y > ybar + 3*sd)
```

would count the number of elements in `y` that were farther than `3*sd` from `ybar` on either side. The right-hand side can be expressed more concisely as `sum(abs(y-ybar) > 3*sd)`.

The function `xor` computes (element-wise) the exclusive or of its two arguments.

The functions `any` and `all` are useful to collapse a logical vector.

Sometimes one needs to test if two objects are 'the same'. That can be made precise in various ways, and two are provided in S. Logical function `identical` tests for exact equality. Function `all.equal` makes a suitable approximate test (for example, it allows for rounding error in numeric components) and either returns `TRUE` or a character vector describing the difference(s). Thus to test for 'equality' one uses (see page 58 for `&&`)

```
res <- all.equal(obj1, obj2)
if(!(is.logical(res) && res)) warning("objects differ")
```

Operator precedence

The formal precedence of operators is given in Table 3.1. However, as usual it is better to use parentheses to group expressions rather than rely on remembering these rules. They can be found on-line from `help(Syntax)`.

Table 3.1: Precedence of operators, from highest to lowest.

$	list element extraction
@	slot extraction
[[[	vector and list element extraction
^	exponentiation
-	unary minus
:	sequence generation
%% %/% %*%	and other special operators %...%
* /	multiply and divide
+ -	addition, subtraction
< > <= >= == !=	comparison operators
!	logical negation
& | && ||	logical operators (& && above | || in R)
~	formula
<- -> _ =	assignment

Generating regular sequences

There are several ways in S to generate sequences of numbers. For example, 1:30 is the vector c(1, 2, ..., 29, 30). The colon operator has a high precedence within an expression, so 2*1:15 is the vector c(2, 4, 6, ..., 28, 30). Put n <- 10 and compare the sequences 1:n-1 and 1:(n-1).

A construction such as 10:1 may be used to generate a sequence in reverse order.

The function seq is a more general facility for generating sequences. It has five arguments, only some of which may be specified in any one call. The first two arguments, named from and to, if given, specify the beginning and end of the sequence, and if these are the only two arguments the result is the same as the colon operator. That is, seq(2, 10) and seq(from = 2, to = 10) give the same vector as 2:10.

The third and fourth arguments to seq are named by and length, and specify a step size and a length for the sequence. If by is not given, the default by = 1 is used. For example,

```
s3 <- seq(-5, 5, by = 0.2)
s4 <- seq(length = 51, from = -5, by = 0.2)
```

generate in both s3 and s4 the vector $(-5.0, -4.8, -4.6, \ldots, 4.6, 4.8, 5.0)$.

The fifth argument is named along and has a vector as its value. If it is the only argument given it creates a sequence 1, 2, ..., length(*vector*), or the empty sequence if the value is empty. (This makes seq(along = x) preferable to 1:length(x) in most circumstances.) If specified rather than to or length its length determines the length of the result.

A companion function is rep which can be used to repeat an object in various ways. The simplest forms are

```
s5 <- rep(x, times = 5)  # repeat whole vector
s5 <- rep(x, each = 5)   # repeat element-by-element
```

which will put five copies of x end-to-end in s5, or make five consecutive copies of each element.

A times = v argument may specify a vector of the same length as the first argument, x. In this case the elements of v must be non-negative integers, and the result is a vector obtained by repeating each element in x a number of times as specified by the corresponding element of v. Some examples will make the process clear:

```
x <- 1:4         # puts c(1,2,3,4)              into x
i <- rep(2, 4)   # puts c(2,2,2,2)              into i
y <- rep(x, 2)   # puts c(1,2,3,4,1,2,3,4)      into y
z <- rep(x, i)   # puts c(1,1,2,2,3,3,4,4)      into z
w <- rep(x, x)   # puts c(1,2,2,3,3,3,4,4,4,4)  into w
```

As a more useful example, consider a two-way experimental layout with four row classes, three column classes and two observations in each of the twelve cells. The observations themselves are held in a vector y of length 24 with column classes stacked above each other, and row classes in sequence within each column class. Our problem is to generate two indicator vectors of length 24 that will give the row and column class, respectively, of each observation. Since the three column classes are the first, middle and last eight observations each, the column indicator is easy. The row indicator requires two calls to rep:

```
> ( colc <- rep(1:3, each = 8) )
> [1] 1 1 1 1 1 1 1 1 2 2 2 2 2 2 2 2 3 3 3 3 3 3 3 3
> ( rowc <- rep(rep(1:4, each = 2), 3) )
> [1] 1 1 2 2 3 3 4 4 1 1 2 2 3 3 4 4 1 1 2 2 3 3 4 4
```

These can also be generated arithmetically using the ceiling function

```
> 1 + (ceiling(1:24/8) - 1) %% 3 -> colc; colc
> [1] 1 1 1 1 1 1 1 1 2 2 2 2 2 2 2 2 3 3 3 3 3 3 3 3
> 1 + (ceiling(1:24/2) - 1) %% 4 -> rowc; rowc
> [1] 1 1 2 2 3 3 4 4 1 1 2 2 3 3 4 4 1 1 2 2 3 3 4 4
```

In general the expression 1 + (ceiling(1:n/r) - 1) %% m generates a sequence of length n consisting of the numbers 1, 2, ..., m each repeated r times. This is often a useful idiom (for which R has a function gl).

3.4 Character Vector Operations

The form of character vectors can be unexpected and should be carefully appreciated. Unlike say C, they are vectors of character strings, not of characters, and most operations are performed separately on each component.

Note that "" is a legal character string with no characters in it, known as the empty string. This should be contrasted with character(0) which is the empty character vector. As vectors, "" has length 1 and character(0) has length 0.

S follows C conventions in entering character strings, so special characters need to be escaped. Thus \ is entered as \\, and \b \n \r \t (backspace, newline, tab, carriage return) are available, as well as octal values such as \176. These are escaped when printing by `print` but not when output by `cat`.

There are several functions for operating on character vectors. The function `nchar` gives (as a vector) the number of characters in each element of its character vector argument. The function `paste` takes an arbitrary number of arguments, coerces them to strings or character vectors if necessary and joins them, element by element, as character vectors. For example,

```
> paste(c("X", "Y"), 1:4)
[1] "X 1" "Y 2" "X 3" "Y 4"
```

Any short arguments are recycled in the usual way. By default the joined elements are separated by a blank; this may be changed by using the argument `sep`, often the empty string:

```
> paste(c("X", "Y"), 1:4, sep = "")
[1] "X1" "Y2" "X3" "Y4"
```

Another argument, `collapse`, allows the result to be concatenated into a single string. It prescribes another character string to be inserted between the components during concatenation. If it is NULL, the default, or `character(0)`, no such global concatenation takes place. For example,

```
> paste(c("X", "Y"), 1:4, sep = "", collapse = " + ")
[1] "X1 + Y2 + X3 + Y4"
```

Substrings of the strings of a character vector may be extracted (element-by-element) using the `substring` function. It has three arguments

```
substring(text, first, last = 1000000)
```

where `text` is the character vector, `first` is a vector of first character positions to be selected and `last` is a vector of character positions for the last character to be selected. If `first` or `last` are shorter vectors than `text` they are recycled. For example, the dataset `state.name` is a character vector of length 50 containing the names of the states of the United States of America in alphabetic order. To extract the first four letters in the names of the last seven states:

```
> # R: data(state)
> substring(state.name[44:50], 1, 4)
[1] "Utah" "Verm" "Virg" "Wash" "West" "Wisc" "Wyom"
```

Note the use of the index vector [44:50] to select the last seven states.

The function `abbreviate` provides a more general mechanism for generating abbreviations. In this example it gives in S-PLUS (R is slightly different)

```
> as.vector(abbreviate(state.name[44:50]))
[1] "Utah" "Vrmn" "Vrgn" "Wshn" "WsVr" "Wscn" "Wymn"
> as.vector(abbreviate(state.name[44:50], use.classes = F))
[1] "Utah" "Verm" "Virg" "Wash" "WVir" "Wisc" "Wyom"
```

We used `as.vector` to suppress the names attribute that contains the unabbreviated names!

Simple matching is done by the function `match`. This matches (exactly) each element of its first argument against the elements of its second argument, and returns a vector of indices into the second argument or the `nomatch` argument (which defaults to `NA`) if there is no match.

Missing values

S-PLUS has no support for missing values in character vectors; factors should be   S+
used instead. There is a class `"string"` described in Chambers (1998) but not
fully implemented in S-PLUS.

R does have support for missing values in character vectors. The value printed   R
as `NA` *may* represent a missing value, but it can also represent a character string,
e.g. North America. For example,

```
> (x <- c("a", NA, "NA"))
[1] "a" NA   "NA"
> is.na(x)
[1] FALSE   TRUE FALSE
```

R does try to differentiate them where possible either by omitting quotes (as here)
or printing the missing value as `<NA>`.

Regular expressions

Regular expressions are powerful ways to match character patterns familiar to
users of such tools as `sed`, `grep`, `awk` and `perl`. For example, ' . ' matches any
character (use '\ . ' to match ' . ') and ' .* ' matches zero or more occurrences of
any character, that is, any character string. The beginning is matched by ^ and the
end by $.

The function `grep` searches for patterns given by a regular expression in a
vector of character strings, and returns the indices of the strings in which a match
is found.

Function `regexpr` also matches one regular expression to a character vector,
returning more details of the match, the first matching position and the match
length. For example,

```
> grep("na$", state.name)
[1]   3 14 18 26 33 40
> regexpr("na$", state.name)
 [1] -1 -1   6 -1 -1 -1 -1 -1 -1 -1 -1 -1 -1   6 -1 -1 -1   8 -1
[20] -1 -1 -1 -1 -1 -1   6 -1 -1 -1 -1 -1 -1 13 -1 -1 -1 -1 -1
[39] -1 13 -1 -1 -1 -1 -1 -1 -1 -1 -1 -1
attr(, "match.length"):
 [1] -1 -1   2 -1 -1 -1 -1 -1 -1 -1 -1 -1   2 -1 -1 -1   2 -1
[20] -1 -1 -1 -1 -1 -1   2 -1 -1 -1 -1 -1 -1   2 -1 -1 -1 -1 -1
[39] -1   2 -1 -1 -1 -1 -1 -1 -1 -1 -1 -1
> state.name[regexpr("na$", state.name)> 0]
[1] "Arizona"        "Indiana"        "Louisiana"
[4] "Montana"        "North Carolina" "South Carolina"
```

The functions `regMatch` and `regMatchPos` of S-PLUS have a very similar role, but encode the answer somewhat differently. They can match multiple regular expressions, recycling arguments as needed.

3.5 Formatting and Printing

Function `print` does not allow much control over the layout of a report. The function `cat` is similar to `paste` with argument `collapse = ""` in that it coerces its arguments to character strings and concatenates them. However, instead of returning a character string result it prints out the result in the session window or optionally on an external file. For example, to print out today's date on our UNIX system:[11]

```
> d <- date()
> cat("Today's date is:", substring(d, 1, 10),
                          substring(d, 25, 28), "\n")
Today's date is: Sun Jan  6 2002
```

Note that an explicit newline (`"\n"`) is needed. Function `cat`, unlike `print`, interprets escaped characters (see page 52).

Other arguments to `cat` allow the output to be broken into lines of specified length, and optionally labelled:

```
> cat(1, 2, 3, 4, 5, 6, fill = 8, labels = letters)
a 1 2
c 3 4
e 5 6
```

and `fill = T` fills to the current output width.

Function `cat` effectively uses `as(, "character")` to convert an object, and so converts numbers in S-PLUS at full precision, as in

```
> cat(powers.of.pi, "\n")
0.101321183642338 0.318309886183791 1 3.14159265358979 ....
```

Often it is best to manage that conversion ourselves using `format`, which provides the most general way to prepare data for output. It coerces data to character strings in a common format.

```
> format(powers.of.pi)
        -2         -1          0          1          2
  "0.10132"  "0.31831"  "1.00000"  "3.14159"  "9.86960"
> cat(format(powers.of.pi), "\n", sep="  ")
0.1013212  0.3183099  1.0000000  3.1415927  9.8696044
```

For example, the S-PLUS print function `print.summary.lm` for summaries of linear regressions contains the lines

---

[11] The format for others may well differ.

```
cat("\nCoefficients:\n")
print(format(round(x$coef, digits = digits)), quote = F)
cat("\nResidual standard error:",
    format(signif(x$sigma, digits)), "on", rdf,
    "degrees of freedom\n")
cat("Multiple R-Squared:", format(signif(x$r.squared, digits)),
    "\n")
cat("F-statistic:", format(signif(x$fstatistic[1], digits)),
    "on", x$fstatistic[2], "and", x$fstatistic[3],
    "degrees of freedom, the p-value is", format(signif(1 -
    pf(x$fstatistic[1], x$fstatistic[2], x$fstatistic[3]),
    digits)), "\n")
```

Note the use of `signif` and `round` to specify the accuracy required. (For `round` the number of digits is specified, whereas for `signif` it is the number of significant digits.)

There is a tendency to output values such as `0.6870000000000001`, even after rounding to (here) three digits. (Not from `print`, but from `write`, `cat`, `paste`, `as.character` and so on.) Use `format` to avoid this.

By default the accuracy of printed and formatted values is controlled by the `options` parameter `digits`, which defaults to 7.

3.6   Calling Conventions for Functions

Functions may have their arguments *specified* or *unspecified* when the function is defined. (We saw how to write simple functions on page 4.)

When the arguments are unspecified there may be an arbitrary number of them. They are shown as ... when the function is defined or printed. Examples of functions with unspecified arguments include the concatenation function `c(...)` and the parallel maximum and minimum functions `pmax(...)` and `pmin(...)`.

Where the arguments are specified there are two conventions for supplying values for the arguments when the function is called:

1. arguments may be specified in the same order in which they occur in the function definition, in which case the values are supplied in order, and

2. arguments may be specified as `name = value`, when the order in which the arguments appear is irrelevant. The name may be abbreviated providing it partially matches just one named argument.

It is important to note that these two conventions may be mixed. A call to a function may begin with specifying the arguments in positional form but specify some later arguments in the named form. For example, the two calls

```
t.test(x1, y1, var.equal = F, conf.level = 0.99)
t.test(conf.level = 0.99, var.equal = F, x1, y1)
```

are equivalent.

Functions with named arguments also have the option of specifying *default values* for those arguments, in which case if a value is not specified when the function is called the default value is used. For example, the function `t.test` has in S-PLUS an argument list defined as

```
t.test <- function(x, y = NULL, alternative = "two.sided",
        mu = 0, paired = F, var.equal = T, conf.level = 0.95)
```

so that our previous calls can also be specified as

```
t.test(x1, y1, , , , F, 0.99)
```

and in all cases the default values for `alternative`, `mu` and `paired` are used. Using the positional form and omitting values, as in this last example, is rather prone to error, so the named form is preferred except for the first couple of arguments.

Some functions (for example `paste`) have both unspecified and specified arguments, in which case the specified arguments occurring after the ... argument on the definition must be named exactly if they are to be matched at all.

The argument names and any default values for an S function can be found from the on-line help, by printing the function itself or succinctly using the `args` function. For example, in S-PLUS,

```
> args(hist)  ## look at hist.default in R
function(x, nclass = "Sturges", breaks, plot = TRUE, probability
        = FALSE, include.lowest = T, ...,
        xlab = deparse(substitute(x)))
NULL
```

shows the arguments, their order and those default values that are specified for the `hist` function for plotting histograms. (The return value from `args` always ends with `NULL`.) Note that even when no default value is specified the argument itself may not need to be specified. If no value is given for `breaks` when the `hist` function is called, default values are calculated within the function. Unspecified arguments are passed on to a plotting function called from within `hist`.

Functions are considered in much greater detail in Venables and Ripley (2000).

3.7 Model Formulae

Model formulae were introduced into S as a compact way to specify linear models, but have since been adopted for so many diverse purposes that they are now best regarded as an integral part of the S language. The various uses of model formulae all have individual features that are treated in the appropriate chapter, based on the common features described here.

A formula is of the general form

```
response ~ expression
```

where the left-hand side, `response`, may in some uses be absent and the right-hand side, `expression`, is a collection of terms joined by operators usually resembling an arithmetical expression. The meaning of the right-hand side is context dependent. For example, in non-linear regression it is an arithmetical expression and all operators have their usual arithmetical meaning. In linear and generalized linear modelling it specifies the form of the model matrix and the operators have a different meaning. In Trellis graphics it is used to specify the abscissa variable for a plot, but a vertical bar, │, operator is allowed to indicate conditioning variables.

It is conventional (but not quite universal) that a function that interprets a formula also has arguments `weights`, `data`, `subset` and `na.action`. Then the formula is interpreted in the *context* of the argument `data` which must be a list, usually a data frame; the objects named on either side of the formula are looked for first in `data` and then searched for in the usual way.[12] The `weights` and `subset` arguments are also interpreted in the context of the data frame.

We have seen a few formulae in Chapter 1, all for linear models, where the response is the dependent variable and the right-hand side specifies the explanatory variables. We had

```
fm <- lm(y ~ x,  data = dum)
abline(lm(time ~ dist))
fm <-  aov(Speed ~ Run + Expt)
fm0 <- update(fm, . ~ . - Run)
```

Notice that in these cases `+` indicates inclusion, not addition, and `-` exclusion. In most cases we had already attached the data frame, so we did not specify it via a data argument. The function `update` is a very useful way to change the call to functions using model formulae; it reissues the call having updated the formula (and any other arguments specified when it is called). The formula term " . " has a special meaning in a call to `update`; it means 'what is there already' and may be used on either side of the `~`.

It is implicit in this description that the objects referred to in the formula are of the same length, or constants that can be replicated to that length; they should be thought of as all being measured on the same set of cases. Other arguments allow that set of cases to be altered; `subset` is an expression evaluated in the context of `data` that should evaluate to a valid indexing vector (of types 1, 2 or 4 on pages 27 and 28). The `na.action` argument specifies what is to be done when missing values are found by specifying a function to be applied to the data frame of all the data needed to process the formula. The default action in S-PLUS is usually `na.fail`, which reports an error and stops, but some functions have more accommodating defaults. (R defaults to `na.omit` which drops every case   R containing a missing value.)

Further details of model formulae are given in later chapters. Many of these involve special handling of factors and functions appearing on the right-hand side of a formula.

---

[12] For S-PLUS this as described in Section 3.1. R has been experimenting with looking in the *environment* of `formula` which usually means starting the search where `formula` was defined.

3.8   Control Structures

Control structures are the commands that make decisions or execute loops.

The `if` statement has the form

```
if (condition)   true.branch    else    false.branch
```

First the expression `condition` is evaluated. A `NA` condition is an error. If the result is true (or numeric and non-zero) the value of the `if` statement is that of the expression `true.branch`, otherwise that of the expression `false.branch`. The `else` part is optional and omitting it is equivalent to using "`else NULL`". If `condition` has a vector value only the first component is used and a warning is issued. The `if` function can be extended over several lines,[13] and the statements may be compound statements enclosed in braces `{ }`.

Two additional logical operators, `&&` and `||`, are useful with `if` statements. Unlike `&` and `|`, which operate component-wise on vectors, these operate on scalar logical expressions. With `&&` the right-hand expression is only evaluated if the left-hand one is true, and with `||` only if it is false. This conditional evaluation property can be used as a safety feature, as on page 49.

We saw the vector function `ifelse` on page 30. That does evaluate both its arguments.

Loops: The `for`, `while` and `repeat` statements

A `for` loop allows a statement to be iterated as a variable assumes values in a specified sequence. The statement has the form

```
for(variable in sequence) statement
```

where `in` is a keyword, `variable` is the loop variable and `sequence` is the vector of values it assumes as the loop proceeds. This is often of the form `1:10` or `seq(along = x)` but it may be a list, in which case `variable` assumes the value of each component in turn. The `statement` part will often be a grouped statement and hence enclosed within braces, `{ }`.

The `while` and `repeat` loops do not make use of a loop variable:

```
while (condition) statement
repeat statement
```

In both cases the commands in the body of the loop are repeated. For a `while` loop the normal exit occurs when `condition` becomes false; the `repeat` statement continues indefinitely unless exited by a `break` statement.

A `next` statement within the body of a `for`, `while` or `repeat` loop causes a jump to the beginning of the next iteration. The `break` statement causes an immediate exit from the loop.

---

[13] A little care is needed when entering `if` ... `else` statements to ensure that the input is not syntactically complete before the `else` clause, and braces can help with achieving this.

A single-parameter maximum-likelihood example

For a simple example with a statistical context we estimate the parameter λ of the zero-truncated Poisson distribution by maximum likelihood. The probability distribution is specified by

$$\Pr(Y = y) = \frac{e^{-\lambda} \lambda^y}{(1 - e^{-\lambda}) \, y!} \qquad y = 1, 2, \ldots$$

and corresponds to observing only non-zero values of a Poisson count. The mean is $E(Y) = \lambda/(1 - e^{-\lambda})$. The maximum likelihood estimate $\hat{\lambda}$ is found by equating the sample mean to its expectation $\bar{y} = \hat{\lambda}/(1 - e^{-\hat{\lambda}})$. If this equation is written as $\hat{\lambda} = \bar{y} \, (1 - e^{-\hat{\lambda}})$, Newton's method leads to the iteration scheme

$$\hat{\lambda}_{m+1} = \hat{\lambda}_m - \frac{\hat{\lambda}_m - \bar{y} \, (1 - e^{-\hat{\lambda}_m})}{1 - \bar{y} \, e^{-\hat{\lambda}_m}}$$

First we generate our artificial sample from a distribution with $\lambda = 1$.

```
> yp <- rpois(50, lambda = 1) # full Poisson sample of size 50
> table(yp)
  0  1  2 3 5
 21 12 14 2 1
> y <- yp[yp > 0]              # truncate the zeros; n = 29
```

We use a termination condition based both on convergence of the process and an iteration count limit, for safety. An obvious starting value is $\hat{\lambda}_0 = \bar{y}$.

```
> ybar <- mean(y); ybar
[1] 1.7586
> lam <- ybar
> it <- 0                          # iteration count
> del <- 1                         # iterative adjustment
> while (abs(del) > 0.0001 && (it <- it + 1) < 10) {
      del <- (lam - ybar*(1 - exp(-lam)))/(1 - ybar*exp(-lam))
      lam <- lam - del
      cat(it, lam, "\n") }
1 1.32394312696735
2 1.26142504977282
3 1.25956434178259
4 1.25956261931933
```

To generate output from a loop in progress, an explicit call to a function such as `print` or `cat` has to be used. For tracing output `cat` is usually convenient since it can combine several items. Numbers are coerced to character in full precision; using `format(lam)` in place of `lam` is the simplest way to reduce the number of significant digits to the `options` default.

3.9   Array and Matrix Operations

Arrays may be used in ordinary arithmetic expressions and the result is an array
formed by element-by-element operations on the data vector. The `dim` attributes
of operands generally need to be the same, and this becomes the dimension vector
of the result. So if `A`, `B` and `C` are all arrays of the same dimensions,

```
D <- 2*A*B + C + 1
```

makes `D` a similar array with its data vector the result of the evident element-by-
element operations. The precise rules concerning mixed array and vector calculations are complex (Venables and Ripley, 2000, p. 27).

Elementary matrix operations

We have seen that a matrix is merely a data vector with a `dim` attribute specifying a double index. However, S contains many operators and functions for
matrices; for example `t(X)` is the transpose function. The functions[14] `nrow(A)`
and `ncol(A)` give the number of rows and columns in the matrix `A`. There are
functions `row` and `col` that can be applied to matrices to produce a matrix of
the same size filled with the row or column number. Thus to produce the upper
triangle of a square matrix `A` we can use

```
A[col(A) >= row(A)]
```

This uses a logical vector index and so returns the upper triangle in column-major
order. For the lower triangle we can use `<=` or the function `lower.tri`. A few S
functions want the lower triangle of a symmetric matrix in row-major order; note
that this is the upper triangle in column-major order.

Matrices can be built up from other vectors and matrices by the functions
`cbind` and `rbind`; see page 31.

The operator `%*%` is used for matrix multiplication. Vectors that occur in
matrix multiplications are promoted either to row or to column vectors, whichever
is multiplicatively coherent. Note that if `A` and `B` are square matrices of the same
size, then `A * B` is the matrix of element-by-element products whereas `A %*% B`
is the matrix product. If `x` is a vector, then

```
x %*% A %*% x
```

is a quadratic form $x^T A x$, where x is the column vector and ^{T} denotes transpose.

Note that `x %*% x` seems to be ambiguous, as it could mean either $x^T x$ or
$x x^T$. A more precise definition of `%*%` is that of an inner product rather than a
matrix product, so in this case $x^T x$ is the result. (For $x x^T$ use `x %o% x`; see
below.)

The function `crossprod` forms 'crossproducts', meaning that

```
XT.y <- crossprod(X, y)
```

---

[14]Note that the names are singular; it is all too easy to write `nrows` !

calculates $X^T y$. This matrix could be calculated as `t(X) %*% y` but using `crossprod` is more efficient. If the second argument is omitted it is taken to be the same as the first. Thus `crossprod(X)` calculates the matrix $X^T X$.

An important operation on arrays is the *outer product*. If `a` and `b` are two numeric arrays, their outer product is an array whose dimension vector is obtained by concatenating their two dimension vectors (order is important), and whose data vector is obtained by forming all possible products of elements of the data vector of `a` with those of `b`. The outer product is formed by the operator `%o%`:

```
ab <- a %o% b
```

or by the function `outer`:

```
ab <- outer(a, b, "*")
ab <- outer(a, b)        # as "*" is the default.
```

Multiplication may be replaced by an arbitrary function of two variables (or its name as a character string). For example, if we wished to evaluate the function

$$f(x, y) = \frac{\cos(y)}{1 + x^2}$$

over a regular grid of values with x- and y-coordinates defined by the S vectors `x` and `y`, respectively, we could use

```
z <- outer(x, y, function(x, y) cos(y)/(1 + x^2))
```

using an anonymous function.

The function `diag` either creates a diagonal matrix from a vector argument or extracts as a vector the diagonal of a matrix argument. Used on the assignment side of an expression it allows the diagonal of a matrix to be replaced.[15] For example, to form a covariance matrix in multinomial fitting we could use

```
> p <- dbinom(0:4, size = 4, prob = 1/3)  # an example
> CC <- -(p %o% p)
> diag(CC) <- p + diag(CC)
> structure(3^8 * CC, dimnames = list(0:4, 0:4))  # convenience
       0    1     2     3    4
0   1040 -512  -384  -128  -16
1   -512 1568  -768  -256  -32
2   -384 -768  1368  -192  -24
3   -128 -256  -192   584   -8
4    -16  -32   -24    -8   80
```

In addition `diag(n)` for a positive integer `n` generates an $n \times n$ identity matrix. This is an exception to the behaviour for vector arguments; it is safer to use `diag(x, length(x))` which will give a diagonal matrix with diagonal `x` for a vector of any length, even one.

---

[15] This is one of the few places where the recycling rule is disabled; the replacement must be a scalar or of the correct length.

More functions operating on matrices

The standard operations of linear algebra are either available as functions or can easily be programmed, making S a flexible matrix manipulation language (if rather slower than specialized matrix languages).

The function `solve` inverts matrices and solves systems of linear equations; `solve(A)` inverts A and `solve(A, b)` solves A %*% x = b. (If the system is over-determined, the least-squares fit is found, but matrices of less than full rank give an error.)

The function `chol` returns the Choleski decomposition $A = U^T U$ of a non-negative definite symmetric matrix. (Note that there is another convention, in which the lower-triangular form $A = LL^T$ with $L = U^T$ is used.) Function `backsolve` solves upper triangular systems of matrices, and is often used in conjunction with `chol`. (R has `forwardsolve`, the analogue for lower-triangular matrices.)

Eigenvalues and eigenvectors

The function `eigen` calculates the eigenvalues and eigenvectors of a square matrix. The result is a list of two components, `values` and `vectors`. If we need only the eigenvalues we can use:

```
eigen(Sm, only.values = T)$values
```

Real symmetric matrices have real eigenvalues, and the calculation for this case can be much simpler and more stable. Argument `symmetric` may be used to specify whether a matrix is (to be regarded as) symmetric. The default value is T if the matrix exactly equals its transpose, otherwise F.

Singular value decomposition and generalized inverses

An $n \times p$ matrix X has a *singular value decomposition* (SVD) of the form

$$X = U\Lambda V^T$$

where U and V are $n \times \min(n, p)$ and $p \times \min(n, p)$ matrices of orthonormal columns, and Λ is a diagonal matrix. Conventionally the diagonal elements of Λ are ordered in decreasing order; the number of non-zero elements is the rank of X. A proof of its existence can be found in Golub and Van Loan (1989), and a discussion of the *statistical* value of SVDs in Thisted (1988), Gentle (1998) and Monahan (2001).

The function `svd` takes a matrix argument M and calculates the singular value decomposition. The components of the result are u and v, the orthonormal matrices and d, a vector of singular values. If either U or V is not required its calculation can be avoided by the argument nu = 0 or nv = 0.

The QR decomposition

A faster decomposition to calculate than the SVD is the QR decomposition, defined as

$$M = Q R$$

where, if M is $n \times p$, Q is an $n \times n$ matrix of orthonormal columns (that is, an orthogonal matrix) and R is an $n \times p$ matrix with zero elements apart from the first p rows that form an upper triangular matrix (see Golub and Van Loan, 1989, §5.2). The function qr(M) implements the algorithm detailed in Golub & Van Loan, and hence returns the result in a somewhat inconvenient form to use directly. The result is a list that can be used by other tools. For example,

```
M.qr <- qr(M)              # QR decomposition
Q <- qr.Q(M.qr)            # Extract  a Q (n x p) matrix
R <- qr.R(M.qr)            # Extract an R (p x p) matrix
y.res <- qr.resid(M.qr, y) # Project onto error space
```

The last command finds the residual vector after projecting the vector y onto the column space of M. Other tools that use the result of qr include qr.fitted for fitted values and qr.coef for regression coefficients.

Note that by default qr.R only extracts the first p rows of the matrix R and qr.Q only the first p columns of Q, which form an orthonormal basis for the column space of M if M is of maximal rank. To find the complete form of Q we need to call qr.Q with an extra argument complete = T. The columns of Q beyond the rth, where $r \leqslant p$ is the rank of M, form an orthonormal basis for the null space or *kernel* of M, which is occasionally useful for computations. A simple function to extract it is

```
Null <- function(M) {
   tmp <- qr(M)
   set <- if(tmp$rank == 0) 1:ncol(M) else -(1:tmp$rank)
   qr.Q(tmp, complete = T)[, set, drop = F]
}
```

Determinant and trace

There are several ways to write determinant functions. Often it is known in advance that a determinant will be non-negative or that its sign is not needed, in which case methods to calculate the absolute value of the determinant suffice. In this case it may be calculated as the product of the singular values, or slightly faster but possibly less accurately from the QR-decomposition as

```
absdet <- function(M) abs(prod(diag(qr(M)$qr)))
```

If the sign is unknown and important, the determinant may be calculated as the product of the eigenvalues. These will in general be complex and the result may have complex roundoff error even though the exact result is known to be real, so a simple function to perform the calculation is

```
det <- function(M) Re(prod(eigen(M, only.values = T)$values))
```

(R has a det function covering both methods.)   R

The following trace function is so simple the only reason for having it might be to make code using it more readable.

```
tr <- function(M) sum(diag(M))
```

More linear algebra

An S environment will never be competitive on speed with specialized matrix-manipulation systems, but for most statistical problems linear algebra is a small part of the task. Where it is not, more advanced facilities are available in some environments.

S-PLUS has a library section Matrix on some platforms[16] and R has a similar contributed package[17] Matrix on CRAN. These use objects of class "Matrix" to represent rectangular matrices with specializations to triangular, symmetric/Hermitian, orthogonal, diagonal matrices and so on. Functions are provided for determinant (det), norm (norm) and condition number (rcond) as well as methods for %*%, eigen, solve, svd and t.

R   R has functions La.eigen and La.svd based like Matrix on LAPACK routines (Anderson and ten others, 1999). These are stabler and often much faster than eigen and svd. On many platforms R can be built to take advantage of enhanced BLAS routines which can speed up linear algebra considerably.

Vectorized calculations

Users coming to S from other languages are often slow to take advantage of the power of S to do vectorized calculations, that is, calculations that operate on entire vectors rather than on individual components in sequence. This often leads to unnecessary loops. For example, consider calculating the Pearson chi-squared statistic for testing independence in a two-way contingency table. This is defined as

$$X_P^2 = \sum_{i=1}^{r}\sum_{j=1}^{s} \frac{(f_{ij} - e_{ij})^2}{e_{ij}}, \qquad e_{ij} = \frac{f_{i.}f_{.j}}{f_{..}}$$

Two nested for loops may seem to be necessary, but in fact no explicit loops are needed. If the frequencies f_{ij} are held as a matrix the most efficient calculation in S uses matrix operations:

```
fi. <- f %*% rep(1, ncol(f))
f.j <- rep(1, nrow(f)) %*% f
e <- (fi. %*% f.j)/sum(fi.)
X2p <- sum((f - e)^2/e)
```

Explicit loops in S should be regarded as potentially expensive in time and memory use and ways of avoiding them should be considered. (Note that this will be impossible with genuinely iterative calculations such as our Newton scheme on page 59.)

---

[16] But not on Windows.
[17] By Douglas Bates and Saikat DebRoy; not for classic MacOS.

The functions apply *and* sweep

The function apply allows functions to operate on an array using sections successively. For example, consider the dataset iris [18] which is a $50 \times 4 \times 3$ array of four observations on 50 specimens of each of three species. Suppose we want the means for each variable by species; we can use apply.

The arguments of apply are

1. the name of the array, X;

2. an integer vector MARGIN giving the indices defining the sections of the array to which the function is to be separately applied. It is helpful to note that if the function applied has a scalar result, the result of apply is an array with dim(X)[MARGIN] as its dimension vector;

3. the function, or the name of the function FUN to be applied separately to each section;

4. any additional arguments needed by the function as it is applied to each section.

Thus we need to use

```
> apply(iris, c(2, 3), mean)
          Setosa Versicolor Virginica
Sepal L.   5.006     5.936     6.588
Sepal W.   3.428     2.770     2.974
Petal L.   1.462     4.260     5.552
Petal W.   0.246     1.326     2.026
> apply(iris, c(2, 3), mean, trim = 0.1)
          Setosa Versicolor Virginica
Sepal L. 5.0025     5.9375    6.5725
Sepal W. 3.4150     2.7800    2.9625
Petal L. 1.4600     4.2925    5.5100
Petal W. 0.2375     1.3250    2.0325
```

where we also show how arguments can be passed to the function, in this case to give a trimmed mean. If we want the overall means we can use

```
> apply(iris, 2, mean)
 Sepal L. Sepal W. Petal L. Petal W.
   5.8433   3.0573    3.758   1.1993
```

Note how dimensions have been dropped to give a vector. If the result of FUN is itself a vector of length d, say, then the result of apply is an array with dimension vector c(d, dim(X)[MARGIN]), with single-element dimensions dropped. Also note that matrix results are reduced to vectors; if we ask for the covariance matrix for each species,

```
ir.var <- apply(iris, 3, var)
```

we get a 16×3 matrix. We can add back the correct dimensions, but in so doing we lose the dimnames. We can restore both by

---

[18] iris3 in R, so use data(iris3) and the appropriate substitutions below.

```
ir.var <- array(ir.var, dim = dim(iris)[c(2, 2, 3)],
                dimnames = dimnames(iris)[c(2, 2, 3)])
```

The function `apply` is often useful to replace explicit[19] loops. Note too that for *linear* computations it is rather inefficient. We can form the means by matrix multiplication:

```
> matrix(rep(1/50, 50) %*% matrix(iris, nrow = 50), nrow = 4,
          dimnames = dimnames(iris)[-1])
          a  Setosa Versicolor Virginica
Sepal L.  5.006     5.936      6.588
Sepal W.  3.428     2.770      2.974
Petal L.  1.462     4.260      5.552
Petal W.  0.246     1.326      2.026
```

which will be very much faster on larger examples, but is much less transparent. We can also make use of functions such as `colMeans` and `rowMeans` (and in S+ S-PLUS `colVars`, `rowVars` and `colStdev`) if these will do what we want. For example,

```
colMeans(iris)
```

averages across the first dimension.

The function `aperm` is often useful with array/matrix arithmetic of this sort. It permutes the indices, so that `aperm(iris, c(2, 3, 1))` is a $4 \times 3 \times 50$ array. (Note that the matrix transpose operation is a special case.) We can get the overall means of each measurement by

```
colMeans(aperm(iris, c(1, 3, 2)), dims = 2)
```

Function `sweep`

The functions `apply` and `sweep` are often used together. For example, having found the means of the `iris` data, we may want to remove them by subtraction or perhaps division. We can use sweep in each case:

```
ir.means <- colMeans(iris)
sweep(iris, c(2, 3), ir.means)
log(sweep(iris, c(2, 3), ir.means, "/"))
```

Of course, we could have subtracted the log means in the second case.

3.10   Introduction to Classes and Methods

The primary purpose of the S programming environment is to construct and manipulate objects. These objects may be fairly simple, such as numeric vectors, factors, arrays or data frames, or reasonably complex such as an object conveying the results of a model-fitting process. The manipulations fall naturally into broad categories such as plotting, printing, summarizing and so forth. They may also

---

[19]There is an internal loop in R.

involve several objects, for example performing an arithmetic operation on two objects to construct a third.

Since S is intended to be an extensible environment new kinds of object are designed by users to fill new needs, but it will usually be convenient to manipulate them using familiar functions such as `plot`, `print` and `summary`. For the new kind of object the standard manipulations will usually have an obvious purpose, even though the precise action required differs at least in detail from any previous action in this category.

Consider the `summary` function. If a user designs a new kind of object called, say, a "`newfit`" object, it will often be useful to make available a method of summarizing such objects so that the important features are easy to appreciate and the less important details are suppressed. One way of doing this is to write a new function, say `summary.newfit`, which could be used to perform the particular kind of printing action needed. If `myobj` is a particular `newfit` object it could then be summarized using

```
> summary.newfit(myobj)
```

We could also write functions with names `plot.newfit`, `residuals.newfit`, `coefficients.newfit` for the particular actions appropriate for `newfit` objects for plotting, summarizing, extracting residuals, extracting coefficients and so on. It would be most convenient if the user could just type `summary(myobj)`, and this is one important idea behind *object-oriented programming*. To make it work we need to have a standard method by which the evaluator may recognise the different classes of object being presented to it. In S this is done by giving the object a `class`. There is a `class` replacement function available to set the class, which would normally be done when the object was created. For example,

```
> class(myobj) <- "newfit"
```

Functions like `summary`, `plot` and `residuals` are called *generic* functions. They have the property of adapting their action to match the class of object presented to them. The specific implementations such as `summary.lm` are known as *method* functions. It may not be completely clear if a function *is* generic, but S-PLUS has the test `isGeneric`.

At this point we need to emphasize that current S-PLUS, earlier versions of S-PLUS and R handle classes and generic functions internally quite differently, and the details are given in Venables and Ripley (2000). However, there is fairly extensive backwards compatibility, and most of S-PLUS uses the compatibility features. You may notice when listing functions that many classes in S-PLUS are set by `oldClass` not `class`.

Method dispatch

Generic functions use the class of their first few arguments to decide which method function to use. In the simple cases the generic function will consist of a call to UseMethod as in

```
> summary
function(object, ...) UseMethod("summary")
```

Then a call to `methods` will list all the method functions currently available, for example

```
> methods(summary)
        splus              splus                  splus
 "summary.agnes"  "summary.aov"  "summary.aovlist"
             splus                splus
 "summary.bootstrap"  "summary.censorReg"
                   splus                splus
 "summary.censorRegList"  "summary.clara"
                   splus                       main
 "summary.compare.fits"  "summary.connection"
  ....
```

in S-PLUS 6.0. The method for the closest matching class is used; the details of 'closest matching' are complex and given in Venables and Ripley (2000). However, the underlying idea is simple. Objects from linear models have class `"lm"`, those from generalized linear models class `"glm"` and those from analysis of variance models class `"aov"`. Each of the last two classes is defined to *inherit* from class `"lm"`, so if there is no specific method function for the class, that for `"lm"` will be tried. This can greatly reduce the number of method functions needed. Furthermore, method functions can build on those from simpler classes; for example, `predict.glm` works by manipulating the results of a call to `predict.lm`.

S+   Function `plot` in S-PLUS works in both the current class system and in back-compatibility mode. To see all its methods one needs

```
> showMethods("plot")
     Database          x          y
[1,] "splus"   "ANY"          "ANY"
[2,] "splus"   "ANY"          "missing"
[3,] "splus"   "timeSeries"   "ANY"
[4,] "splus"   "signalSeries" "ANY"
> methods("plot")
        nlme3              nlme3        splus          splus
 "plot.ACF"  "plot.Variogram"  "plot.acf"  "plot.agnes"

          splus              nlme3              splus
 "plot.arima"  "plot.augPred"  "plot.censorReg"
  ....
```

Method dispatch is very general; the meaning of arithmetic and logical operators such as $-$ and $\&$ may depend on the classes of the operands.

Chapter 4

Graphics

Both S-PLUS and R provide comprehensive graphics facilities for static two-dimensional plots, from simple facilities for producing common diagnostic plots by plot(*object*) to fine control over publication-quality graphs. In consequence, the number of graphics parameters is huge. In this chapter, we build up the complexity gradually. Most readers will not need the material in Section 4.4, and indeed the material there is not used elsewhere in this book. However, we *have* needed to make use of it, especially in matching existing graphical styles.

Some graphical ideas are best explored in their statistical context, so that, for example, histograms are covered in Chapter 5, survival curves in Chapter 13, biplots in Chapter 11 and time-series graphics in Chapter 14. Table 4.1 gives an overview of the high-level graphics commands with page references.

There are many books on graphical design. Cleveland (1993) discusses most of the methods of this chapter and the detailed design choices (such as the aspect ratios of plots and the presence of grids) that can affect the perception of graphical displays. As these are to some extent a matter of personal preference and this is also a guide to S, we have kept to the default choices. Spence (2001) and Wilkinson (1999) and the classics of Tufte (1983, 1990, 1997) discuss the visual exploration of data.

Trellis graphics (Becker *et al.*, 1996) are a later addition to S with a somewhat different style and philosophy to the basic plotting functions. We describe the basic functions first, then the Trellis functions in Section 4.5. R has a variant on Trellis in its lattice package. The Windows version of S-PLUS has a very different (and less powerful) style of object-oriented editable graphics which we do not cover. One feature we do find useful is the ability to interactively change the viewpoint in perspective plots (see page 422). The rgl package[1] for R under Windows provides similar facilities.

There are quite a few small differences in the R graphics model, and the description here tries to be completely accurate only for S-PLUS 6.

---

[1] Available at http://www.stats.uwo.ca/faculty/murdoch/software/.

Table 4.1: High-level plotting functions. Page references are given to the most complete description in the text. Those marked by † have alternatives in Trellis.

Function	Page	Description
abline	74	Add lines to the current plot in slope-intercept form.
axis	80	Add an axis to the plot.
barplot†	72	Bar graphs.
biplot	312	Represent rows and columns of a data matrix.
brush spin	75	Dynamic graphics. not R.
contour†	76	Contour plot. The Trellis equivalent is contourplot.
dotchart†		Produce a dot chart.
eqscplot	75	Plot with geometrically equal scales (our library).
faces		Chernoff's faces plot of multivariate data.
frame	78	Advance to next figure region.
hist	112	Histograms. We prefer our function truehist.
hist2d	130	Two-dimensional histogram calculations.
identify locator	80	Interact with an existing plot.
image† image.legend	76	High-density image plot functions. The Trellis version is levelplot.
interaction.plot		Interaction plot for a two-factor experiment.
legend	81	Add a legend to the current plot.
matplot	88	Multiple plots specified by the columns of a matrix.
mtext	81	Add text in the margins.
pairs†	75	All pairwise plots between multiple variables. The Trellis version is splom.
par	83	Set or ask about graphics parameters.
persp† perspp persp.setup	76	Three-dimensional perspective plot functions. Similar Trellis functions are called wireframe and cloud.
pie†		Produce a pie chart.
plot		Generic plotting function.
polygon		Add polygon(s) to the present plot, possibly filled.
points lines	73	Add points or lines to the current plot.
qqplot qqnorm	108	Quantile-quantile and normal Q-Q plots.
rect		(R only) Add rectangles, possibly filled.
scatter.smooth	230	Scatterplot with a smooth curve.
segments arrows	88	Draw line segments or arrows on the current plot.
stars		Star plots of multivariate data.
symbols		Draw variable-sized symbols on a plot.
text	73	Add text symbols to the current plot.
title	79	Add title(s).

Table 4.2: Some of the graphical devices available.

S-PLUS:

`motif`	UNIX: X11–windows systems.
`graphsheet`	Windows, screen, printer, bitmaps.
`win.printer`	Windows, a wrapper for a `graphsheet`.
`postscript`	PostScript printers.
`hplj`	UNIX: Hewlett-Packard LaserJet printers.
`hpgl`	Hewlett-Packard HP-GL plotters.
`pdf.graph`	Adobe's PDF format.
`wmf.graph`	Windows metafiles.
`java.graph`	Java device.

R:

`X11`	UNIX: X11–windows systems.
`windows`	Windows, screen, printer, metafiles.
`macintosh`	classic MacOS screen device.
`postscript`	PostScript printers.
`pdf`	PDF files.
`xfig`	files for XFig.
`png`	PNG bitmap graphics.
`jpeg`	JPEG bitmap graphics.
`bitmap`	several bitmap formats *via* GhostScript.

4.1 Graphics Devices

Before any plotting commands can be used, a graphics device must be opened to receive graphical output. Most commonly this is a window on the screen of a workstation or a printer. A list of supported devices on the current hardware with some indication of their capabilities is available from the on-line help system by ?Devices. (Note the capital letter.)

A graphics device is opened by giving the command in Table 4.2, possibly with parameters giving the size and position of the window; for example, using S-PLUS on UNIX,

```
motif("-geometry 600x400-0+0")
```

opens a small graphics window initially positioned in the top right-hand corner of the screen. All current S environments will automatically open a graphics device if one is needed, but we often choose to open the device ourselves and so take advantage of the ability to customize it.

To make a device request permission before each new plot to clear the screen use either par(ask = T) (which affects just the current device) or dev.ask(ask = T) (not R: applies to every device). R

All open graphics devices may be closed using graphics.off(); quitting the S session does this automatically.

UK deaths from lung disease

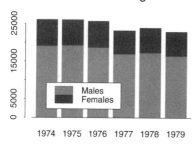
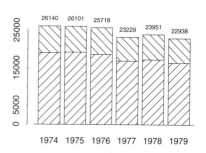

Figure 4.1: Two different styles of bar chart showing the annual UK deaths from certain lung diseases. In each case the lower block is for males, the upper block for females.

It is possible to have several graphical devices open at once. By default the most recently opened one is used, but `dev.set` can be used to change the current device (by number). The function `dev.list` lists currently active devices, and `dev.off` closes the current device, or one specified by number. There are also commands `dev.cur`, `dev.next` and `dev.prev` which return the number of the current, next or previous device on the list. The `dev.copy` function copies the current plot to the specified device (by default the next device on the list).

Note that for some devices little or no output will appear on a file until `dev.off` or `graphics.off` is called.

Many of the graphics devices on windowing systems have menus of choices, for example, to make hardcopies and to alter the colour scheme in use. The S-PLUS `motif` device has a Copy option on its Graph menu that allows a (smaller) copy of the current plot to be copied to a new window, perhaps for comparison with later plots. (The copy window can be dismissed by the Delete item on its Graph menu.)

There are some special considerations for users of `graphsheet` devices on S-PLUS for Windows: see page 451.

4.2 Basic Plotting Functions

The function `plot` is a generic function that, when applied to many types of S objects, will give one or more plots. Many of the plots appropriate to univariate data such as boxplots and histograms are considered in Chapter 5.

Bar charts

The function to display barcharts is `barplot`. This has many options (described in the on-line help), but some simple uses are shown in Figure 4.1. (Many of the details are covered in Section 4.3.)

```
# R: data(mdeaths); data(fdeaths); library(ts)
lung.deaths <- aggregate(ts.union(mdeaths, fdeaths), 1)
```

```
barplot(t(lung.deaths), names = dimnames(lung.deaths)[[1]],
        main = "UK deaths from lung disease")
legend(locator(1), c("Males", "Females"), fill = c(2, 3))
loc <- barplot(t(lung.deaths), names = dimnames(lung.deaths)[[1]],
               angle = c(45, 135), density = 10, col = 1)
total <- rowSums(lung.deaths)
text(loc, total + par("cxy")[2], total, cex = 0.7) #R: xpd = T
```

Line and scatterplots

The default plot function takes arguments `x` and `y`, vectors of the same length, or a matrix with two columns, or a list (or data frame) with components `x` and `y` and produces a simple scatterplot. The axes, scales, titles and plotting symbols are all chosen automatically, but can be overridden with additional graphical parameters that can be included as named arguments in the call. The most commonly used ones are:

`type = "c"`	Type of plot desired. Values for c are: p for points only (the default), l for lines only, b for both points and lines (the lines miss the points), s, S for step functions (s specifies the level of the step at the left end, S at the right end), o for overlaid points and lines, h for high-density vertical line plotting, and n for no plotting (but axes are still found and set).
`axes = L`	If F all axes are suppressed (default T, axes are automatically constructed).
`xlab = "string"` `ylab = "string"`	Give labels for the x- and/or y-axes (default: the names, including suffices, of the x- and y-coordinate vectors).
`sub = "string"` `main = "string"`	sub specifies a title to appear under the x-axis label and main a title for the top of the plot in larger letters (default: both empty).
`xlim = c(lo ,hi)` `ylim = c(lo, hi)`	Approximate minimum and maximum values for x- and/or y-axis settings. These values are normally automatically rounded to make them 'pretty' for axis labelling.

The functions `points`, `lines`, `text` and `abline` can be used to add to a plot, possibly one created with `type = "n"`. Brief summaries are:

`points(x,y,...)`	Add points to an existing plot (possibly using a different plotting character). The plotting character is set by `pch=` and the size of the character by `cex=` or `mkh=`.
`lines(x,y,...)`	Add lines to an existing plot. The line type is set by `lty=` and width by `lwd=`. The `type` options may be used.
`text(x,y,labels,...)`	Add text to a plot at points given by `x,y`. `labels` is an integer or character vector; `labels[i]` is plotted at point $(x[i],y[i])$. The default is `seq(along=x)`.

1	2	3	4	5	6	7	8	9	10	11	12	13	14	15	16	17	18
○	△	+	×	◇	▽	⊠	✳	⬦	⊕	⍓	⊞	⊠	◳	■	●	▲	◆

19	20	21	22	23	24	25	26	27
▼	⊹	✳	−	│	⊠	♂	♀	⊠

19	20	21	22	23	24	25
●	•	⊙	◻	◇	△	▽

Figure 4.2: Plotting symbols or marks, specified by pch = n. Those on the left of the second row are only available on graphsheet devices, and those on the right of the second row only in R (where the fill colour for 21 to 25 has been taken as light grey).

abline(a, b, ...)	Draw a line in intercept and slope form, (a,b), across an
abline(h = c, ...)	existing plot. h = c may be used to specify y-coordinates
abline(v = c, ...)	for the heights of horizontal lines to go across a plot, and v
abline(*lmobject*, ...)	= c similarly for the x-coordinates for vertical lines. The
	coefficients of a suitable *lmobject* are used.

These are the most commonly used graphics functions; we have shown examples of their use in Chapter 1, and show many more later. (There are also functions arrows and symbols that we do not use in this book.) The plotting characters available for plot and points can be characters of the form pch = "o" or numbered from 0 to 27, which uses the marks shown in Figure 4.2.

Size of text and symbols

Confusingly, the size of plotting characters is selected in one of two very different ways. For plotting characters (by pch = "o") or text (by text), the parameter cex (for 'character expansion') is used. This defaults to the global setting (which

S+   defaults to 1), and rescales the character by that factor. In S-PLUS for a mark set by pch = n, the size is controlled by the mkh parameter which gives the height of the symbol *in inches*. (This will be clear for printers; for screen devices the default device region is about 8 in × 6 in and this is not changed by resizing the window.) However, if mkh = 0 (the default, and always in R) the size is then controlled by cex and the default size of each symbol is approximately that of 0. Care is needed in changing cex on a call to plot, as this may[2] also change the size of the axis labels. It is better to use, for example,

```
plot(x, y, type = "n")                    # axes only
points(x, y, pch = 4, mkh = 0, cex = 0.7)  # add the points
```

If cex is used to change the size of all the text on a plot, it will normally be desirable to set mex to the same value to change the interline spacing. An alternative to specifying cex is csi, which specifies the absolute size of a character (in inches). (There is no msi.)

The default text size can be changed for some graphics devices, for example, by argument pointsize for the postscript, win.printer, windows and macintosh devices.

---

[2]In R these are controlled by cex.axis ; there are also cex.main , cex.sub and cex.lab , the last for the axis titles.

Equally scaled plots

There are many plots, for example, in multivariate analysis, that represent distances in the plane and for which it is essential to have a scaling of the axes that is geometrically accurate. This can be done in many ways, but most easily by our function `eqscplot` which behaves as the default plot function but shrinks the scale on one axis until geometrical accuracy is attained.

Warning: when screen devices (except a `graphsheet`) are resized the S-PLUS process is not informed, so `eqscplot` can only work for the original window shape.

R has an argument `asp` that can be given to many high-level plotting functions and fixes scales so that x units are `asp` times as large as y units, even across window resizing.

Multivariate plots

The plots we have seen so far deal with one or two variables. To view more we have several possibilities. A *scatterplot matrix* or *pairs* plot shows a matrix of scatterplots for each pair of variables, as we saw in Figure 1.2, which was produced by `splom(~ hills)`. Enhanced versions of such plots are a *forte* of Trellis graphics, so we do not discuss how to make them in the base graphics system.

Dynamic graphics

S-PLUS has limited facilities for dynamic plots; R has none. Both can work with XGobi and GGobi (see page 302) to add dynamic brushing, selecting and rotating.

The S-PLUS function `brush` allows interaction with the (lower half) of a scatterplot matrix. An example is shown in Figure 1.3 on page 9. As it is much easier to understand these by using them, we suggest you try

```
brush(hills)
```

and experiment.

Points can be highlighted (marked with a symbol) by moving the brush (a rectangular window) over them with button 1 held down. When a point is highlighted, it is shown highlighted in all the displays. Highlighting is removed by brushing with button 2 held down. It is also possible to add or remove points by clicking with button 1 in the scrolling list of row names.

One of four possible (device-dependent) marking symbols can be selected by clicking button 1 on the appropriate one in the display box on the right. The marking is by default persistent, but this can be changed to 'transient' in which only points under the brush are labelled (and button 1 is held down). It is also possible to select marking by row label as well as symbol.

The brush size can be altered under UNIX by picking up a corner of the brush in the `brush size` box with the mouse button 1 and dragging to the required size. Under Windows, move the brush to the background of the main `brush` window, hold down the left mouse button and drag the brush to the required size.

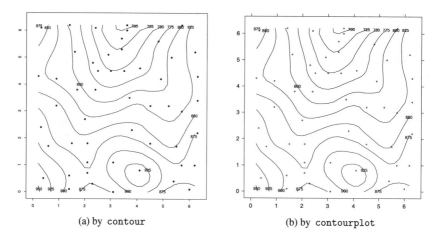

(a) by contour   (b) by contourplot

Figure 4.3: Contour plots of loess smoothing of the topo dataset. Note the differences in the axes and the way points are depicted.

The plot produced by brush will also show a three-dimensional plot (unless spin = F), and this can be produced on its own by spin. Under UNIX clicking with mouse button 1 will select three of the variables for the x-, y- and z-axes. The plot can be spun in several directions and resized by clicking in the appropriate box. The speed box contains a vertical line or slider indicating the current setting.

Plots from brush and spin can only be terminated by clicking with the mouse button 1 on the quit box or button.

Obviously brush and spin are available only on suitable screen devices, including motif and graphsheet. Hardcopy is possible only by directly printing the window used, not by copying the plot to a printer graphics device.

Plots of surfaces

The functions contour, persp and image allow the display of a function defined on a two-dimensional regular grid. Their Trellis equivalents give more elegant output, so we do not discuss them in detail. The function contour allows more control than contourplot. We anticipate an example from Chapter 15 of plotting a smooth topographic surface for Figure 4.3, and contrast it with contourplot.

```
topo.loess <- loess(z ~ x * y, topo, degree = 2, span = 0.25)
topo.mar <- list(x = seq(0, 6.5, 0.2), y = seq(0, 6.5, 0.2))
topo.lo <- predict(topo.loess, expand.grid(topo.mar))
par(pty = "s")          # square plot
contour(topo.mar$x, topo.mar$y, topo.lo, xlab = "", ylab = "",
    levels = seq(700,1000,25), cex = 0.7)
points(topo$x, topo$y)
par(pty =  "m")
```

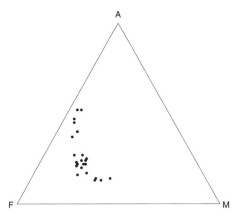

Figure 4.4: A ternary plot of the compositions of 23 rocks from Aitchison (1986).

```
contourplot(z ~ x * y, mat2tr(topo.lo), aspect = 1,
    at = seq(700, 1000, 25), xlab = "", ylab = "",
    panel = function(x, y, subscripts, ...) {
        panel.contourplot(x, y, subscripts, ...)
        panel.xyplot(topo$x,topo$y, cex = 0.5)
    }
)
```

This generates values of the surface on a regular 33×33 grid generated by
`expand.grid`. Our MASS library provides the functions `con2tr` and `mat2tr`
to convert objects designed for input to `contour` and matrices as produced by
`predict.loess` into data frames suitable for the Trellis 3D plotting routines.

The S-PLUS for Windows GUI and R under Windows have ways to visualize
such surfaces interactively; see pages 69 and 422.

Making new types of plots

The basic components described so far can be used to create new types of plot as
the need arises.

Ternary plots are used for compositional data (Aitchison, 1986) where there
are three components whose proportions add to one. These are represented by a
point in an equilateral triangle, where the distances to the sides add to a constant.
These are implemented in a function `ternary` which is given on the help page
of the MASS dataset `Skye`; see Figure 4.4.

4.3 Enhancing Plots

In this section we cover a number of ways that are commonly used to enhance
plots, without reaching the level of detail of Section 4.4.

Some plots (such as Figure 1.1) are square, whereas others are rectangular. The shape is selected by the graphics parameter `pty`. Setting `par(pty = "s")` selects a square plotting region, whereas `par(pty = "m")` selects a maximally sized (and therefore usually non-square) region.

Multiple figures on one plot

We have already seen several examples of plotting two or more figures on a single device surface, apart from scatterplot matrices. The graphics parameters `mfrow` and `mfcol` subdivide the plotting region into an array of figure regions. They differ in the order in which the regions are filled. Thus

```
par(mfrow = c(2, 3))
par(mfcol = c(2, 3))
```

both select a 2×3 array of figures, but with the first they are filled along rows, and with the second down columns. A new figure region is selected for each new plot, and figure regions can be skipped by using `frame`.

All but two[3] of the multi-figure plots in this book were produced with `mfrow`. Most of the side-by-side plots were produced with `par(mfrow = c(2, 2))`, but using only the first two figure regions.

The `split.screen` function provides an alternative and more flexible way of generating multiple displays on a graphics device. An initial call such as

```
split.screen(figs = c(3, 2))
```

subdivides the current device surface into a 3×2 array of *screens*. The screens created in this example are numbered 1 to 6, *by rows*, and the original device surface is known as screen 0. The current screen is then screen 1 in the upper left corner, and plotting output will fill the screen as it would a figure. Unlike multi-figure displays, the next plot will use the same screen unless another is specified using the `screen` function. For example, the command

```
screen(3)
```

causes screen 3 to become the next current screen.

On screen devices the function `prompt.screen` may be used to define a screen layout interactively. The command

```
split.screen(prompt.screen())
```

allows the user to define a screen layout by clicking mouse button 1 on diagonally opposite corners. In our experience this requires a steady hand, although there is a `delta` argument to `prompt.screen` that can be used to help in aligning screen edges. Alternatively, if the `figs` argument to `split.screen` is specified as an $N \times 4$ matrix, this divides the plot into N screens (possibly overlapping) whose corners are specified by giving (xl, xu, yl, yu) as the row of the matrix (where the whole region is $(0, 1, 0, 1)$).

---

[3]Figures 6.2 and 6.3, where the `fig` parameter was used.

The `split.screen` function may be used to subdivide the current screen recursively, thus leading to irregular arrangements. In this case the screen numbering sequence continues from where it had reached.

Split-screen mode is terminated by a call to `close.screen(all = T)`; individual screens can be shut by `close.screen(n)`.

The function `subplot`[5] provides a third way to subdivide the device surface. This has call `subplot(fun, ...)` which adds the graphics output of `fun` to an existing plot. The size and position can be determined in many ways (see the on-line help); if all but the first argument is missing a call to `locator` is used to ask the user to click on any two opposite corners of the plot region.[6]

Use of the `fig` parameter to `par` provides an even more flexible way to subdivide a plot; see Section 4.4 and Figure 6.2 on page 153.

With multiple figures it is normally necessary to reduce the size of the text. If either the number of rows or columns set by `mfrow` or `mfcol` is three or more, the text size is halved by setting `cex = 0.5` (and `mex = 0.5`; see Section 4.4). This may produce characters that are too small and some resetting may be appropriate. (On the other hand, for a 2×2 layout the characters will usually be too large.) For all other methods of subdividing the plot surface the user will have to make an appropriate adjustment to `cex` and `mex` or to the default text size (for example by changing `pointsize` on the `postscript` and other devices).

Adding information

The basic plots produced by `plot` often need additional information added to give context, particularly if they are not going to be used with a caption. We have already seen the use of `xlab`, `ylab`, `main` and `sub` with scatterplots. These arguments can all be used with the function `title` to add titles to existing plots. The first argument is `main`, so

```
title("A Useful Plot?")
```

adds a main title to the current plot.

Further points and lines are added by the `points` and `lines` functions. We have seen how plot symbols can be selected with `pch=`. The line type is selected by `lty=`. This is device-specific, but usually includes solid lines (1) and a variety of dotted, dashed and dash-dot lines. Line width is selected by `lwd=`, with standard width being 1, and the effect being device-dependent.

Using colour

The colour model of S-PLUS graphics is quite complex. Colours are referred to as numbers, and set by the parameter `col`. Sometimes only one colour is allowed (e.g., `points`) and sometimes `col` can be a vector giving a colour for each plot item (e.g., `text`). There will always be at least two colours, 0 (the background, useful for erasing by over-plotting) and 1. However, how many colours there are and what they appear as is set by the device. Furthermore, there are separate

---

[5]Not in R, which has another approach called `layout` .

[6]Not the figure region; Figure 4.5 on page 81 shows the distinction.

colour groups, and what they are is device-specific. For example, `motif` devices have separate colour spaces for lines (including symbols), text, polygons (including histograms, bar charts and pie charts) and images, and `graphsheet` devices have two spaces, one for lines and text, the other for polygons and images. Thus the colours can appear completely differently when a graph is copied from device to device, in particular on screen and on a hardcopy. It is usually a good idea to design a colour scheme for each device.

It is necessary to read the device help page thoroughly (and for `postscript`, also that for `ps.options.send`).

R has a different and more coherent colour model involving named colours, user-settable palettes and even transparency. See the help topics `palette` and `colors` for more details.

Identifying points interactively

The function `identify` has a similar calling sequence to `text`. The first two arguments give the x- and y-coordinates of points on a plot and the third argument gives a vector of labels for each point. (The first two arguments may be replaced by a single list argument with two of its components named x and y, or by a two-column matrix.) The labels may be a character string vector or a numeric vector (which is coerced to character). Then clicking with mouse button 1 near a point on the plot causes its label to be plotted; labelling all points or clicking anywhere in the plot with button 2 terminates the process.[7] (The precise position of the click determines the label position, in particular to left or right of the point.) We saw an example in Figure 1.4 on page 10. The function returns a vector of index numbers of the points that were labelled.

In Chapter 1 we used the `locator` function to add new points to a plot. This function is most often used in the form `locator(1)` to return the (x, y) coordinates of a single button click to place a label or legend, but can also be used to return the coordinates of a series of points, terminated by clicking with mouse button 2.

Adding further axes and grids

It is sometimes useful to add further axis scales to a plot, as in Figure 8.1 on page 212 which has scales for both kilograms and pounds. This is done by the function `axis`. There we used

```
attach(wtloss)
oldpar <- par() # R: oldpar <- par(no.readonly = TRUE)
# alter margin 4; others are default
par(mar = c(5.1, 4.1, 4.1, 4.1))
plot(Days, Weight, type = "p", ylab = "Weight (kg)")
Wt.lbs <- pretty(range(Weight*2.205))
axis(side = 4, at = Wt.lbs/2.205, lab = Wt.lbs, srt = 90)
mtext("Weight (lb)", side = 4, line = 3)
detach()
```

---

[7]With R on a Macintosh (which only has one mouse button) click outside the plot window to terminate `locator` or `identify`.

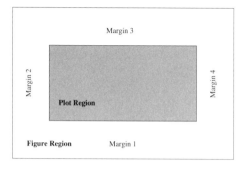

Figure 4.5: Anatomy of a graphics figure.

```
par(oldpar)
```

This adds an axis on side 4 (labelled clockwise from the bottom; see Figure 4.5)
with labels rotated by $90°$ (srt = 90 , not needed in R where unlike S-PLUS the   R
rotation is controlled by the setting of las) and then uses mtext to add a label
'underneath' that axis. Other parameters are explained in Section 4.4. Please read
the on-line documentation very carefully to determine which graphics parameters
are used in which circumstances.

Grids can be added by using axis with long tick marks, setting parameter
tck = 1 (yes, obviously). For example, a dotted grid is created by

```
axis(1, tck = 1, lty = 2); axis(2, tck = 1, lty = 2)
```

and the location of the grid lines can be specified using at=.

Adding legends

Legends are added by the function legend. Since it can label many types of variation such as line type and width, plot symbol, colour, and fill type, its description
is very complex. All calls are of the form

```
legend(x, y, legend, ...)
```

where x and y give either the upper left corner of the legend box or both upper
left and lower right corners. These are often most conveniently specified on-
screen by using locator(1) or locator(2). Argument legend is a character
vector giving the labels for each variation. Most of the remaining arguments are
vectors of the same length as legend giving the appropriate coding for each
variation, by lty=, lwd=, col=, fill=, angle= and density=. Argument
pch is a single character string concatenating the symbols; for numerical pch in
S-PLUS use the vector argument marks .

By default the legend is contained in a box; the drawing of this box can be
suppressed by argument bty = "n".

The Trellis function key provides a more flexible approach to constructing
legends, and can be used with basic plots. (See page 104 for further details.)

Non-English labels

Non-native English speakers will often want to include characters from their other languages in labels. For Western European languages written in ISO-latin1 encoding this will normally work; it does for all the R devices and for motif and graphsheet devices in S-PLUS. To use such characters with the postscript device under S-PLUS, set

```
ps.options(setfont = ps.setfont.latin1)
```

If you are unable to enter the characters from the keyboard, octal escapes of the form "\341" (which encodes á) can be used.

R's postscript device allows arbitrary encodings via its encoding parameter, and S-PLUS's ps.setfont.latin1 could be modified to use a different encoding such as ISO-latin2.

Mathematics in labels

Users frequently wish to include the odd subscript, superscript and mathematical symbol in labels. There is no general solution, but for the S-PLUS postscript driver[8] Alan Zaslavsky's package postscriptfonts adds these features. We can label Figure 7.3 (on page 209) by λ (from font 13, the PostScript symbol font).

```
library(postscriptfonts)
x <- 0:100
plik <- function(lambda)
   sum(dpois(x, lambda) * 2 * ( (lambda - x) +
       x * log(pmax(1, x)/lambda)))
lambda <- c(1e-8, 0.05, seq(0.1, 5, 0.1))
plot(lambda, sapply(lambda, plik), type = "l",
      ylim = c(0, 1.4), ylab = "", xlab = "")
abline(h = 1, lty = 3)
mixed.mtext(texts = "l", side = 1, line = 3, font = 13) # xlab
mixed.mtext(texts = "E~f13~d~.l~f1~.(deviance)", adj = 0.5,
             side = 2, line = 3, font = 13)              # ylab
```

R has rather general facilities to label with mathematics: see ?plotmath and Murrell and Ihaka (2000). Here we could use (on most devices, including on-screen)

```
plot(lambda, sapply(lambda, plik), type = "l", ylim = c(0, 1.4),
      xlab = expression(lambda),
      ylab = expression(paste(E[lambda], "(deviance)")))
```

4.4   Fine Control of Graphics

The graphics process is controlled by *graphics parameters*, which are set for each graphics device. Each time a new device is opened these parameters for that

---

[8]Under UNIX or Windows.

device are reset to their default values. Graphics parameters may be set, or their current values queried, using the par function. If the arguments to par are of the name = value form the graphics parameter name is set to value, if possible, and other graphics parameters may be reset to ensure consistency. The value returned is a list giving the previous parameter settings. Instead of supplying the arguments as name = value pairs, par may also be given a single list argument with named components.

If the arguments to par are quoted character strings, "name", the current value of graphics parameter name is returned. If more than one quoted string is supplied the value is a list of the requested parameter values, with named components. The call par() with no arguments returns a list of all the graphics parameters.

Some of the many graphics parameters are given in Tables 4.3 and 4.4 (on pages 84 and 87). Those in Table 4.4 can also be supplied as arguments to high-level plot functions, when they apply just to the figure produced by that call. (The layout parameters are ignored by the high-level plot functions.)

The figure region and layout parameters

When a device is opened it makes available a rectangular surface, the *device region*, on which one or more plots may appear. Each plot occupies a rectangular section of the device surface called a *figure*. A figure consists of a rectangular *plot region* surrounded by a *margin* on each side. The margins or sides are numbered one to four, clockwise starting from the bottom. The plot region and margins together make up the *figure region*, as in Figure 4.5 on page 81. The device surface, figure region and plot region have their vertical sides parallel and hence their horizontal sides also parallel.

The size and position of figure and plot regions on a device surface are controlled by *layout parameters*, most of which are listed in Table 4.3. Lengths may be set in either absolute or relative units. Absolute lengths are in *inches*, whereas relative lengths are in *text lines* (so relative to the current font size).

Margin sizes are set using mar for text lines or mai for inches. These are four-component vectors giving the sizes of the lower, left, upper and right margins in the appropriate units. Changing one causes a consistent change in the other; changing mex will change mai but not mar.

Positions may be specified in relative units using the unit square as a coordinate system for which some enclosing region, such as the device surface or the figure region, is the unit square. The fig parameter is a vector of length four specifying the current figure as a fraction of the device surface. The first two components give the lower and upper x-limits and the second two give the y-limits. Thus to put a point plot in the left-hand side of the display and a Q-Q plot on the right-hand side we could use:

```
postscript(file = "twoplot.ps")    # open a postscript device
par(fig = c(0, 2/3, 0, 1))         # set a figure on the left
plot(x, y)                         # point plot
par(fig = c(2/3, 1, 0, 1))         # set a figure on the right
```

Table 4.3: Some graphics layout parameters with example settings.

`din, fin, pin`	Absolute device size, figure size and plot region size in inches. `fin = c(6, 4)`
`fig`	Define the figure region as a fraction of the device region. `fig = c(0, 0.5, 0,1)`
`font`	Small positive integer determining a text font for characters and hence an interline spacing. For S-PLUS's `postscript` device one of the standard PostScript fonts given by `ps.options("fonts")`. In R font 1 is plain, font 2 italic, font 3 bold, font 4 bold italic and font 5 is the symbol font. `font = 3`
`mai, mar`	The four margin sizes, in inches (`mai`), or in text line units (`mar`, that is, *relative* to the current font size). Note that `mar` need not be an integer. `mar = c(3, 3, 1, 1) + 0.1`
`mex`	Number of text lines per interline spacing. `mex = 0.7`
`mfg`	Define a position within a specified multi-figure display. `mfg = c(2, 2, 3, 2)`
`mfrow, mfcol`	Define a multi-figure display. `mfrow = c(2, 2)`
`new`	Logical value indicating whether the current figure has been used. `new = T`
`oma, omi, omd`	Define outer margins in text lines or inches, or by defining the size of the array of figures as a fraction of the device region. `oma = c(0, 0, 4, 0)`
`plt`	Define the plot region as a fraction of the figure region. `plt = c(0.1, 0.9, 0.1, 0.9)`
`pty`	Plot type, or shape of plotting region, `"s"` or `"m"`
`uin`	(not R) Return inches per user coordinate for x and y.
`usr`	Limits for the plot region in user coordinates. `usr = c(0.5, 1.5, 0.75, 10.25)`

```
qqnorm(resid(obj))                    # diagnostic plot
dev.off()
```

The left-hand figure occupies $2/3$ of the device surface and the right-hand figure $1/3$. For regular arrays of figures it is simpler to use `mfrow` or `split.screen`.

Positions in the plot region may also be specified in absolute *user coordinates*. Initially user coordinates and relative coordinates coincide, but any high-level plotting function changes the user coordinates so that the x- and y-coordinates range from their minimum to maximum values as given by the plot axes. The graphics parameter `usr` is a vector of length four giving the lower and upper x- and y-limits for the user coordinate system. Initially its setting is `usr = c(0,1,0,1)`. Consider another simple example:

```
> motif()                    # open a device
> par("usr")                 # usr coordinates
[1] 0 1 0 1
```

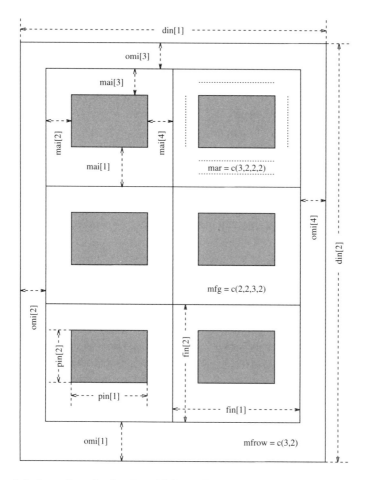

Figure 4.6: An outline of a 3×2 multi-figure display with outer margins showing some graphics parameters. The current figure is at position $(2, 2)$ and the display is being filled by rows. In this figure "fin[1]" is used as a shorthand for par("fin")[1], and so on.

```
> x <- 1:20
> y <- x + rnorm(x)          # generate some data
> plot(x, y)                 # produce a scatterplot
> par("usr")                 # user coordinates now match the plot
[1]  0.2400 20.7600  1.2146 21.9235
```

Any attempt to plot outside the user coordinate limits causes a warning message unless the general graphics parameter xpd is set to T.

Figure 4.6 shows some of the layout parameters for a multi-figure layout. Such an array of figures may occupy the entire device surface, or it may have *outer margins*, which are useful for annotations that refer to the entire array. Outer margins are set with the parameter oma (in text lines) or omi (in inches). Alternatively omd may be used to set the region containing the array of figures in a similar way to which `fig` is used to set one figure. This implicitly determines the outer margins as the complementary region. In contrast to what happens with the margin parameters `mar` and `mai`, a change to `mex` will leave the outer margin size, `omi`, constant but adjust the number of text lines, `oma`.

Text may be put in the outer margins by using `mtext` with parameter `outer = T`.

Common axes for figures

There are at least two ways to ensure that several plots share a common axis or axes.

1. Use the same `xlim` or `ylim` (or both) setting on each plot and ensure that the parameters governing the way axes are formed, such as `lab`, `las`, `xaxs` and allies, do not change.

2. Set up the desired axis system with the first plot and then use `par` to set the low-level parameter `xaxs = "d"`, `yaxs = "d"` or both as appropriate. This ensures that the axis or axes are not changed by further high-level plot commands on the same device.

An example: A Q-Q normal plot with envelope

In Chapter 5 we recommend assessing distributional form by quantile-quantile plots. A simple way to do this is to plot the sorted values against quantile approximations to the expected normal order statistics and draw a line through the 25 and 75 percentiles to guide the eye, performed for the variable `Infant.Mortality` of the Swiss provinces data (on fertility and socio-economic factors on Swiss provinces in about 1888) by

```
## in R just use data(swiss)
swiss <- data.frame(Fertility = swiss.fertility, swiss.x)
attach(swiss)
qqnorm(Infant.Mortality)
qqline(Infant.Mortality)
```

The reader should check the result and compare it with the style of Figure 4.7.

Another suggestion to assess departures is to compare the sample Q-Q plot with the envelope obtained from a number of other Q-Q plots from generated normal samples. This is discussed in (Atkinson, 1985, §4.2) and is based on an idea of Ripley (see Ripley, 1981, Chapter 8). The idea is simple. We generate a number of other samples of the same size from a normal distribution and scale all samples to mean 0 and variance 1 to remove dependence on location and scale parameters. Each sample is then sorted. For each order statistic the maximum and minimum values for the generated samples form the upper and lower envelopes. The envelopes are plotted on the Q-Q plot of the scaled original sample and form a guide to what constitutes serious deviations from the expected behaviour under normality. Following Atkinson our calculation uses 19 generated normal samples.

We begin by calculating the envelope and the x-points for the Q-Q plot.

Table 4.4: Some of the more commonly used general and high-level graphics parameters with example settings.

Text:

adj	Text justification. 0 = left justify, 1 = right justify, 0.5 = centre.
cex	Character expansion. cex = 2
csi	Height of font (inches). csi = 0.11
font	Font number: device-dependent.
srt	String rotation in degrees. srt = 90
cin cxy	Character width and height in inches and usr coordinates (for information, not settable).

Symbols:

col	Colour for symbol, line or region. col = 2
lty	Line type: solid, dashed, dotted, etc. lty = 2
lwd	Line width, usually as a multiple of default width. lwd = 2
mkh	Mark height (inches). Ignored in R. mkh = 0.05
pch	Plotting character or mark. pch = "*" or pch = 4 for marks. (See page 74.)

Axes:

bty	Box type, as "o", "l", "7", "c", "n".
exp	(not R) Notation for exponential labels. exp = 1
lab	Tick marks and labels. lab = c(3, 7, 4)
las	Label orientation. 0 = parallel to axis, 1 = horizontal, 2 = vertical.
log	Control log axis scales. log = "y"
mgp	Axis location. mgp = c(3, 1, 0)
tck	Tick mark length as signed fraction of the plot region dimension. tck = -0.01
xaxp yaxp	Tick mark limits and frequency. xaxp = c(2, 10, 4)
xaxs yaxs	Style of axis limits. xaxs = "i"
xaxt yaxt	Axis type. "n" (null), "s" (standard), "t" (time) or "l" (log).

High Level:

ann	(R only) Should titles and axis labels be plotted?
ask	Prompt before going on to next plot? ask = F
axes	Print axes? axes = F
main	Main title. main = "Figure 1"
sub	Subtitle. sub = "23-Jun-2002"
type	Type of plot. type = "n"
xlab ylab	Axis labels. ylab = "Speed in km/sec"
xlim ylim	Axis limits. xlim = c(0, 25)
xpd	May points or lines go outside the plot region? xpd = T

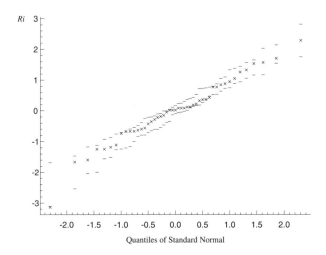

Figure 4.7: The Swiss fertility data. A Q-Q normal plot with envelope for infant mortality.

```
samp <- cbind(Infant.Mortality, matrix(rnorm(47*19), 47, 19))
samp <- apply(scale(samp), 2, sort)
rs <- samp[, 1]
xs <- qqnorm(rs, plot = F)$x
env <- t(apply(samp[, -1], 1, range))
```

As an exercise in building a plot with specific requirements we now present the envelope and Q-Q plot in a style very similar to Atkinson's. To ensure that the Q-Q plot has a y-axis large enough to take the envelope we could calculate the y-limits as before, or alternatively use a matrix plot with `type = "n"` for the envelope at this stage. The axes and their labels are also suppressed for the present:

```
matplot(xs, cbind(rs, env), type = "pnn",
        pch = 4, mkh = 0.06, axes = F, xlab = "", ylab = "")
```

The argument setting `type = "pnn"` specifies that the first column (`rs`) is to produce a point plot and the remaining two (`env`) no plot at all, but the axes will allow for them. Setting `pch = 4` specifies a 'cross' style plotting symbol (see Figure 4.2) similar to Atkinson's, and `mkh = 0.06` establishes a suitable size for the plotting symbol.

Atkinson uses small horizontal bars to represent the envelope. We can now calculate a half length for these bars so that they do not overlap and do not extend beyond the plot region. Then we can add the envelope bars using `segments`:

```
xyul <- par("usr")
smidge <- min(diff(c(xyul[1], xs, xyul[2])))/2
segments(xs - smidge, env[, 1], xs + smidge, env[, 1])
segments(xs - smidge, env[, 2], xs + smidge, env[, 2])
```

Atkinson's axis style differs from the default S style in several ways. There are many more tick intervals; the ticks are inside the plot region rather than outside;

there are more labelled ticks; and the labelled ticks are longer than the unlabelled.
From experience ticks along the x-axis at 0.1 intervals with labelled ticks at 0.5
intervals seems about right, but this is usually too close on the y-axis. The axes
require four calls to the axis function:

```
xul <- trunc(10*xyul[1:2])/10
axis(1, at=seq(xul[1], xul[2], by = 0.1), labels = F, tck = 0.01)
xi <- trunc(xyul[1:2])
axis(1, at = seq(xi[1], xi[2], by = 0.5), tck = 0.02)
yul <- trunc(5*xyul[3:4])/5
axis(2, at = seq(yul[1], yul[2], by = 0.2), labels = F, tck= 0.01)
yi <- trunc(xyul[3:4])
axis(2, at = yi[1]:yi[2], tck = 0.02)
```

Finally we add the L-box, put the x-axis title at the centre and the y-axis title
at the top:

```
box(bty = "l")              # lower case "L"
# S: ps.options()$fonts
mtext("Quantiles of Standard Normal", side=1, line=2.5, font=3)
# S: mtext("Ri", side = 2, line = 2, at = yul[2], font = 10)
# R: mtext(expression(R[i]), side = 2, line = 2, at = yul[2])
```

where in S-PLUS fonts 3 and 10 are Times-Roman and Times-Italic on the device
used (postscript under UNIX), found from the list given by ps.options().
    The final plot is shown in Figure 4.7 on page 88.

4.5   Trellis Graphics

Trellis graphics were developed to provide a consistent graphical 'style' and to
extend conditioning plots; the style is a development of that used in Cleveland
(1993).
    Trellis is very prescriptive, and changing the display style is not always an
easy matter.
    It may be helpful to understand that Trellis is written entirely in the S lan-
guage, as calls to the basic plotting routines. Two consequences are that it can be
slow and memory-intensive, and that it takes over many of the graphics parameters for its own purposes. (Global settings of graphics parameters are usually not
used, the outer margin parameters omi being a notable exception.) Computation
of a Trellis plot is done in two passes: once when a Trellis object is produced, and
once when that object is printed (producing the actual plot).
    The trellis library contains a large number of examples: use

```
?trellis.examples
```

to obtain an up-to-date list. These are all functions that can be called to plot the
example, and listed to see how the effect was achieved.
    R has a similar system in its package lattice; however, that is built on a
different underlying graphics model called grid and mixes (even) less well with
traditional S graphics. This runs most of the examples shown here, but the output
will not be identical.

Trellis graphical devices

The `trellis.device` graphical device is provided by the `trellis` library. It is perhaps more accurate to call it a meta-device, for it uses one of the underlying graphical devices,[9] but customizes the parameters of the device to use the Trellis style, and in particular its colour schemes.

Trellis graphics is intended to be used on a `trellis.device` device, and may give incorrect results on other devices.

Trellis devices by default use colour for screen windows and greylevels for printer devices. The settings for a particular device can be seen by running the command `show.settings()`. These settings are not the same for all colour screens, nor for all printer devices. Trellis colour schemes have a mid-grey background on colour screens (but not colour printers). If a Trellis plot is used without a graphics device already in use, a suitable Trellis device is started.

Trellis model formulae

Trellis graphics functions make use of the language for model formulae described in Section 3.7. The Trellis code for handling model formulae to produce a data matrix from a data frame (specified by the `data` argument) allows the argument `subset` to select a subset of the rows of the data frame, as one of the first three forms of indexing vector described on page 27. (Character vector indices are not allowed.)

There are a number of inconsistencies in the use of the formula language. There is no `na.action` argument, and missing values are handled inconsistently; generally rows with NAs are omitted, but `splom` fails if there are missing values. Surprisingly, `splom` uses a formula, but does not accept a `data` argument.

Trellis uses an extension to the model formula language, the operator ' | ' which can be read as 'given'. Thus if a is a factor, `lhs ~ rhs | a` will produce a plot for each level of a of the subset of the data for which a has that level (so estimating the conditional distribution given a). Conditioning on two or more factors gives a plot for each combination of the factors, and is specified by an interaction, for example, `| a*b`. For the extension of conditioning to continuous variates via what are known as *shingles*, see page 101.

Trellis plot objects can be kept, and `update` can be used to change them, for example, to add a title or change the axis labels, before re-plotting by printing (often automatically as the result of the call to `update`).

Basic Trellis plots

As Table 4.5 shows, the basic styles of plot exist in Trellis, but with different names and different default styles. Their usage is best seen by considering how to produce some figures in the Trellis style.

Figure 1.2 (page 9) was produced by `splom(~ hills)`. Trellis plots of scatterplot matrices read from bottom to top (as do all multi-panel Trellis displays,

---

[9]Currently `motif`, `postscript`, `graphsheet`, `win.printer`, `pdf.graph`, `wmf.graph` and `java.graph` where these are available.

Table 4.5: Trellis plotting functions. Page references are given to the most complete description in the text.

Function	Page	Description
xyplot	94	Scatterplots.
bwplot	92	Boxplots.
stripplot	98	Display univariate data against a numerical variable.
dotplot		ditto in another style,
histogram		'Histogram', actually a frequency plot.
densityplot		Kernel density estimates.
barchart		Horizontal bar charts.
piechart		Pie chart.
splom	90	Scatterplot matrices.
contourplot	76	Contour plot of a surface on a regular grid.
levelplot	94	Pseudo-colour plot of a surface on a regular grid.
wireframe	94	Perspective plot of a surface evaluated on a regular grid.
cloud	104	A perspective plot of a cloud of points.
key	104	Add a legend.
color.key	94	Add a color key (as used by levelplot).
trellis.par.get	93	Save Trellis parameters.
trellis.par.set	93	Reset Trellis parameters.
equal.count	102	Compute a shingle.

like graphs rather than matrices, despite the meaning of the name splom). By default the panels in a splom plot are square.

Figure 4.8 is a Trellis version of Figure 1.4. Note that the y-axis numbering is horizontal by default (equivalent to the option par(las = 1)), and that points are plotted by open circles rather than filled circles or stars. It is not possible to add to a Trellis plot,[10] so the Trellis call has to include all the desired elements. This is done by writing a *panel function*, in this case

```
# R: library(lqs)
xyplot(time ~ dist, data = hills,
    panel = function(x, y, ...) {
        panel.xyplot(x, y, ...)
        panel.lmline(x, y, type = "l")
        panel.abline(lqs(y ~ x), lty = 3)
        identify(x, y, row.names(hills))
    }
```

[10]As the user coordinate system is not retained; but the plot call can be updated by update and re-plotted.

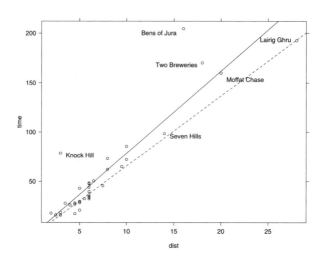

Figure 4.8: A Trellis version of Figure 1.4 (page 10).

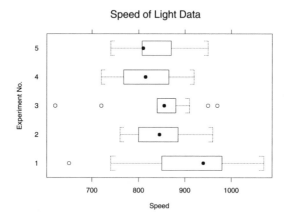

Figure 4.9: A Trellis version of Figure 1.5 (page 11).

)

Figure 4.9 is a Trellis version of Figure 1.5. Boxplots are known as box-and-whisker plots, and are displayed horizontally. This figure was produced by

```
bwplot(Expt ~ Speed, data = michelson, ylab = "Experiment No.")
title("Speed of Light Data")
```

Note the counter-intuitive way the formula is used. This plot corresponds to a one-way layout splitting Speed by experiment, so it is tempting to use Speed as the response. It may help to remember that the formula is of the y ~ x form for the x- and y-axes of the plot. (The same ordering is used for all the univariate plot functions.)

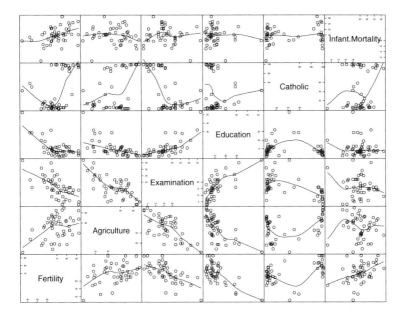

Figure 4.10: A Trellis scatterplot matrix display of the Swiss provinces data.

Figure 4.10 is an enhanced scatterplot matrix, again using a panel function to add to the basic display. Now we see the power of panel functions, as the basic plot commands can easily be applied to multi-panel displays. The aspect = "fill" command allows the array of plots to fill the space; by default the panels are square as in Figure 1.2.

```
splom(~ swiss, aspect = "fill",
   panel = function(x, y, ...) {
       panel.xyplot(x, y, ...); panel.loess(x, y, ...)
   }
)
```

Most Trellis graphics functions have a groups parameter, which we can illustrate on the stormer data used in Section 8.4 (see Figure 4.11).

```
sps <- trellis.par.get("superpose.symbol")
sps$pch <- 1:7
trellis.par.set("superpose.symbol", sps)
xyplot(Time ~ Viscosity, data = stormer, groups = Wt,
   panel = panel.superpose, type = "b",
   key = list(columns = 3,
       text = list(paste(c("Weight:   ", "", ""),
                         unique(stormer$Wt), "gms")),
       points = Rows(sps, 1:3)
       )
)
```

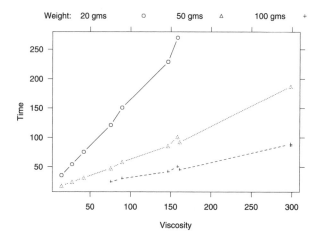

Figure 4.11: A Trellis plot of the stormer data.

Here we have changed the default plotting symbols (which differ by device) to the first seven pch characters shown in Figure 4.2 on page 74. (We could just use the argument pch = 1:7 to xyplot, but then specifying the key becomes much more complicated.)

Figure 4.12 shows further Trellis plots of the smooth surface shown in Figure 4.3. Once again panel functions are needed to add the points. The aspect = 1 parameter ensures a square plot. The drape = T parameter to wireframe is optional, producing the superimposed greylevel (or pseudo-colour) plot.

```
topo.plt <- expand.grid(topo.mar)
topo.plt$pred <- as.vector(predict(topo.loess, topo.plt))
levelplot(pred ~ x * y, topo.plt, aspect = 1,
    at = seq(690, 960, 10), xlab = "", ylab = "",
    panel = function(x, y, subscripts, ...) {
        panel.levelplot(x, y, subscripts, ...)
        panel.xyplot(topo$x,topo$y, cex = 0.5, col = 1)
    }
)
wireframe(pred ~ x * y, topo.plt, aspect = c(1, 0.5),
    drape = T, screen = list(z = -150, x = -60),
    colorkey = list(space="right", height=0.6))
```

(The arguments given by colorkey refer to the color.key function.) There is no simple way to add the points to the perspective display.

Trellises of plots

In multivariate analysis we necessarily look at several variables at once, and we explore here several ways to do so. We can produce a scatterplot matrix of the first three principal components of the crabs data (see page 302) by

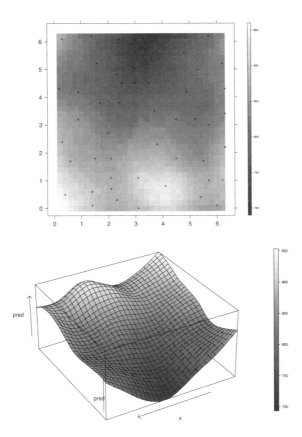

Figure 4.12: Trellis `levelplot` and `wireframe` plots of a `loess` smoothing of the topo dataset.

```
lcrabs.pc <- predict(princomp(log(crabs[,4:8])))
crabs.grp <- c("B", "b", "O", "o")[rep(1:4, each = 50)]
splom(~ lcrabs.pc[, 1:3], groups = crabs.grp,
    panel = panel.superpose,
    key = list(text = list(c("Blue male", "Blue female",
                             "Orange Male", "Orange female")),
        points = Rows(trellis.par.get("superpose.symbol"), 1:4),
        columns = 4)
    )
```

A 'black and white' version of this plot is shown in Figure 4.13. On a 'colour' device the groups are distinguished by colour and are all plotted with the same symbol (o).

However, it might be clearer to display these results as a trellis of `splom` plots, by

```
sex <- crabs$sex; levels(sex) <- c("Female", "Male")
sp <- crabs$sp; levels(sp) <- c("Blue", "Orange")
```

Blue male ○   Blue female +   Orange Male >   Orange female s

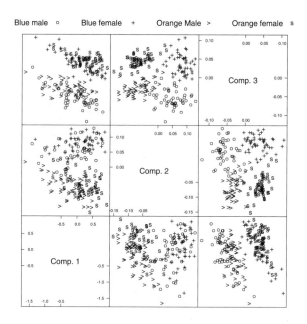

Figure 4.13: A scatterplot matrix of the first three principal components of the `crabs` data.

```
splom(~ lcrabs.pc[, 1:3] | sp*sex, cex = 0.5, pscales = 0)
```

as shown in Figure 4.14. Notice how this is the easiest method to code. It is at the core of the paradigm of Trellis, which is to display many plots of subsets of the data in some meaningful layout.

Now consider data from a multi-factor study, Quine's data on school absences discussed in Sections 6.6 and 7.4. It will help to set up more informative factor labels, as the factor names are not given (by default) in trellises of plots.

```
Quine <- quine
levels(Quine$Eth) <- c("Aboriginal", "Non-aboriginal")
levels(Quine$Sex) <- c("Female", "Male")
levels(Quine$Age) <- c("primary", "first form",
                       "second form", "third form")
levels(Quine$Lrn) <- c("Average learner", "Slow learner")
bwplot(Age ~ Days | Sex*Lrn*Eth, data = Quine)
```

This gives an array of eight boxplots, which by default takes up two pages. On a screen device there will be no pause between the pages unless the argument ask = T is set for par. It is more convenient to see all the panels on one page, which we can do by asking for a different layout (Figure 4.15). We also suppress the colouring of the strip labels by using style = 1; there are currently six preset styles.

```
bwplot(Age ~ Days | Sex*Lrn*Eth, data = Quine, layout = c(4, 2),
       strip = function(...) strip.default(..., style = 1))
```

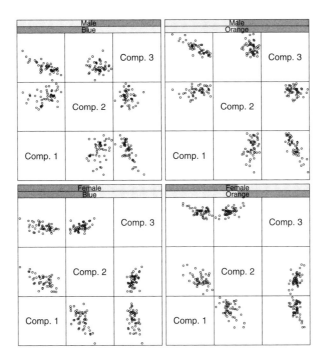

Figure 4.14: A multi-panel version of Figure 4.13.

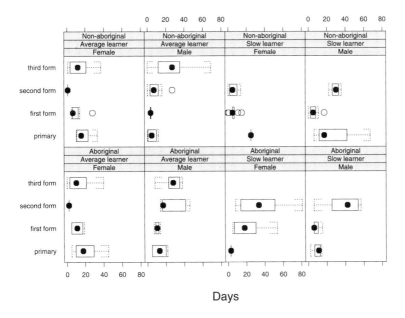

Figure 4.15: A multi-panel boxplot of Quine's school attendance data.

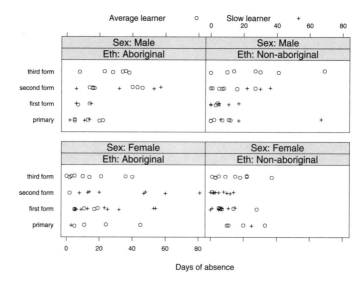

Figure 4.16: A `stripplot` of Quine's school attendance data.

A `stripplot` allows us to look at the actual data. We jitter the points slightly to avoid overplotting.

```
stripplot(Age ~ Days | Sex*Lrn*Eth, data = Quine,
          jitter = T, layout = c(4, 2))

stripplot(Age ~ Days | Eth*Sex, data = Quine,
    groups = Lrn, jitter = T,
    panel = function(x, y, subscripts, jitter.data = F, ...) {
        y <- as.numeric(y)      # only needed in R
        if(jitter.data)  y <- jitter(y)
        panel.superpose(x, y, subscripts, ...)
    },
    xlab = "Days of absence",
    between = list(y = 1), par.strip.text = list(cex = 0.7),
    key = list(columns = 2, text = list(levels(Quine$Lrn)),
        points = Rows(trellis.par.get("superpose.symbol"), 1:2)
        ),
    strip = function(...)
            strip.default(..., strip.names = c(T, T), style = 1)
)
```

The second form of plot, shown in Figure 4.16, uses different symbols to distinguish one of the factors. We include the factor name in the strip labels, using a custom `strip` function.

The Trellis function `dotplot` is very similar to `stripplot`; its panel function includes horizontal lines at each level. Function `stripplot` uses the styles of `xyplot` whereas `dotplot` has its own set of defaults; for example, the default plotting symbol is a filled rather than open circle.

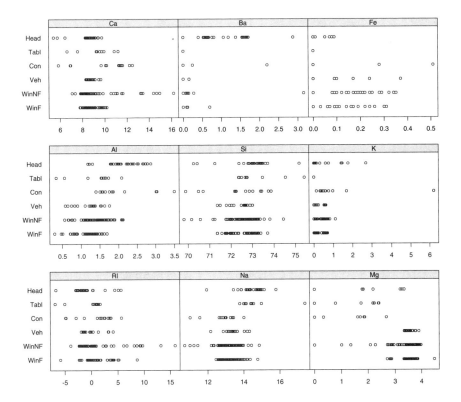

Figure 4.17: Plot by `stripplot` of the forensic glass dataset `fgl`.

As a third example, consider our dataset `fgl`. This has 10 measurements on 214 fragments of glass from forensic testing, the measurements being of the refractive index and composition (percent weight of oxides of Na, Mg, Al, Si, K, Ca, Ba and Fe). The fragments have been classified by six sources. We can look at the types for each measurement by

```
fgl0 <- fgl[ , -10] # omit type.
fgl.df <- data.frame(type = rep(fgl$type, 9),
    y = as.vector(as.matrix(fgl0)),
    meas = factor(rep(1:9, each = 214), labels = names(fgl0)))
stripplot(type ~ y | meas, data = fgl.df,
    scales = list(x = "free"), xlab = "", cex = 0.5,
    strip = function(...) strip.default(style = 1, ...))
```

Layout of a trellis

A trellis of plots is generated as a sequence of plots that are then arranged in rows, columns and pages. The sequence is determined by the order in which the conditioning factors are given: the first varying fastest. The order of the levels of the factor is that of its `levels` attribute.

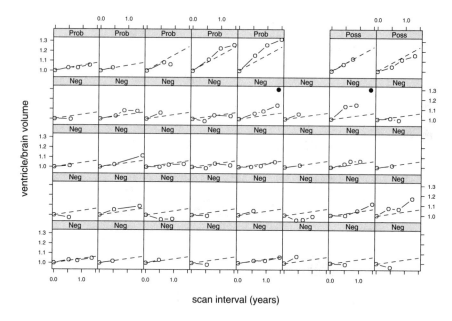

Figure 4.18: The presentation of results from a study of 39 subjects. In a real application this could be larger and so less dominated by the labels. Colour could be used to distringuish groups, too.

How the sequence of plots is displayed on the page(s) is controlled by an algorithm that tries to optimize the use of the space available, but it can be controlled by the `layout` parameters. A specification `layout = c(c, r, p)` asks for c columns, r rows and p pages. (Note the unusual ordering.) Using $c = 0$ allows the algorithm to choose the number of columns; p is used to produce only the first p pages of a many-page trellis.

If the number of levels of a factor is large and not easily divisible (for example, seven), we may find a better layout by leaving some of the cells of the trellis empty using the `skip` argument. Figure 4.18 shows another use of `layout` and `skip`.

The `between` parameter can be used to specify gaps in the trellis layout, as in Figure 4.16. It is a list with x and y components, numeric vectors that specify the gaps in units of character height. The `page` parameter can be used to invoke a function (with argument n, the page number) to label each page. The default page function does nothing.

Subscripts and groups

The `subscripts` argument of the panel function is supplied if the Trellis function is called with argument `subscripts = T`.[11] Then its value is a numeric vector of indices of cases (normally rows of `data`) that have been passed to that panel.

---

[11] And it seems sometimes even if it is not.

Figure 4.18 shows the use of Trellis to present the results of a real study. There were 39 subjects in three groups (marked on the strips), each being brain-scanned 2–4 times over up to 18 months. The plot shows the data and for each patient a dashed line showing the mean rate of change for the alloted group. Two patients whose panels are marked with a dot were later shown to have been incorrectly allocated to the 'normals' group.

Note how we arrange the layout to separate the groups. We make use of the `subscripts` argument to the panel function to identify the subject; vector `pr3` holds a set of predictions at the origin and after 1.5 years from a linear mixed-effects model.

```
xyplot(ratio ~ scant | subject, data = A5,
       xlab = "scan interval (years)",
       ylab = "ventricle/brain volume",
       subscripts = T, ID = A5$ID,
       strip = function(factor, ...)
          strip.default(..., factor.levels = labs, style = 1),
       layout = c(8, 5, 1),
       skip = c(rep(F, 37), rep(T, 1), rep(F, 1)),
       panel = function(x, y, subscripts, ID) {
          panel.xyplot(x, y, type = "b", cex = 0.5)
          which <- unique(ID[subscripts])
          panel.xyplot(c(0, 1.5), pr3[names(pr3) == which],
                       type = "l", lty = 3)
          if(which == 303 || which == 341) points(1.4, 1.3)
       })
```

Note how other arguments, here `ID`, are passed to the panel function as additional arguments. One special extra argument is `groups` which is interpreted by `panel.superpose`, as in Figure 4.11.

Conditioning plots and shingles

The idea of a trellis of plots conditioning on combinations of one or more factors can be extended to conditioning on real-valued variables, in what are known as conditioning plots or *coplots*. Two variables are plotted against each other in a series of plots with the values of further variable(s) restricted to a series of possibly overlapping ranges. This needs an extension of the concept of a factor known as a *shingle*.[12]

Suppose we wished to examine the relationship between `Fertility` and `Education` in the Swiss fertility data as the variable `Catholic` ranges from predominantly non-Catholic to mainly Catholic provinces. We add a smooth fit to each panel (and `span` controls the smoothness: see page 423).

```
Cath <- equal.count(swiss$Catholic, number = 6, overlap = 0.25)
xyplot(Fertility ~ Education | Cath, data = swiss,
    span = 1, layout = c(6, 1), aspect = 1,
    panel = function(x, y, span) {
```

---

[12]In American usage this is a rectangular wooden tile laid partially overlapping on roofs or walls.

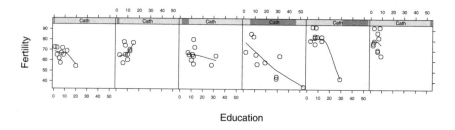

Figure 4.19: A conditioning plot for the Swiss provinces data.

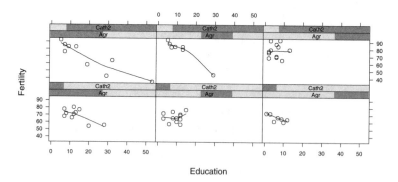

Figure 4.20: Another conditioning plot with two conditioning shingles. The upper row shows the predominantly Catholic provinces.

```
        panel.xyplot(x, y); panel.loess(x, y, span)
    }
)
```

The result is shown in Figure 4.19, with the strips continuing to show the (now overlapping) coverage for each panel. Fertility generally falls as education rises and rises as the proportion of Catholics in the population rises. Note that the level of education is lower in predominantly Catholic provinces.

The function `equal.count` is used to construct a shingle with suitable ranges for the conditioning intervals.

Conditioning plots may also have more than one conditioning variable. Let us condition on Catholic and agriculture simultaneously. Since the dataset is small it seems prudent to limit the number of panels to six in all.

```
Cath2 <- equal.count(swiss$Catholic, number = 2, overlap = 0)
Agr <- equal.count(swiss$Agric, number = 3, overlap = 0.25)
xyplot(Fertility ~ Education | Agr * Cath2, data = swiss,
    span = 1, aspect = "xy",
    panel = function(x, y, span) {
        panel.xyplot(x, y); panel.loess(x, y, span)
    }
)
```

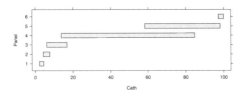

Figure 4.21: A plot of the shingle `Cath`.

The result is shown in Figure 4.20. In general, the fertility rises with the proportion of Catholics and agriculture and falls with education. There is no convincing evidence of substantial interaction.

Shingles have levels, and can be printed and plotted:

```
> Cath
Data:
 [1] 10.0  84.8  93.4  33.8   5.2  90.6  92.9  97.2  97.7  91.4
    ....
Intervals:
   min    max count
   2.2    4.5    10
   4.2    7.7    10
    ....
Overlap between adjacent intervals:
[1] 3 2 3 2 3
> levels(Cath)
   min    max
   2.2    4.5
    ....
> plot(Cath, aspect = 0.3)
```

Multiple displays per page

Recall from page 89 that a Trellis object is plotted by printing it. The method `print.trellis` has optional arguments `position`, `split` and `more`. The argument `more` should be set to T for all but the last part of a figure. The position of individual parts on the device surface can be set by either `split` or `position`. A `split` argument is of the form $c(x, y, nx, ny)$ for four integers. The second pair gives a division into a $nx \times ny$ layout, just like the `mfrow` and `mfcol` arguments to `par`. The first pair gives the rectangle to be used within that layout, with origin at the bottom left.

A `position` argument is of the form c(*xmin, ymin, xmax, ymax*) giving the corners of the rectangle within which to plot the object. (This is a different order from `split.screen`.) The coordinate system for this rectangle is $[0, 1]$ for both axes, but the limits can be chosen outside this range.

The `print.trellis` works by manipulating the graphics parameter `omi`, so the outer margin settings are preserved. However, none of the basic methods (page 78) of subdividing the device surface will work, and if a trellis print

fails omi is not reset. (Using par(omi = rep(0, 4), new = F) will reset
the usual defaults.)

Fine control

Detailed control of Trellis plots may be accomplished by a series of arguments
described in the help page for trellis.args, with variants for the wireframe
and cloud perspective plots under trellis.3d.args.

We have seen some of the uses of panel functions. Some care is needed with
computations inside panel functions that use any data (or user-defined objects or
functions) other than their arguments. First, the computations will occur inside a
deeply nested set of function calls, so care is needed to ensure that the data are
visible, often best done by passing the data as extra arguments. Second, those
computations will be done at the time the result is printed (that is, plotted) and so
the data need to be in the desired state at plot time, not just when the trellis object
is created.

If non-default panel functions are used, we may want these to help control
the coordinate system of the plots, for example, to use a fitted curve to decide
the aspect ratio of the panels. This is the purpose of the prepanel argument,
and there are prepanel functions corresponding to the densityplot, lmline,
loess, qq, qqmath and qqmathline panel functions. These will ensure that
the whole of the fitted curve is visible, and they may affect the choice of aspect
ratio.

The parameter aspect controls the aspect ratio of the panels. A numerical
value (most usefully one) sets the ratio, "fill" adjusts the aspect ratio to fill
the space available and "xy" attempts to bank the fitted curves to $\pm45°$. (See
Figure 4.20.)

The scales argument determines how the x and y axes are drawn. It
is a list of components of name = value form, and components x and y
may themselves be lists. The default relation = "same" ensures that the
axes on each panel are identical. With relation = "sliced" the same numbers of data units are used, but the origin may vary by panel, whereas with
relation = "free" the axes are drawn to accommodate just the data for that
panel. One can also specify most of the parameters of the axis function, and
also log = T to obtain a $\log_{10}$ scale or even log = 2 for a $\log_2$ scale.

The function splom has an argument varnames which sets the names of the
variables plotted on the diagonal. The argument pscales determines how the
axes are plotted; set pscales = 0 to omit them.

Keys

The function key is a replacement for legend, and can also be used as an argument to Trellis functions. If used in this way, the Trellis routines allocate space
for the key, and repeat it on each page if the trellis extends to multiple pages.

The call of key specifies the location of the key by the arguments x, y and
corner. By default corner = c(0, 1), when the coordinate (x, y) specifies
the upper left corner of the key. Any other coordinate of the key can be specified

by setting `corner`, but the size of the key is computed from its contents. (If the argument `plot = F`, the function returns a two-element vector of the computed width and height, which can be used to allocate space.) When `key` is used as an argument to a Trellis function, the position is normally specified not by `x` and `y` but by the argument `space` which defaults to `"top"`.

Most of the remaining arguments to `key` will specify the contents of the key. The (optional) arguments `points`, `lines`, `text` and `rectangles` (for `barchart`) will each specify a column of the key in the order in which they appear. Each argument must be a *list* giving the graphics parameters to be used (and for `text`, the first argument must be the character vector to be plotted). (The function `trellis.par.get` is useful to retrieve the actual settings used for graphics parameters.)

The third group of arguments to `key` fine-tunes its appearance—should it be transparent (`transparent = T`), the presence of a border (specified by giving the border colour as argument `border`), the spacing between columns (`between.columns` in units of character width), the background colour, the font(s) used, the existence of a title and so on. Consult the on-line help for the current details. The argument `columns` specifies the number of columns in the key—we used this in Figures 4.11 and 4.13.

Perspective plots

The argument `aspect` is a vector of two values for the perspective plots, giving the ratio of the y and z sizes to the x size; its effect can be seen in Figure 4.12.

The arguments `distance`, `perspective` and `screen` control the perspective view used. If `perspective = T` (the default), the `distance` argument (default 0.2) controls the extent of the perspective, although not on a physical distance scale as 1 corresponds to viewing from infinity. The `screen` argument (default `list(z = 40, x = -60)`) is a list giving the rotations (in degrees) to be applied to the specified axis in turn. The initial coordinate system has x pointing right, z up and y into the page.

The argument `zoom` (default 1) may be used to scale the final plot, and the argument `par.box` controls how the lines forming the enclosing box are plotted.

Chapter 5

Univariate Statistics

In this chapter we cover a number of topics from classical univariate statistics plus some modern versions.

5.1 Probability Distributions

In this section we confine attention to *univariate* distributions, that is, the distributions of random variables X taking values in the real line $\mathbb{R}$.

The standard distributions used in statistics are defined by probability density functions (for continuous distributions) or probability functions $P(X = n)$ (for discrete distributions). We refer to both as *densities* since the probability functions can be viewed mathematically as densities (with respect to counting measure). The *cumulative distribution function* or CDF is $F(x) = P(X \leqslant x)$ which is expressed in terms of the density by a sum for discrete distributions and as an integral for continuous distributions. The *quantile function* $Q(u) = F^{-1}(u)$ is the inverse of the CDF where this exists. Thus the quantile function gives the percentage points of the distribution. (For discrete distributions the quantile is the smallest integer m such that $F(m) \geqslant u$.)

S has built-in functions to compute the density, cumulative distribution function and quantile function for many standard distributions. In many cases the functions cannot be written in terms of standard mathematical functions, and the built-in functions are very accurate approximations. (But this is also true for the numerical calculation of cosines, for example.)

The first letter of the name of the S function indicates the function so, for example, `dnorm`, `pnorm`, `qnorm` are respectively, the density, CDF and quantile functions for the normal distribution. The rest of the function name is an abbreviation of the distribution name. The distributions available[1] are listed in Table 5.1.

The first argument of the function is always the observation value q (for quantile) for the densities and CDF functions, and the probability p for quantile functions. Additional arguments specify the parameters, with defaults for 'standard' versions of the distributions where appropriate. Precise descriptions of the parameters are given in the on-line help pages.

---

[1]More are available in the R package `SuppDists` by Bob Wheeler.

Table 5.1: S function names and parameters for standard probability distributions.

Distribution	S name	Parameters
beta	beta	shape1, shape2
binomial	binom	size, prob
Cauchy	cauchy	location, scale
chi-squared	chisq	df
exponential	exp	rate
F	f	df1, df2
gamma	gamma	shape, rate
geometric	geom	prob
hypergeometric	hyper	m, n, k
log-normal	lnorm	meanlog, sdlog
logistic	logis	location, scale
negative binomial	nbinom	size, prob
normal	norm	mean, sd
Poisson	pois	lambda
T	t	df
uniform	unif	min, max
Weibull	weibull	shape, scale
Wilcoxon	wilcox	m, n

These functions can be used to replace statistical tables. For example, the 5% critical value for a (two-sided) t test on 11 degrees of freedom is given by qt(0.975, 11), and the P value associated with a Poisson(25)-distributed count of 32 is given by (by convention) 1 - ppois(31, 25). The functions can be given vector arguments to calculate several P values or quantiles.

Q-Q Plots

One of the best ways to compare the distribution of a sample x with a distribution is to use a Q-Q plot. The normal probability plot is the best-known example. For a sample x the quantile function is the inverse of the empirical CDF; that is,

$$\text{quantile}\,(p) = \min\,\{z \mid \text{proportion } p \text{ of the data} \leqslant z\,\}$$

The function quantile calculates the quantiles of a single set of data. The function qqplot(x, y, ...) plots the quantile functions of two samples x and y against each other, and so compares their distributions. The function qqnorm(x) replaces one of the samples by the quantiles of a standard normal distribution. This idea can be applied quite generally. For example, to test a sample against a t_9 distribution we might use

```
plot( qt(ppoints(x), 9), sort(x) )
```

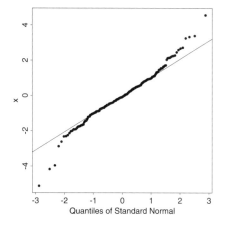

Figure 5.1: Normal probability plot of 250 simulated points from the t_9 distribution.

where the function `ppoints` computes the appropriate set of probabilities for the
plot. These values are $(i - 1/2)/n$,[2] and are generated in increasing order.   R
    The function `qqline` helps assess how straight a `qqnorm` plot is by plotting
a straight line through the upper and lower quartiles. To illustrate this we generate 250 points from a t distribution and compare them to a normal distribution
(Figure 5.1).

```
x <- rt(250, df = 9)
par(pty = "s")
qqnorm(x); qqline(x)
```

The greater spread of the extreme quantiles for the data is indicative of a long-
tailed distribution.
    This method is most often applied to residuals from a fitted model. Some
people prefer the roles of the axes to be reversed, so that the data go on the x-axis
and the theoretical values on the y-axis; this is achieved by giving `qqnorm` by
the argument `datax = T`.

Fitting univariate distributions

Users sometimes find it surprising that S does not come with facilities[3] to fit the
standard univariate distributions to a data vector. We decided to write a general-
purpose maximum-likelihood fitting routine that can be applied to any of the univariate distributions in Table 5.1. Of course, in many cases[4] the maximum likelihood estimators are known in closed form. Function `fitdistr` in MASS fits one
or more continuous parameters with or without constraints. For example,

---

[2] In R and in earlier S-PLUS versions, $(i - 3/8)/(n + 1/4)$ for $n \leqslant 10$.
[3] We understand that these are available in the extra-cost Environmental Statistics module for
S-PLUS.
[4] Including binomial, exponential geometric, log-normal, normal, Poisson and uniform.

```
> x <- rgamma(100, shape = 5, rate = 0.1)
> fitdistr(x, "gamma")
   shape        rate
  4.9614       0.099018
 (0.66552 )  (0.014112)
> x2 <- rt(250, df = 9)
> fitdistr(x2, "t", df = 9)
     m          s
  0.12754     1.0963
 (0.075919)  (0.058261)
> fitdistr(x2, "t")
     m          s         df
  0.12326     1.0502     6.2594
 (0.075343)  (0.078349) (2.9609  )
```

Note that fitdistr does not protect you for what may be rather ill-advised attempts, including estimating ν for a t distribution (Lange *et al.*, 1989; Fernandez and Steel, 1999). The estimated standard errors in parentheses are from the observed information matrix.

Some lateral thinking gives other ways to fit some distributions. Function survReg[5] discussed in Chapter 13 fits parametric models of the form (13.10), which reduce in the special case of no covariates and no censoring to

$$\ell(T) \sim \mu + \sigma \epsilon$$

The dist argument specifies the distribution of ϵ as one of weibull (the default), exponential, rayleigh, lognormal or loglogistic, all with a log transformation ℓ, or extreme, logistic, gaussian or t with an identity transformation.

Multivariate distributions

S-PLUS supplies functions dmvnorm, pmvnorm and rmvnorm, but pmvnorm applies only to bivariate distributions (except in the trivial case of independent components). These specify the distribution via the mean and either of the variance matrix or the correlation matrix and the standard deviations of the components.

Our library section MASS has function mvrnorm that also allows producing samples with specified empirical mean vector and covariance matrix.

The R package mvtnorm by Torsten Hothorn has functions pmvnorm and pmvt for normal, central and non-central t distributions in two or more dimensions, as well as dmvnorm and rmvnorm.

5.2   Generating Random Data

There are S functions to generate independent random samples from all the probability distributions listed in Table 5.1. These have prefix r and first argument n,

---

[5] survreg in package survival in R.

the size of the sample required. For most of the functions the parameters can be specified as vectors, allowing the samples to be non-identically distributed. For example, we can generate 100 samples from the contaminated normal distribution in which a sample is from $N(0, 1)$ with probability 0.95 and otherwise from $N(0, 9)$, by

```
contam <- rnorm( 100, 0, (1 + 2*rbinom(100, 1, 0.05)) )
```

The function `sample` resamples from a data vector, with or without replacement. It has a number of quite different forms. Here `n` is an integer, `x` is a data vector and `p` is a probability distribution on `1, ..., length(x)`:

`sample(n)`	select a random permutation from $1, \ldots, n$
`sample(x)`	randomly permute `x`, for `length(x) > 1`.
`sample(x, replace = T)`	a bootstrap sample
`sample(x, n)`	sample n items from `x` without replacement
`sample(x, n, replace = T)`	sample n items from `x` with replacement
`sample(x, n, replace = T, prob=p)`	probability sample of n items from `x`.

The last of these provides a way to sample from an arbitrary (finite) discrete distribution; set `x` to the vector of values and `prob` to the corresponding probabilities.

The numbers produced by these functions are of course *pseudo*-random rather than genuinely random. The state of the generator is controlled by a set of integers stored in the S object `.Random.seed`. Whenever `sample` or an 'r' function is called, `.Random.seed` is read in, used to initialize the generator, and then its current value is written back out at the end of the function call. If there is no `.Random.seed` in the current working database, one is created with default values (a fixed value in S-PLUS, from the current time in R).

To re-run a simulation call the function `set.seed` initially. This selects from one of at least 1024 pre-selected seeds.

Details of the pseudo-random number generators are given in our on-line complements.[6]

5.3   Data Summaries

Standard univariate summaries such as `mean`, `median` and `var` are available. The `summary` function returns the mean, quartiles and the number of missing values, if non-zero.

The `var` function will take a data matrix and give the variance-covariance matrix, and `cor` computes the correlations, either from two vectors or a data matrix. The function `cov.wt` returns the means and variance[7] matrix and optionally the correlation matrix of a data matrix. As its name implies, the rows of the data matrix can be weighted in these summaries.

---

[6] See page 461 for where to obtain these.

[7] `cov.wt` does not adjust its divisor for the estimation of the mean, using divisor n even when unweighted.

The function `quantile` computes quantiles at specified probabilities, by default $(0, 0.25, 0.5, 0.75, 1)$ giving a "five number summary" of the data vector. This function linearly interpolates, so if $x_{(1)}, \ldots, x_{(n)}$ is the ordered sample,

$$\texttt{quantile(x, p)} = [1 - (p(n-1) - \lfloor p(n-1) \rfloor)] \, x_{1+\lfloor p(n-1) \rfloor}$$
$$+ [p(n-1) - \lfloor p(n-1) \rfloor] \, x_{2+\lfloor p(n-1) \rfloor}$$

where $\lfloor \, \rfloor$ denotes the 'floor' or integer part of. (This differs from the definitions of a quantile given earlier for use with a Q-Q plot. Hyndman and Fan (1996) give several other definitions as used in other systems, and supply S code to implement them.) There are also standard functions `max`, `min` and `range`.

R   The functions `mean` and `cor` (not R) will compute trimmed summaries using the argument `trim`. More sophisticated robust summaries are discussed in Section 5.5.

These functions differ in the way they handle missing values. Functions `mean`, `median`, `max`, `min`, `range` and `quantile` have an argument `na.rm` that defaults to false, but can be used to remove missing values. The functions `var` and `cor` allow several options for the handling of missing values.

Histograms and stem-and-leaf plots

The standard histogram function is `hist(x, ...)` which plots a conventional histogram. More control is available via the extra arguments; `probability = T` gives a plot of unit total area rather than of cell counts.

The argument `nclass` of `hist` suggests the number of bins, and `breaks` specifies the breakpoints between bins. One problem is that `nclass` is only a suggestion, and it is often exceeded. Another is that the definition of the bins that is of the form $[x_0, x_1], (x_1, x_2], \ldots$, not the convention most people prefer.[8]

The default for `nclass` is $\lceil \log_2 n + 1 \rceil$. This is known as Sturges' formula, corresponding to a bin width of $\text{range}(x)/(\log_2 n + 1)$, based on a histogram of a normal distribution (Scott, 1992, p. 48). Note that outliers may inflate the range dramatically and so increase the bin width in the centre of the distribution. Two rules based on compromises between the bias and variance of the histogram for a reference normal distribution are to choose bin width as

$$h = 3.5 \hat{\sigma} n^{-1/3} \tag{5.1}$$
$$h = 2 R n^{-1/3} \tag{5.2}$$

due to Scott (1979) and Freedman and Diaconis (1981), respectively. Here $\hat{\sigma}$ is the estimated standard deviation and R the inter-quartile range. The Freedman–Diaconis formula is immune to outliers, and chooses rather smaller bins than the Scott formula. These are available as `nclass = "fd"` and `nclass = "scott"`.

These are not always satisfactory, as Figure 5.2 shows. (The suggested numbers of bins are 8, 5, 25, 7, 42 and 35; the numbers actually used differ in different

---

[8]The special treatment of x_0 is governed by the switch `include.lowest` which defaults to T. In R argument `right = FALSE` can be used to select left-closed and right-open bins.

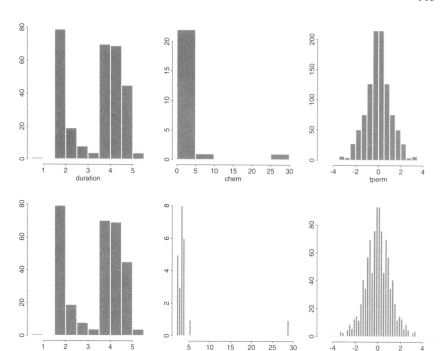

Figure 5.2: Histograms drawn by hist with bin widths chosen by the Scott rule (5.1) (top row) and the Freedman–Diaconis rule (5.2) (bottom row) for datasets geyser$duration, chem and tperm. Computed in S-PLUS 6.0.

implementations.) Column duration of data frame geyser gives the duration (in minutes) of 299 eruptions of the Old Faithful geyser in the Yellowstone National Park (from Azzalini and Bowman, 1990); chem is discussed later in this section and tperm in Section 5.7.

The beauty of S is that it is easy to write one's own function to plot a histogram. Our function is called truehist (in MASS). The primary control is by specifying the bin width h, but suggesting the number of bins by argument nbins will give a 'pretty' choice of h. Function truehist is used in Figures 5.5 and later.

A *stem-and-leaf* plot is an enhanced histogram. The data are divided into bins, but the 'height' is replaced by the next digits in order. We apply this to swiss.fertility, the standardized fertility measure for each of 47 French-speaking provinces of Switzerland at about 1888. The output here is from S-PLUS; R's is somewhat different.

```
> # R: data(swiss); swiss.fertility <- swiss[, 1]
> stem(swiss.fertility)

N = 47    Median = 70.4
Quartiles = 64.4, 79.3
```

```
Decimal point is 1 place to the right of the colon

   3 : 5
   4 : 35
   5 : 46778
   6 : 024455555678899
   7 : 00222345677899
   8 : 0233467
   9 : 222
```

Apart from giving a visual picture of the data, this gives more detail. The actual data, in sorted order, are 35, 43, 45, 54, ... and this can be read from the plot. Sometimes the pattern of numbers (all odd? many 0s and 5s?) gives clues. Quantiles can be computed (roughly) from the plot. If there are outliers, they are marked separately:

```
> stem(chem)

N = 24    Median = 3.385
Quartiles = 2.75, 3.7

Decimal point is at the colon

   2 : 22445789
   3 : 00144445677778
   4 :
   5 : 3

High: 28.95
```

(This dataset on measurements of copper in flour from the Analytical Methods Committee (1989a) is discussed further in Section 5.5.) Sometimes the result is less successful, and manual override is needed:

```
> stem(abbey)

N = 31    Median = 11
Quartiles = 8, 16

Decimal point is at the colon

    5 : 2
    6 : 59
    7 : 0004
    8 : 00005
    . . . .
   26 :
   27 :
   28 : 0
```

```
High:    34  125

> stem(abbey, scale = -1) ## use scale = 0.4 in R

N = 31    Median = 11
Quartiles = 8, 16

Decimal point is 1 place to the right of the colon

    0 : 56777778888899
    1 : 011224444
    1 : 6778
    2 : 4
    2 : 8

High:    34  125
```

Here the `scale` argument sets the backbone to be 10s rather than units. In S-PLUS the `nl` argument controls the number of rows per backbone unit as 2, 5 or 10. The details of the design of stem-and-leaf plots are discussed by Mosteller and Tukey (1977), Velleman and Hoaglin (1981) and Hoaglin, Mosteller and Tukey (1983).

Boxplots

A *boxplot* is a way to look at the overall shape of a set of data. The central box shows the data between the 'hinges' (roughly quartiles), with the median represented by a line. 'Whiskers' go out to the extremes of the data, and very extreme points are shown by themselves.

```
par(mfrow = c(1, 2))   # Figure 5.3
boxplot(chem, sub = "chem", range = 0.5)
boxplot(abbey, sub = "abbey")
par(mfrow = c(1, 1))
bwplot(type ~ y | meas, data = fgl.df, scales = list(x="free"),
    strip = function(...) strip.default(..., style=1), xlab = "")
```

Note how these plots are dominated by the outliers.

There is a bewildering variety of optional parameters to `boxplot` documented in the on-line help page. It is possible to plot boxplots for groups side by side (see Figure 14.16 on page 408) but the Trellis function `bwplot` (see Figure 5.4 on page 116 and Figure 4.9 on page 92) will probably be preferred.

5.4   Classical Univariate Statistics

S-PLUS and R each have a section on classical statistics. The same functions are used to perform tests and to calculate confidence intervals.

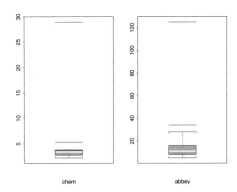

Figure 5.3: Boxplots for the `chem` and `abbey` data.

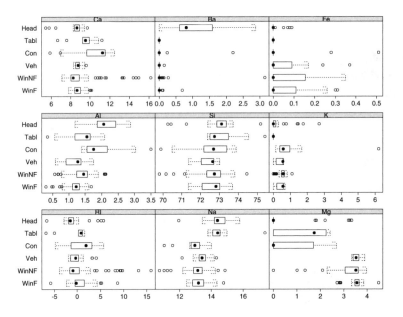

Figure 5.4: Boxplots by type for the `fgl` dataset.

Table 5.2 shows the amounts of shoe wear in an experiment reported by Box, Hunter and Hunter (1978). There were two materials (A and B) that were randomly assigned to the left and right shoes of 10 boys. We use these data to illustrate one-sample and paired and unpaired two-sample tests. (The rather voluminous output has been edited.)

First we test for a mean of 10 and give a confidence interval for the mean, both for material A.

```
> attach(shoes)
> t.test(A, mu = 10)
```

Table 5.2: Data on shoe wear from Box, Hunter and Hunter (1978).

boy	*A*		*B*	
1	13.2	(L)	14.0	(R)
2	8.2	(L)	8.8	(R)
3	10.9	(R)	11.2	(L)
4	14.3	(L)	14.2	(R)
5	10.7	(R)	11.8	(L)
6	6.6	(L)	6.4	(R)
7	9.5	(L)	9.8	(R)
8	10.8	(L)	11.3	(R)
9	8.8	(R)	9.3	(L)
10	13.3	(L)	13.6	(R)

```
        One-sample t-Test

data:  A
t = 0.8127, df = 9, p-value = 0.4373
alternative hypothesis: true mean is not equal to 10
95 percent confidence interval:
  8.8764 12.3836
sample estimates:
 mean of x
     10.63

> t.test(A)$conf.int
[1]   8.8764 12.3836
attr(, "conf.level"):
[1] 0.95

> wilcox.test(A, mu = 10)

        Exact Wilcoxon signed-rank test

data:  A
signed-rank statistic V = 34, n = 10, p-value = 0.5566
alternative hypothesis: true mu is not equal to 10
```

Next we consider two-sample paired and unpaired tests, the latter assuming equal variances or not. Note that we are using this example for illustrative purposes; only the paired analyses are really appropriate.

```
> var.test(A, B)

        F test for variance equality

data:  A and B
```

```
F = 0.9474, num df = 9, denom df = 9, p-value = 0.9372
95 percent confidence interval:
 0.23532 3.81420
sample estimates:
 variance of x variance of y
         6.009        6.3427


> t.test(A, B, var.equal = T)  # default in S-PLUS

        Standard Two-Sample t-Test

data:  A and B
t = -0.3689, df = 18, p-value = 0.7165
95 percent confidence interval:
 -2.7449  1.9249
sample estimates:
 mean of x mean of y
     10.63     11.04


> t.test(A, B, var.equal = F) # default in R

        Welch Modified Two-Sample t-Test

data:  A and B
t = -0.3689, df = 17.987, p-value = 0.7165
95 percent confidence interval:
 -2.745  1.925
    ....


> wilcox.test(A, B)

        Wilcoxon rank-sum test

data:  A and B
rank-sum normal statistic with correction Z = -0.5293,
   p-value = 0.5966


> t.test(A, B, paired = T)

        Paired t-Test

data:  A and B
t = -3.3489, df = 9, p-value = 0.0085
95 percent confidence interval:
 -0.68695 -0.13305
sample estimates:
 mean of x - y
        -0.41


> wilcox.test(A, B, paired = T)
```

```
           Wilcoxon signed-rank test

data:   A and B
signed-rank normal statistic with correction Z = -2.4495,
    p-value = 0.0143
```

The sample size is rather small, and one might wonder about the validity of the t-distribution. An alternative for a randomized experiment such as this is to base inference on the permutation distribution of d = B-A. Figure 5.5 shows that the agreement is very good. The computation of the permutations is discussed in Section 5.7.

The full list of classical tests in S-PLUS is:

```
binom.test       chisq.test       cor.test         fisher.test
friedman.test    kruskal.test     mantelhaen.test  mcnemar.test
prop.test        t.test           var.test         wilcox.test
chisq.gof        ks.gof
```

Many of these have alternative methods—for cor.test there are methods "pearson", "kendall" and "spearman". We have already seen one- and two-sample versions of t.test and wilcox.test, and var.test which compares the variances of two samples. The function cor.test tests for non-zero correlation between two samples, either classically or via ranks. R has all the .test functions, and many more univariate tests.[9]   R

Functions chisq.gof and ks.gof (not in R) compute chi-square and   R
Kolmogorov–Smirnov tests of goodness-of-fit. Function cdf.compare plots comparisons of cumulative distributions such as the right-hand panel of Figure 5.5.

```
par(mfrow = c(1, 2))
truehist(tperm, xlab = "diff")
x <- seq(-4, 4, 0.1)
lines(x, dt(x, 9))
# S: cdf.compare(tperm, distribution = "t", df = 9)
# R: alternative in the scripts
legend(-5, 1.05, c("Permutation dsn","t_9 cdf"), lty = c(1, 3))
```

5.5   Robust Summaries

Outliers are sample values that cause surprise in relation to the majority of the sample. This is not a pejorative term; outliers may be correct, but they should always be checked for transcription errors. They can play havoc with standard statistical methods, and many *robust* and *resistant* methods have been developed since 1960 to be less sensitive to outliers.

---

[9]Use library(help = ctest) to see the current list. For some exact distributions see package exactRankTests by Torsten Hothorn.

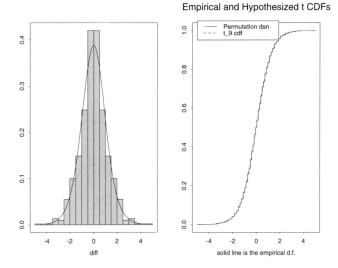

Figure 5.5: Histogram and empirical CDF of the permutation distribution of the paired *t*–test in the shoes example. The density and CDF of t_9 are shown overlaid.

The sample mean $\overline{y}$ can be upset completely by a single outlier; if any data value $y_i \rightarrow \pm\infty$, then $\overline{y} \rightarrow \pm\infty$. This contrasts with the sample median, which is little affected by moving any single value to $\pm\infty$. We say that the median is *resistant* to *gross errors* whereas the mean is not. In fact the median will tolerate up to 50% gross errors before it can be made arbitrarily large; we say its *break-down point* is 50% whereas that for the mean is 0%. Although the mean is the optimal estimator of the location of the normal distribution, it can be substantially sub-optimal for distributions close to the normal. Robust methods aim to have high efficiency in a neighbourhood of the assumed statistical model.

There are several books on robust statistics. Huber (1981) is rather theoretical, Hampel *et al.* (1986) and Staudte and Sheather (1990) less so. Rousseeuw and Leroy (1987) is principally concerned with regression, but is very practical. Robust and resistant methods have long been one of the strengths of S.

Why will it not suffice to screen data and remove outliers? There are several aspects to consider.

1. Users, even expert statisticians, do not always screen the data.

2. The sharp decision to keep or reject an observation is wasteful. We can do better by down-weighting dubious observations than by rejecting them, although we may wish to reject completely wrong observations.

3. It can be difficult or even impossible to spot outliers in multivariate or highly structured data.

4. Rejecting outliers affects the distribution theory, which ought to be adjusted. In particular, variances will be underestimated from the 'cleaned' data.

For a fixed underlying distribution, we define the *relative efficiency* of an estimator $\tilde{\theta}$ relative to another estimator $\hat{\theta}$ by

$$RE(\tilde{\theta};\hat{\theta}) = \frac{\text{variance of } \hat{\theta}}{\text{variance of } \tilde{\theta}}$$

since $\hat{\theta}$ needs only RE times as many observations as $\tilde{\theta}$ for the same precision, approximately. The asymptotic relative efficiency (ARE) is the limit of the RE as the sample size $n \to \infty$. (It may be defined more widely via asymptotic variances.) If $\hat{\theta}$ is not mentioned, it is assumed to be the optimal estimator. There is a difficulty with biased estimators whose variance can be small or zero. One solution is to use the mean-square error, another to rescale by $\theta/E(\hat{\theta})$. Iglewicz (1983) suggests using var $(\log \hat{\theta})$ (which is scale-free) for estimators of scale.

We can apply the concept of ARE to the mean and median. At the normal distribution $ARE(\text{median}; \text{mean}) = 2/\pi \approx 64\%$. For longer-tailed distributions the median does better; for the t distribution with five degrees of freedom (which is often a better model of error distributions than the normal) $ARE(\text{median}; \text{mean}) \approx 96\%$.

The following example from Tukey (1960) is more dramatic. Suppose we have n observations $Y_i \sim N(\mu, \sigma^2), i = 1, \ldots, n$ and we want to estimate σ^2. Consider $\hat{\sigma}^2 = s^2$ and $\tilde{\sigma}^2 = d^2\pi/2$ where

$$d = \frac{1}{n}\sum_i |Y_i - \overline{Y}|$$

and the constant is chosen since for the normal $d \to \sqrt{2/\pi}\,\sigma$. The $ARE(\tilde{\sigma}^2; s^2)$ = 0.876. Now suppose that each Y_i is from $N(\mu, \sigma^2)$ with probability $1 - \epsilon$ and from $N(\mu, 9\sigma^2)$ with probability ϵ. (Note that both the overall variance and the variance of the uncontaminated observations are proportional to σ^2.) We have

ϵ (%)	$ARE(\tilde{\sigma}^2; s^2)$
0	0.876
0.1	0.948
0.2	1.016
1	1.44
5	2.04

Since the mixture distribution with $\epsilon = 1\%$ is indistinguishable from normality for all practical purposes, the optimality of s^2 is very fragile. We say it lacks *robustness of efficiency*.

There are better estimators of σ than $d\sqrt{\pi/2}$ (which has breakdown point 0%). Two alternatives are proportional to

$$IQR = X_{(3n/4)} - X_{(n/4)}$$
$$MAD = \underset{i}{\text{median}}\,\{|Y_i - \underset{j}{\text{median}}\,(Y_j)|\}$$

(Order statistics are linearly interpolated where necessary.) At the normal,

$$MAD \rightarrow \text{median}\{|Y - \mu|\} \approx 0.6745\sigma$$
$$IQR \rightarrow \sigma\left[\Phi^{-1}(0.75) - \Phi^{-1}(0.25)\right] \approx 1.35\sigma$$

(We refer to $MAD/0.6745$ as the MAD estimator, calculated by function mad.) Both are not very efficient but are very resistant to outliers in the data. The MAD estimator has ARE 37% at the normal (Staudte and Sheather, 1990, p. 123).

Consider n independent observations Y_i from a location family with pdf $f(y - \mu)$ for a function f symmetric about zero, so it is clear that μ is the centre (median, mean if it exists) of the distribution of Y_i. We also think of the distribution as being not too far from the normal. There are a number of obvious estimators of μ, including the sample mean, the sample median, and the MLE.

The *trimmed mean* is the mean of the central $1 - 2\alpha$ part of the distribution, so αn observations are removed from each end. This is implemented by the function mean with the argument trim specifying α. Obviously, trim = 0 gives the mean and trim = 0.5 gives the median (although it is easier to use the function median). (If αn is not an integer, the integer part is used.)

Most of the location estimators we consider are *M-estimators*. The name derives from 'MLE-like' estimators. If we have density f, we can define $\rho = -\log f$. Then the MLE would solve

$$\min_{\mu} \sum_i -\log f(y_i - \mu) = \min_{\mu} \sum_i \rho(y_i - \mu)$$

and this makes sense for functions ρ not corresponding to pdfs. Let $\psi = \rho'$ if this exists. Then we will have $\sum_i \psi(y_i - \hat{\mu}) = 0$ or $\sum_i w_i (y_i - \hat{\mu}) = 0$ where $w_i = \psi(y_i - \hat{\mu})/(y_i - \hat{\mu})$. This suggests an iterative method of solution, updating the weights at each iteration.

Examples of M-estimators

The mean corresponds to $\rho(x) = x^2$, and the median to $\rho(x) = |x|$. (For even n any median will solve the problem.) The function

$$\psi(x) = \begin{cases} x & |x| < c \\ 0 & \text{otherwise} \end{cases}$$

corresponds to *metric trimming* and large outliers have no influence at all. The function

$$\psi(x) = \begin{cases} -c & x < -c \\ x & |x| < c \\ c & x > c \end{cases}$$

is known as *metric Winsorizing*[10] and brings in extreme observations to $\mu \pm c$. The corresponding $-\log f$ is

$$\rho(x) = \begin{cases} x^2 & \text{if } |x| < c \\ c(2|x| - c) & \text{otherwise} \end{cases}$$

---

[10] A term attributed by Dixon (1960) to Charles P. Winsor.

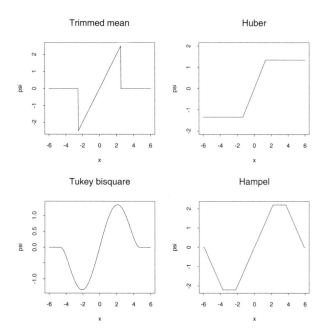

Figure 5.6: The ψ-functions for four common M-estimators.

and corresponds to a density with a Gaussian centre and double-exponential tails. This estimator is due to Huber. Note that its limit as $c \to 0$ is the median, and as $c \to \infty$ the limit is the mean. The value $c = 1.345$ gives 95% efficiency at the normal.

Tukey's *biweight* has

$$\psi(t) = t\left[1 - \left(\frac{t}{R}\right)^2\right]_+^2$$

where $[\,]_+$ denotes the positive part of. This implements 'soft' trimming. The value $R = 4.685$ gives 95% efficiency at the normal.

Hampel's ψ has several linear pieces,

$$\psi(x) = \text{sgn}(x)\begin{cases} |x| & 0 < |x| < a \\ a & a < |x| < b \\ a(c - |x|)/(c - b) & b < |x| < c \\ 0 & c < |x| \end{cases}$$

for example, with $a = 2.2s, b = 3.7s, c = 5.9s$. Figure 5.6 illustrates these functions.

There is a scaling problem with the last four choices, since they depend on a scale factor (c, R or s). We can apply the estimator to rescaled results, that is,

$$\min_\mu \sum_i \rho\left(\frac{y_i - \mu}{s}\right)$$

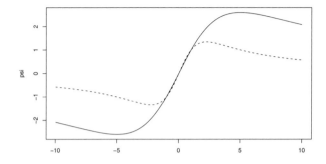

Figure 5.7: Function ψ for maximum-likelihood fitting of t_{25} (solid) and t_5 (dashed).

for a scale factor s, for example the MAD estimator. Alternatively, we can estimate s in a similar way. The MLE for density $s^{-1}f((x-\mu)/s)$ gives rise to the equation

$$\sum_i \psi\left(\frac{y_i-\mu}{s}\right)\left(\frac{y_i-\mu}{s}\right) = n$$

which is not resistant (and is biased at the normal). We modify this to

$$\sum_i \chi\left(\frac{y_i-\mu}{s}\right) = (n-1)\gamma$$

for bounded χ, where γ is chosen for consistency at the normal distribution, so $\gamma = E\chi(N)$. The main example is "Huber's proposal 2" with

$$\chi(x) = \psi(x)^2 = \min(|x|, c)^2 \tag{5.3}$$

In very small samples we need to take account of the variability of $\hat{\mu}$ in performing the Winsorizing.

If the location μ is known we can apply these estimators with $n-1$ replaced by n to estimate the scale s alone.

It is interesting to compare these estimators with maximum-likelihood estimation for a real long-tailed distribution, say t_ν. Figure 5.7 shows the functions $\psi = (-\log f)'$ for $\nu = 25, 5$, both of which show mildly re-descending form.

S-PLUS provides functions `location.m` for location M-estimation and `mad`, and `scale.tau` does scale estimation. Our library section MASS supplies functions `huber` and `hubers` for the Huber M-estimator with MAD and "proposal 2" scale respectively, with default $c = 1.5$.

Examples

We give two datasets taken from analytical chemistry (Abbey, 1988; Analytical Methods Committee, 1989a,b). The dataset `abbey` contains 31 determinations of nickel content ($\mu g\ g^{-1}$) in SY-3, a Canadian syenite rock. Dataset `chem` contains 24 determinations of copper ($\mu g\ g^{-1}$) in wholemeal flour; these data are part of a larger study that suggests $\mu = 3.68$.

```
> sort(chem)
 [1]   2.20  2.20  2.40  2.40  2.50  2.70  2.80  2.90  3.03
[10]   3.03  3.10  3.37  3.40  3.40  3.40  3.50  3.60  3.70
[19]   3.70  3.70  3.70  3.77  5.28 28.95
> mean(chem)
[1] 4.2804
> median(chem)
[1] 3.385
> # S: location.m(chem)
[1] 3.1452
    ....
> # S: location.m(chem, psi.fun = "huber")
[1] 3.2132
    ....
> mad(chem)
[1] 0.52632
> # S: scale.tau(chem)
[1] 0.639
> # S: scale.tau(chem, center = 3.68)
[1] 0.91578
> unlist(huber(chem))
     mu       s
 3.2067 0.52632
> unlist(hubers(chem))
     mu       s
 3.2055 0.67365
> fitdistr(chem, "t", list(m = 3, s = 0.5), df = 5)
      m       s
 3.1854 0.64217
```

The sample is clearly highly asymmetric with one value that appears to be out by
a factor of 10. It was checked and reported as correct by the laboratory. With
such a distribution the various estimators are estimating different aspects of the
distribution and so are not comparable. Only for symmetric distributions do all the
location estimators estimate the same quantity, and although the true distribution
is unknown here, it is unlikely to be symmetric.

```
> sort(abbey)
 [1]    5.2   6.5   6.9   7.0   7.0   7.0   7.4   8.0   8.0
[10]    8.0   8.0   8.5   9.0   9.0  10.0  11.0  11.0  12.0
[19]   12.0  13.7  14.0  14.0  14.0  16.0  17.0  17.0  18.0
[28]   24.0  28.0  34.0 125.0
> mean(abbey)
[1] 16.006
> median(abbey)
[1] 11
> # S: location.m(abbey)
[1] 10.804
> # S: location.m(abbey, psi.fun = "huber")
[1] 11.517
```

```
> unlist(hubers(abbey))
    mu      s
 11.732 5.2585
> unlist(hubers(abbey, k = 2))
    mu      s
 12.351 6.1052
> unlist(hubers(abbey, k = 1))
    mu      s
 11.365 5.5673
> fitdistr(abbey, "t", list(m = 12, s = 5), df = 10)
     m      s
 11.925 7.0383
```

Note how reducing the constant k (representing c) reduces the estimate of location, as this sample (like many in analytical chemistry, where most gross errors stem from contamination) has a long right tail.

5.6 Density Estimation

The non-parametric estimation of probability density functions is a large topic; several books have been devoted to it, notably Silverman (1986), Scott (1992), Wand and Jones (1995) and Simonoff (1996). Bowman and Azzalini (1997) concentrate on providing an introduction to kernel-based methods, providing an easy-to-use package sm.[11]

The histogram with probability = T is of course an estimator of the density function. The histogram depends on the starting point of the grid of bins. The effect can be surprisingly large; see Figure 5.8. The figure also shows that by averaging the histograms we can obtain a much clearer view of the distribution.

This idea of an *average shifted histogram* or *ASH* density estimate is a useful motivation and is discussed in detail in Scott (1992). However, the commonest from of density estimation is a *kernel density estimate* of the form

$$\hat{f}(x) = \frac{1}{nb} \sum_{j=1}^{n} K\left(\frac{x - x_j}{b}\right) \tag{5.4}$$

for a sample $x_1, \ldots, x_n$, a fixed kernel $K()$ and a bandwidth b; the kernel is normally chosen to be a probability density function.

S+   S-PLUS has a function density. The default kernel is the normal (argument window="g" for Gaussian), with alternatives "rectangular", "triangular" and "cosine" (the latter being $(1 + \cos \pi x)/2$ over $[-1, 1]$). The bandwidth width is the length of the non-zero section for the alternatives, and four times the standard deviation for the normal. (Note that these definitions are twice and four times those most commonly used.)

R   R also has a function density, with bandwidth specified as bw,[12] a multiple

---

[11] Available for S-PLUS from http://www.stats.gla.ac.uk/~adrian/sm and http://azzalini.stat.unipd.it/Book_sm, and for R from CRAN.

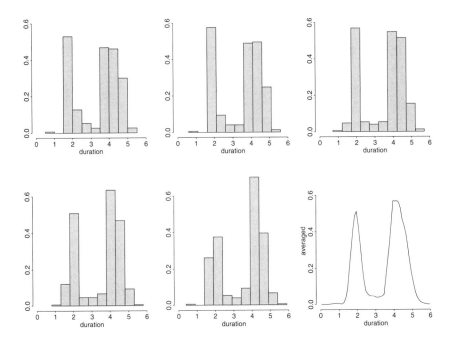

Figure 5.8: Five shifted histograms with bin width 0.5 and the frequency polygon of their average, for the Old Faithful geyser duration data. The code used is in the scripts for this chapter.

of the standard deviation of the kernel specified by kernel.[13] There is a wider range of kernels but once again the normal kernel is the default.

The choice of bandwidth is a compromise between smoothing enough to remove insignificant bumps and not smoothing too much to smear out real peaks. Mathematically there is a compromise between the bias of $\hat{f}(x)$, which increases as b is increased, and the variance which decreases. Theory (see (5.6) on page 129) suggests that the bandwidth should be proportional to $n^{-1/5}$, but the constant of proportionality depends on the unknown density.

The default choice of bandwidth in S-PLUS is not recommendable. There are better-supported rules of thumb such as

$$\hat{b} = 1.06 \min(\hat{\sigma}, R/1.34)n^{-1/5} \qquad (5.5)$$

for the IQR R and the Gaussian kernel of bandwidth the standard deviation (to be quadrupled for use with S-PLUS) (Silverman, 1986, pp. 45–47), invoked by width = "nrd". The default in R is the variant with 1.06 replaced by 0.9 R (Silverman, 1986, (3.31) on p. 48).

---

[12] width can be used for compatibility with S-PLUS.
[13] window can also be used.

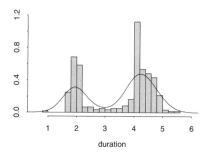 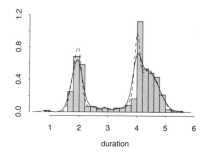

Figure 5.9: Density plots for the Old Faithful duration data. Superimposed on a histogram are the kernel density estimates with a Gaussian kernel and normal reference bandwidth (left) and the Sheather–Jones 'direct plug-in' (right, solid) and 'solve-the-equation' (right, dashed) bandwidth estimates with a Gaussian kernel.

For other kernels there are theoretical scaling factors to give 'equivalent' kernels (Scott, 1992, pp. 141–2), but these are very close to scaling to the same standard deviation for the kernel (and so are not needed in R's parametrization).

We return to the geyser duration data. Here the bandwidth suggested by (5.5) is too large (Figure 5.9). (Scott, 1992, §6.5.1.2) suggests that the value

$$b_{OS} = 1.144\hat{\sigma}n^{-1/5}$$

provides an upper bound on the (Gaussian) bandwidths one would want to consider, again to be quadrupled for use in `density`. The normal reference given by (5.5) is only slightly below this upper bound, and will be too large for densities with multiple modes: see Figure 5.9, the left panel of which was generated by

```
attach(geyser)
truehist(duration, nbins = 15, xlim = c(0.5, 6), ymax = 1.2)
lines(density(duration, width = "nrd"))
```

Note that the data in this example are (negatively) serially correlated, so the theory for independent data must be viewed as only a guide.

Bandwidth selection

Ways to find compromise value(s) of b automatically are discussed by Scott (1992). These are based on estimating the *mean integrated square error*

$$MISE = E\int |\hat{f}(x;b) - f(x)|^2 \, dx = \int E|\hat{f}(x;b) - f(x)|^2 \, dx$$

and choosing the smallest value as a function of b. We can expand $MISE$ as

$$MISE = E\int \hat{f}(x;b)^2 \, dx - 2E\hat{f}(X;b) + \int f(x)^2 \, dx$$

where the third term is constant and so can be dropped.

A currently favoured approach is to make an asymptotic expansion of $MISE$ of the form

$$MISE = \frac{1}{nb} \int K^2 + \frac{1}{4} b^4 \int (f'')^2 \{ \int x^2 K \}^2 + O(1/nb + b^4)$$

so if we neglect the remainder, the optimal bandwidth would be

$$b_{AMISE} = \left[\frac{\int K^2}{n \int (f'')^2 \{ \int x^2 K \}^2} \right]^{1/5} \tag{5.6}$$

The 'direct plug-in' estimators use (5.6), which involves the integral $\int (f'')^2$. This in turn is estimated using the second derivative of a kernel estimator with a different bandwidth, chosen by repeating the process, this time using a reference bandwidth. The 'solve-the-equation' estimators solve (5.6) when the bandwidth for estimating $\int (f'')^2$ is taken as function of b (in practice proportional to $b^{5/7}$). Details are given by Sheather and Jones (1991) and Wand and Jones (1995, §3.6); this is implemented for the Gaussian kernel in functions `bandwidth.sj` (S-PLUS) and `bw.SJ` (R). The right panel of Figure 5.9 was generated by

```
truehist(duration, nbins = 15, xlim = c(0.5, 6), ymax = 1.2)
lines(density(duration, width = "SJ", n = 256), lty = 3)
# R: lines(density(duration, width = "SJ-dpi", n = 256), lty = 1)
# S: lines(density(duration, n = 256,
                width = bandwidth.sj(duration, method = "dpi")),
        lty = 1)
```

There have been a number of comparative studies of bandwidth selection rules (including Park and Turlach, 1992, and Cao, Cuevas and González-Manteiga, 1994) and a review by Jones, Marron and Sheather (1996). 'Second generation' rules such as Sheather–Jones seem to be preferred and indeed to be close to optimal.

As another example, consider our dataset `galaxies` from Roeder (1990), which shows evidence of at least four peaks (Figure 5.10).

```
gal <- galaxies/1000
plot(x = c(0, 40), y = c(0, 0.3), type = "n", bty = "l",
     xlab = "velocity of galaxy (1000km/s)", ylab = "density")
rug(gal)
# S: lines(density(gal, width = bandwidth.sj(gal, method = "dpi"),
                n = 256), lty = 1)
# R: lines(density(gal, width = "SJ-dpi", n = 256), lty = 1)
lines(density(gal, width = "SJ", n = 256), lty = 3)
## see later for explanation
library(logspline)
x <- seq(5, 40, length = 500)
lines(x, dlogspline(x, logspline.fit(gal)), lty = 2)
```

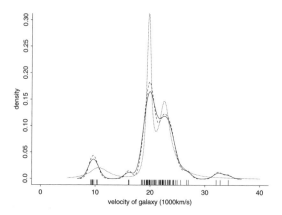

Figure 5.10: Density estimates for the 82 points of the `galaxies` data. The solid and dashed lines are Gaussian kernel density estimates with bandwidths chosen by two variants of the Sheather–Jones method. The dotted line is a logspline estimate.

End effects

Most density estimators will not work well when the density is non-zero at an end of its support, such as the exponential and half-normal densities. (They are designed for continuous densities and this is discontinuity.) One trick is to reflect the density and sample about the endpoint, say, a. Thus we compute the density for the sample `c(x, 2a-x)`, and take double its density on $[a, \infty)$ (or $(-\infty, a]$ for an upper endpoint). This will impose a zero derivative on the estimated density at a, but the end effect will be much less severe. For details and further tricks see Silverman (1986, §3.10). The alternative is to modify the kernel near an endpoint (Wand and Jones, 1995, §2.1), but we know of no S implementation of such *boundary kernels*.

Two-dimensional data

It is often useful to look at densities in two dimensions. Visualizing in more dimensions is difficult, although Scott (1992) provides some examples of visualization of three-dimensional densities.

The dataset on the Old Faithful geyser has two components, `duration`, the duration which we have studied, and `waiting`, the waiting time in minutes until the next eruption. There is also evidence of non-independence of the durations.

S+ S-PLUS provides a function `hist2d` for two-dimensional histograms, but its output is too rough to be useful. We apply two-dimensional kernel analysis directly; this is most straightforward for the normal kernel aligned with axes, that is, with variance $\mathrm{diag}(h_x^2, h_y^2)$. Then the kernel estimate is

$$f(x, y) = \frac{\sum_s \phi\big((x - x_s)/h_x\big)\phi\big((y - y_s)/h_y\big)}{n\, h_x h_y}$$

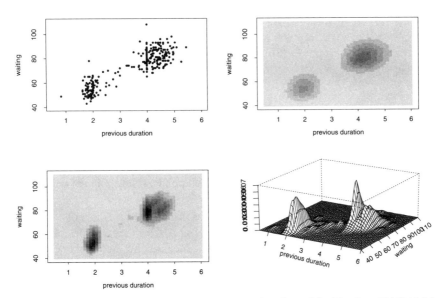

Figure 5.11: Scatter plot and two-dimensional density plots of the bivariate Old Faithful geyser data. Note the effects of the observations of duration rounded to two or four minutes. The code to produce this plot is in the scripts.

which can be evaluated on a grid as XY^T where $X_{is} = \phi\big((gx_i - x_s)/h_x\big)$ and (gx_i) are the grid points, and similarly for Y. Our function kde2d [14] implements this; the results are shown in Figure 5.11.

```
geyser2 <- data.frame(as.data.frame(geyser)[-1, ],
                      pduration = geyser$duration[-299])
attach(geyser2)
par(mfrow = c(2, 2))
plot(pduration, waiting, xlim = c(0.5, 6), ylim = c(40, 110),
     xlab = "previous duration", ylab = "waiting")
f1 <- kde2d(pduration, waiting, n = 50, lims=c(0.5, 6, 40, 110))
image(f1, zlim = c(0, 0.075),
         xlab = "previous duration", ylab = "waiting")
f2 <- kde2d(pduration, waiting, n = 50, lims=c(0.5, 6, 40, 110),
     h = c(width.SJ(duration), width.SJ(waiting)) )
image(f2, zlim = c(0, 0.075),
         xlab = "previous duration", ylab = "waiting")
# S: persp(f2, eye = c(50, -1000, 0.5),
# R: persp(f2,  phi = 30, theta = 20, d = 5,
         xlab = "previous duration", ylab = "waiting", zlab = "")
```

Users of **S-PLUS** or **R** under Windows can explore interactively the fitted density surface, as illustrated on page 422.

An alternative approach using binning and the 2D fast Fourier transform is taken by Wand's function bkde2D in library section KernSmooth.

---

[14]This predates a similar function of the same name on statlib.

Density estimation *via* model fitting

There are several proposals (Simonoff, 1996, pp. 67–70, 90–92) to use a univariate density estimator of the form

$$f(y) = \exp g(y; \theta) \tag{5.7}$$

for a parametric family $g(\cdot; \theta)$ of smooth functions, most often splines. The fit criterion is maximum likelihood, possibly with a smoothness penalty. The advantages of (5.7) are that it automatically provides a non-negative density estimate, and that it may be more natural to consider 'smoothness' on a relative rather than absolute scale.

The library section `logspline` by Charles Kooperberg implements one variant on this theme by Kooperberg and Stone (1992). This uses a cubic spline (see page 229) for g in (5.7), with smoothness controlled by the number of knots selected. There is an AIC-like penalty; the number of the knots is chosen to maximize

$$\sum_{i=1}^{n} g(y_i; \widehat{\theta}) - n \log \int \exp g(y; \widehat{\theta}) \, dy - a \times \text{number of parameters} \tag{5.8}$$

The default value of a is $\log n$ (sometimes known as BIC) but this can be specified by argument `penalty` of `logspline.fit`. The initial knots are selected at quantiles of the data and then deleted one at a time using the Wald criterion for significance. Finally, (5.8) is used to choose one of the knot sequences considered.

Local polynomial fitting

Kernel density estimation can be seen as fitting a locally constant function to the data; other approaches to density estimation use local polynomials (Fan and Gijbels, 1996; Loader, 1999) which have the advantage of being much less sensitive to end effects. There are S implementations in the library sections `KernSmooth` by Wand and `locfit` by Loader.

We compare kernel density and local polynomial estimators for the `galaxies` data. The difference for the same bandwidth[15] is negligible except at the ends.

```
library(KernSmooth)
plot(x = c(0, 40), y = c(0, 0.3), type = "n", bty = "l",
     xlab = "velocity of galaxy (1000km/s)", ylab = "density")
rug(gal)
lines(bkde(gal, bandwidth = dpik(gal)))
lines(locpoly(gal, bandwidth = dpik(gal)), lty = 3)
```

---

[15] dpik is Wand's implementation of "SJ-dpi" .

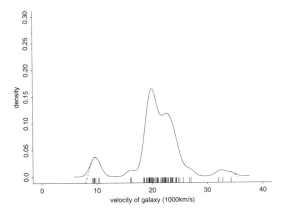

Figure 5.12: Density estimates for `galaxies` by kernel (solid line) and local polynomial (dashed line) methods.

5.7   Bootstrap and Permutation Methods

Several modern methods of what is often called *computer-intensive statistics* make use of extensive repeated calculations to explore the sampling distribution of a parameter estimator $\hat{\theta}$. Suppose we have a random sample $x_1, \ldots, x_n$ drawn independently from one member of a parametric family $\{F_\theta \mid \theta \in \Theta\}$ of distributions. Suppose further that $\hat{\theta} = T(\boldsymbol{x})$ is a *symmetric* function of the sample, that is, does not depend on the sample order.

The *bootstrap* procedure (Efron and Tibshirani, 1993; Davison and Hinkley, 1997) is to take m samples from $\boldsymbol{x}$ *with replacement* and to calculate $\hat{\theta}^*$ for these samples, where conventionally the asterisk is used to denote a bootstrap resample. Note that the new samples consist of an integer number of copies of each of the original data points, and so will normally have ties. Efron's idea was to assess the variability of $\hat{\theta}$ about the unknown true θ by the variability of $\hat{\theta}^*$ about $\hat{\theta}$. In particular, the bias of $\hat{\theta}$ may be estimated by the mean of $\hat{\theta}^* - \hat{\theta}$.

As an example, suppose that we needed to know the median m of the `galaxies` data. The obvious estimator is the sample median, which is 20 833 km/s. How accurate is this estimator? The large-sample theory says that the median is asymptotically normal with mean m and variance $1/4n\,f(m)^2$. But this depends on the unknown density at the median. We can use our best density estimators to estimate $f(m)$, but as we have seen we can find considerable bias and variability if we are unlucky enough to encounter a peak (as in a unimodal symmetric distribution). Let us try the bootstrap:

```
density(gal, n = 1, from = 20.833, to = 20.834, width = "SJ")$y
[1] 0.13009
1/(2 * sqrt(length(gal)) * 0.13)
[1] 0.42474
set.seed(101); m <- 1000; res <- numeric(m)
for (i in 1:m) res[i] <- median(sample(gal, replace = T))
```

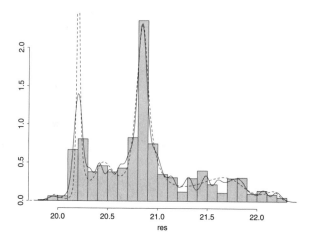

Figure 5.13: Histogram of the bootstrap distribution for the median of the `galaxies` data, with a kernel density estimate (solid) and a logspline density estimate (dashed).

```
mean(res - median(gal))
[1] 0.032258
sqrt(var(res))
[1] 0.50883
```

which took less than a second and confirms the adequacy of the large-sample mean and variance for our example. In this example the bootstrap resampling can be avoided, for the bootstrap distribution of the median can be found analytically (Efron, 1982, Chapter 10; Staudte and Sheather, 1990, p. 84), at least for odd n.

The bootstrap distribution of $\hat{\theta}_i$ about $\hat{\theta}$ is far from normal (Figure 5.13).

```
truehist(res, h = 0.1)
# R: lines(density(res, width = "SJ-dpi", n = 256)
# S: lines(density(res, width = bandwidth.sj(res, method = "dpi"),
                   n = 256))
quantile(res, p = c(0.025, 0.975))
   2.5% 97.5%
 20.175 22.053
x <- seq(19.5, 22.5, length = 500)
lines(x, dlogspline(x, logspline.fit(res)), lty = 3)
```

In larger problems it is important to do bootstrap calculations efficiently, and there are two suites of S functions to do so. S-PLUS has `bootstrap`, and library section[16] `boot` which is very comprehensive.

```
> library(boot)
> set.seed(101)
> gal.boot <- boot(gal, function(x, i) median(x[i]), R = 1000)
> gal.boot
```

---

[16] Written by Angelo Canty to support Davison and Hinkley (1997). Available for S-PLUS and R.

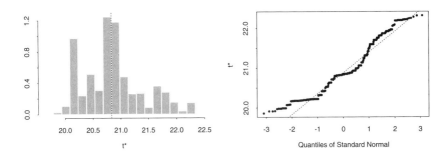

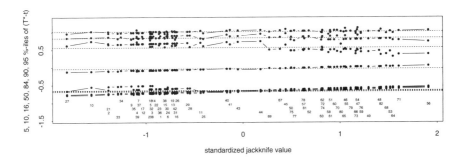

Figure 5.14: Plot for bootstrapped median from the `galaxies` data. The top row shows a histogram (with dashed line the observed value) and a Q-Q plot of the bootstrapped samples. The bottom plot is from `jack.after.boot` and displays the effects of individual observations (here negligible).

```
Bootstrap Statistics :
     original    bias     std. error
t1*   20.834 0.038747     0.52269

boot.ci(gal.boot, conf = c(0.90, 0.95),
        type = c("norm", "basic", "perc", "bca"))

Intervals :
Level       Normal                  Basic
90%   (19.94, 21.65 )     (19.78, 21.48 )
95%   (19.77, 21.82 )     (19.59, 21.50 )

Level     Percentile                BCa
90%   (20.19, 21.89 )     (20.18, 21.87 )
95%   (20.17, 22.07 )     (20.14, 21.96 )

> plot(gal.boot)   # Figure 5.14
```

The `bootstrap` suite of functions has fewer options:

```
> # bootstrap is in S-PLUS only
> gal.bt <- bootstrap(gal, median, seed = 101, B = 1000)
> summary(gal.bt)
   ....

Summary Statistics:
        Observed    Bias  Mean       SE
median   20.83 0.03226 20.87 0.5088

Empirical Percentiles:
        2.5%    5%   95% 97.5%
median 20.18 20.19 21.87 22.05

BCa Confidence Limits:
        2.5%    5%   95% 97.5%
median 20.18 20.19 21.87 22.07
> plot(gal.bt)
> qqnorm(gal.bt)
```

The limits can also be obtained by

```
> limits.emp(gal.bt)
         2.5%     5%     95%   97.5%
median 20.175 20.188 21.867 22.053
> limits.bca(gal.bt)
         2.5%     5%     95%   97.5%
median 20.179 20.193 21.867 22.072
```

One approach to a confidence interval for the parameter θ is to use the quantiles of the bootstrap distributions; this is termed the *percentile confidence interval* and was the original approach suggested by Efron. The bootstrap distribution here is quite asymmetric, and the intervals based on normality are not adequate. The 'basic' intervals are based on the idea that the distribution of $\hat{\theta}_i - \hat{\theta}$ mimics that of $\hat{\theta} - \theta$. If this were so, we would get a $1 - \alpha$ confidence interval as

$$1 - \alpha = P(L \leqslant \hat{\theta} - \theta \leqslant U) \approx P(L \leqslant \hat{\theta}_i - \hat{\theta} \leqslant U)$$

so the interval is $(\hat{\theta} - U, \hat{\theta} - L)$ where $L + \hat{\theta}$ and $U + \hat{\theta}$ are the $\alpha/2$ and $1 - \alpha/2$ points of the bootstrap distribution, say $k_{\alpha/2}$ and $k_{1-\alpha/2}$. Then the basic bootstrap interval is

$$(\hat{\theta} - U, \hat{\theta} - L) = (\hat{\theta} - [k_{1-\alpha/2} - \hat{\theta}], \hat{\theta} - [k_{\alpha/2} - \hat{\theta}]) = (2\hat{\theta} - k_{1-\alpha/2}, 2\hat{\theta} - k_{\alpha/2})$$

which is the percentile interval reflected about the estimate $\hat{\theta}$. In asymmetric problems the basic and percentile intervals will differ considerably (as here), and the basic intervals seem more rational.

The BC_a intervals are an attempt to shift and scale the percentile intervals to compensate for their biases, apparently unsuccessfully in this example. The idea is that if for some unknown increasing transformation g we had $g(\hat{\theta}) - g(\theta) \sim F_0$ for a symmetric distribution F_0, the percentile intervals would be exact. Suppose rather that if $\phi = g(\theta)$,

$$g(\hat{\theta}) - g(\theta) \sim N\big(w\,\sigma(\phi), \sigma^2(\phi)\big) \qquad \text{with } \sigma(\phi) = 1 + a\,\phi$$

Standard calculations (Davison and Hinkley, 1997, p. 204) show that the α confidence limit is given by the $\hat{\alpha}$ percentile of the bootstrap distribution, where

$$\hat{\alpha} = \Phi\left(w + \frac{w + z_\alpha}{1 - a(w + z_\alpha)}\right)$$

and a and w are estimated from the bootstrap samples.

The median (and other sample quantiles) is appreciably affected by discreteness, and it may well be better to sample from a density estimate rather than from the empirical distribution. This is known as the *smoothed bootstrap*. We can do that neatly with `boot`, using a Gaussian bandwidth of 2 and so standard deviation 0.5.

```
sim.gen  <- function(data, mle) {
  n <- length(data)
  data[sample(n, replace = T)]  + mle*rnorm(n)
}
gal.boot2 <- boot(gal, median, R = 1000,
   sim = "parametric", ran.gen = sim.gen, mle = 0.5)
boot.ci(gal.boot2, conf = c(0.90, 0.95),
           type = c("norm","basic","perc"))
Intervals :
Level      Normal             Basic              Percentile
90%  (19.93, 21.48 )   (19.83, 21.36 )   (20.31, 21.83 )
95%  (19.78, 21.63 )   (19.72, 21.50 )   (20.17, 21.95 )
```

The constants a and w in the BC_a interval cannot be estimated by `boot` in this case. The smoothed bootstrap slightly inflates the sample variance (here by 1.2%) and we could rescale the sample if this was appropriate.

For a smaller but simpler example, we return to the differences in shoe wear between materials. There is a fifth type of confidence interval that `boot.ci` can calculate, which needs a variance v^* estimate of the statistic $\hat{\theta}^*$ from each bootstrap sample. Then the confidence interval can be based on the basic confidence intervals for the *studentized* statistics $(\hat{\theta}^* - \hat{\theta})/\sqrt{v^*}$. Theory suggests that the studentized confidence interval may be the most reliable of all the methods we have discussed.

```
> attach(shoes)
> t.test(B - A)
95 percent confidence interval:
 0.13305 0.68695
> shoes.boot <- boot(B - A, function(x,i) mean(x[i]), R = 1000)
> boot.ci(shoes.boot, type = c("norm", "basic", "perc", "bca"))

Intervals :
Level      Normal             Basic
95%  ( 0.1767,  0.6296 )   ( 0.1800,  0.6297 )

Level      Percentile         BCa
95%  ( 0.1903,  0.6400 )   ( 0.1900,  0.6300 )
```

```
mean.fun <- function(d, i) {
  n <- length(i)
  c(mean(d[i]), (n-1)*var(d[i])/n^2)
}
> shoes.boot2 <- boot(B - A, mean.fun, R = 1000)
> boot.ci(shoes.boot2, type = "stud")

Intervals :
Level     Studentized
95%    ( 0.1319,  0.6911 )
```

We have only scratched the surface of bootstrap methods; Davison and Hinkley (1997) provide an excellent practically-oriented account, S software and practical exercises.

Permutation tests

Inference for designed experiments is often based on the distribution over the random choices made during the experimental design, on the belief that this randomness alone will give distributions that can be approximated well by those derived from normal-theory methods. There is considerable empirical evidence that this is so, but with modern computing power we can check it for our own experiment, by selecting a large number of re-labellings of our data and computing the test statistics for the re-labelled experiments.

Consider again the shoe-wear data of Section 5.4. The most obvious way to explore the permutation distribution of the t-test of d = B−A is to select random permutations, but as the permutation distribution has only $2^{10} = 1\,024$ points we can use the exact distribution for Figure 5.5. (The code to generate this efficiently is in the scripts.)

Chapter 6

Linear Statistical Models

Linear models form the core of classical statistics and are still the basis of much of statistical practice; many modern modelling and analytical techniques build on the methodology developed for linear models.

In S most modelling exercises are conducted in a fairly standard way. The dataset is usually held in a single *data frame* object. A primary model is fitted using a *model fitting function*, for which a *formula* specifying the form of the model and the data frame specifying the variables to be used are the basic arguments. The resulting *fitted model object* can be interrogated, analysed and even modified in various ways using generic functions. The important point to note is that the fitted model object carries with it the information that the fitting process has revealed.

Although most modelling exercises conform to this rough paradigm some features of linear models are special. The *formula* for a linear model specifies the response variable and the explanatory variables (or factors) used to model the mean response by a version of the Wilkinson–Rogers notation (Wilkinson and Rogers, 1973) for specifying models that we discuss in Section 6.2.

We begin with an example to give a feel for the process and to present some of the details.

6.1   An Analysis of Covariance Example

The data frame `whiteside` contains a dataset collected in the 1960s by Mr Derek Whiteside of the UK Building Research Station and reported in the collection of small datasets edited by Hand *et al.* (1994, No. 88, p. 69). Whiteside recorded the weekly gas consumption and average external temperature at his own house in south-east England during two 'heating seasons'[1] one before and one after cavity-wall insulation was installed. The object of the exercise was to assess the effect of the insulation on gas consumption.

The variables in data frame `whiteside` are `Insul`, a factor with levels `Before` and `After`, `Temp`, for the weekly average external temperature in degrees Celsius and `Gas`, the weekly gas consumption in 1 000 cubic feet units. We begin by plotting the data in two panels showing separate least-squares lines.

---

[1]We are grateful to Dr Kevin McConway for clarification.

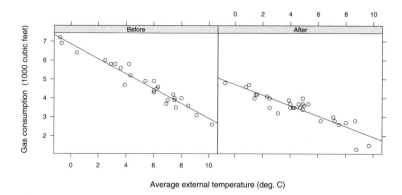

Figure 6.1: Whiteside's data showing the effect of insulation on household gas consumption.

```
xyplot(Gas ~ Temp | Insul, whiteside, panel =
  function(x, y, ...) {
    panel.xyplot(x, y, ...)
    panel.lmline(x, y, ...)
  }, xlab = "Average external temperature (deg. C)",
  ylab = "Gas consumption  (1000 cubic feet)", aspect = "xy",
  strip = function(...) strip.default(..., style = 1))
```

The result is shown in Figure 6.1. Within the range of temperatures given a straight line model appears to be adequate. The plot shows that insulation reduces the gas consumption for equal external temperatures, but it also appears to affect the slope, that is, the rate at which gas consumption increases as external temperature falls.

To explore these issues quantitatively we will need to fit linear models, the primary function for which is `lm`. The main arguments to `lm` are

```
lm(formula, data, weights, subset, na.action)
```

where

> `formula` is the model formula (the only required argument),
>
> `data` is an optional data frame,
>
> `weights` is a vector of positive weights, if non-uniform weights are needed,
>
> `subset` is an index vector specifying a subset of the data to be used (by default all items are used),
>
> `na.action` is a function specifying how missing values are to be handled (by default, missing values are not allowed in S-PLUS but cause cases to be omitted in R.).

R

If the argument `data` is specified, it gives a data frame from which variables are selected ahead of the search path. Working with data frames and using this argument is strongly recommended.

It should be noted that setting `na.action = na.omit` will allow models to be fitted omitting cases that have missing components on a required variable. If any cases are omitted the fitted values and residual vector will no longer match the original observation vector in length; use `na.action = na.exclude` if the fitted values (and so on) should include `NA`s.

Formulae have been discussed in outline in Section 3.7 on page 56. For `lm` the right-hand side specifies the explanatory variables. Operators on the right-hand side of linear model formulae have the special meaning of the Wilkinson–Rogers notation and not their arithmetical meaning.

To fit the separate regressions of gas consumption on temperature as shown in Figure 6.1 we may use

```
gasB <- lm(Gas ~ Temp, data = whiteside, subset = Insul=="Before")
gasA <- update(gasB, subset = Insul=="After")
```

The first line fits a simple linear regression for the 'before' temperatures. The right-hand side of the formula needs only to specify the variable `Temp` since an intercept term (corresponding to a column of unities of the model matrix) is always implicitly included. It may be explicitly included using `1 + Temp`, where the `+` operator implies *inclusion* of a term in the model, not addition.

The function `update` is a convenient way to modify a fitted model. Its first argument is a fitted model object that results from one of the model-fitting functions such as `lm`. The remaining arguments of `update` specify the desired changes to arguments of the call that generated the object. In this case we simply wish to switch subsets from `Insul=="Before"` to `Insul=="After"`; the formula and data frame remain the same. Notice that variables used in the `subset` argument may also come from the data frame and need not be visible on the (global) search path.

Fitted model objects have an appropriate *class*, in this case `"lm"`. Generic functions to perform further operations on the object include

`print` for a simple display,

`summary` for a conventional regression analysis output,

`coef` (or `coefficients`) for extracting the regression coefficients,

`resid` (or `residuals`) for residuals,

`fitted` (or `fitted.values`) for fitted values,

`deviance` for the residual sum of squares,

`anova` for a sequential analysis of variance table, or a comparison of several hierarchical models,

`predict` for predicting means for new data, optionally with standard errors, and

`plot` for diagnostic plots.

Many of these method functions are very simple, merely extracting a component of the fitted model object. The only component likely to be accessed for which no extractor function is supplied[2] is `df.residual`, the residual degrees of freedom.

---

[2]In S-PLUS: there is a `df.residual` function in R.

The output from summary is self-explanatory. Edited results for our fitted models are

```
> summary(gasB)
  ....
Coefficients:
              Value Std. Error t value Pr(>|t|)
(Intercept)   6.854     0.118   57.876    0.000
       Temp  -0.393     0.020  -20.078    0.000

Residual standard error: 0.281 on 24 degrees of freedom

> summary(gasA)
  ....
Coefficients:
              Value Std. Error t value Pr(>|t|)
(Intercept)   4.724     0.130   36.410    0.000
       Temp  -0.278     0.025  -11.036    0.000

Residual standard error: 0.355 on 28 degrees of freedom
```

The difference in residual variances is relatively small, but the formal textbook F-test for equality of variances could easily be done. The sample variances could be extracted in at least two ways, for example

```
varB <- deviance(gasB)/gasB$df.resid    # direct calculation
varB <- summary(gasB)$sigma^2           # alternative
```

It is known this F-test is highly non-robust to non-normality (see, for example, Hampel *et al.* (1986, pp. 55, 188)) so its usefulness here would be doubtful.

To fit both regression models in the same "lm" model object we may use

```
> gasBA <- lm(Gas ~ Insul/Temp - 1, data = whiteside)
> summary(gasBA)
  ....
Coefficients:
                  Value Std. Error t value Pr(>|t|)
    InsulBefore   6.854     0.136   50.409    0.000
     InsulAfter   4.724     0.118   40.000    0.000
InsulBeforeTemp  -0.393     0.022  -17.487    0.000
 InsulAfterTemp  -0.278     0.023  -12.124    0.000

Residual standard error: 0.323 on 52 degrees of freedom
  ....
```

Notice that the estimates are the same but the standard errors are different because they are now based on the pooled estimate of variance.

Terms of the form a/x, where a is a factor, are best thought of as "separate regression models of type 1 + x within the levels of a." In this case an intercept is not needed, since it is replaced by two separate intercepts for the two levels of insulation, and the formula term - 1 removes it.

We can check for curvature in the mean function by fitting separate quadratic rather than linear regressions in the two groups. This may be done as

```
> gasQ <- lm(Gas ~ Insul/(Temp + I(Temp^2)) - 1, data = whiteside)
> summary(gasQ)$coef
                        Value Std. Error  t value    Pr(>|t|)
     InsulBefore    6.7592152  0.1507868  44.8263 0.0000e+00
      InsulAfter    4.4963739  0.1606679  27.9855 0.0000e+00
  InsulBeforeTemp  -0.3176587  0.0629652  -5.0450 6.3623e-06
   InsulAfterTemp  -0.1379016  0.0730580  -1.8876 6.4896e-02
InsulBeforeI(Temp^2) -0.0084726  0.0066247  -1.2789 2.0683e-01
 InsulAfterI(Temp^2) -0.0149795  0.0074471  -2.0114 4.9684e-02
```

The 'identity' function I(...) is used in this context and with data.frame (see page 18). It evaluates its argument with operators having their arithmetical meaning and returns the result. Hence it allows arithmetical operators to be used in linear model formulae, although if any function call is used in a formula their arguments are evaluated in this way.

The separate regression coefficients show that a second-degree term is possibly needed for the After group only, but the evidence is not overwhelming.[3] We retain the separate linear regressions model on the grounds of simplicity.

An even simpler model that might be considered is one with parallel regressions. We can fit this model and test it within the separate regression model using

```
> # R: options(contrasts = c("contr.helmert", "contr.poly"))
> gasPR <- lm(Gas ~ Insul + Temp, data = whiteside)
> anova(gasPR, gasBA)
Analysis of Variance Table
    ....
          Terms Resid. Df    RSS  Test Df Sum of Sq F Value
1    Insul + Temp     53 6.7704
2 Insul/Temp - 1      52 5.4252 1 vs. 2  1   1.3451  12.893

          Pr(F)
1
2 0.00073069
```

When anova is used with two or more nested models it gives an analysis of variance table for those models. In this case it shows that separate slopes are indeed necessary. Note the unusual layout of the analysis of variance table. Here we could conduct this test in a simpler and more informative way. We now fit the model with separate slopes using a different parametrization:

```
> options(contrasts = c("contr.treatment", "contr.poly"))
> gasBA1 <- lm(Gas ~ Insul*Temp, data = whiteside)
> summary(gasBA1)$coef
              Value Std. Error  t value    Pr(>|t|)
(Intercept)  6.85383   0.135964  50.4091 0.0000e+00
      Insul -2.12998   0.180092 -11.8272 2.2204e-16
       Temp -0.39324   0.022487 -17.4874 0.0000e+00
  Insul:Temp 0.11530   0.032112   3.5907 7.3069e-04
```

---

[3]Notice that when the quadratic terms are present first-degree coefficients mean 'the slope of the curve at temperature zero', so a non-significant value does not mean that the linear term is not needed. Removing the non-significant linear term for the 'after' group, for example, would be unjustified.

The call to options is explained more fully in Section 6.2; for now we note that it affects the way regression models are parametrized when factors are used. The formula Insul*Temp expands to 1 + Insul + Temp + Insul:Temp and the corresponding coefficients are, in order, the intercept for the 'before' group, the *difference* in intercepts, the slope for the 'before' group and the *difference* in slopes. Since this last term is significant we conclude that the two separate slopes are required in the model. Indeed note that the F-statistic in the analysis of variance table is the square of the final t-statistic and that the tail areas are identical.

6.2 Model Formulae and Model Matrices

This section contains some rather technical material and might be skimmed at first reading.

A linear model is specified by the response vector y and by the matrix of explanatory variables, or *model matrix*, X. The model formula conveys both pieces of information, the left-hand side providing the response and the right-hand side instructions on how to generate the model matrix according to a particular convention.

A multiple regression with three quantitative determining variables might be specified as y ~ x1 + x2 + x3. This would correspond to a model with a familiar algebraic specification

$$y_i = \beta_0 + \beta_1 x_{i1} + \beta_2 x_{i2} + \beta_3 x_{i3} + \epsilon_i, \qquad i = 1, 2, \ldots, n$$

The model matrix has the partitioned form

$$X = \begin{bmatrix} \mathbf{1} \ x_1 \ x_2 \ x_3 \end{bmatrix}$$

The intercept term (β_0 corresponding to the leading column of ones in X) is implicitly present; its presence may be confirmed by giving a formula such as y ~ 1 + x1 + x2 + x3, but wherever the 1 occurs in the formula the column of ones will always be the first column of the model matrix. It may be omitted and a regression through the origin fitted by giving a - 1 term in the formula, as in y ~ x1 + x2 + x3 - 1.

Factor terms in a model formula are used to specify classifications leading to what are often called analysis of variance models. Suppose a is a factor. An analysis of variance model for the one-way layout defined by a might be written in the algebraic form

$$y_{ij} = \mu + \alpha_j + \epsilon_{ij} \qquad i = 1, 2, \ldots, n_j; \quad j = 1, 2, \ldots, k$$

where there are k classes and the n_j is the size of the jth. Let $n = \sum_j n_j$. This specification is over-parametrized, but we could write the model matrix in the form

$$X = \begin{bmatrix} \mathbf{1} \ X_a \end{bmatrix}$$

where X_a is an $n \times k$ binary incidence (or 'dummy variable') matrix where each row has a single unity in the column of the class to which it belongs.

The redundancy comes from the fact that the columns of X_a add to **1**, making X of rank k rather than $k + 1$. One way to resolve the redundancy is to remove the column of ones. This amounts to setting $\mu = 0$, leading to an algebraic specification of the form

$$y_{ij} = \alpha_j + \epsilon_{ij} \qquad i = 1, 2, \ldots, n_j; \ j = 1, 2, \ldots, k$$

so the α_j parameters are the class means. This formulation may be specified by `y ~ a - 1`.

If we do not break the redundancy by removing the intercept term it must be done some other way, since otherwise the parameters are not identifiable. The way this is done in S is most easily described in terms of the model matrix. The model matrix generated has the form

$$X^\star = \begin{bmatrix} 1 & X_a C_a \end{bmatrix}$$

where C_a, the *contrast matrix* for a, is a $k \times (k - 1)$ matrix chosen so that $X^\star$ has rank k, the number of columns. A necessary (and usually sufficient) condition for this to be the case is that the square matrix $\begin{bmatrix} 1 & C_a \end{bmatrix}$ be non-singular.

The reduced model matrix $X^\star$ in turn defines a linear model, but the parameters are often not directly interpretable and an algebraic formulation of the precise model may be difficult to write down. Nevertheless, the relationship between the newly defined and original (redundant) parameters is clearly given by

$$\alpha = C_a \alpha^\star \tag{6.1}$$

where α are the original α parameters and $\alpha^\star$ are the new.

If c_a is a non-zero vector such that $c_a^T C_a = \mathbf{0}$ it can be seen immediately that using $\alpha^\star$ as parameters amounts to estimating the original parameters, α subject to the *identification constraint* $c_a^T \alpha = 0$ which is usually sufficient to make them unique. Such a vector (or matrix) c_a is called an *annihilator* of C_a or a basis for the orthogonal complement of the range of C_a.

If we fit the one-way layout model using the formula

`y ~ a`

the coefficients we obtain will be estimates of μ and $\alpha^\star$. The corresponding constrained estimates of the α may be obtained by multiplying by the contrasts matrix or by using the function `dummy.coef`. Consider an artificial example:

```
> dat <- data.frame(a = factor(rep(1:3, 3)),
                    y = rnorm(9, rep(2:4, 3), 0.1))
> obj <- lm(y ~ a, dat)
> (alf.star <- coef(obj))
   (Intercept)        a1        a2
        2.9719   0.51452   0.49808
> Ca <- contrasts(dat$a)        # contrast matrix for 'a'
> drop(Ca %*% alf.star[-1])
        1          2          3
  -1.0126   0.016443   0.99615
```

```
> dummy.coef(obj)
$"(Intercept)":
[1] 2.9719

$a:
        1         2        3
  -1.0126 0.016443 0.99615
```

Notice that the estimates of α sum to zero because the contrast matrix used here implies the identification constraint $1^T \alpha = 0$.

Contrast matrices

S+   By default S-PLUS uses so-called *Helmert* contrast matrices for unordered factors and orthogonal polynomial contrast matrices for ordered factors. The forms of these can be deduced from the following artificial example:

```
> N <- factor(Nlevs <- c(0,1,2,4))
> contrasts(N)
  [,1] [,2] [,3]
0  -1   -1   -1
1   1   -1   -1
2   0    2   -1
4   0    0    3
> contrasts(ordered(N))
        .L      .Q        .C
0 -0.67082   0.5 -0.22361
1 -0.22361  -0.5  0.67082
2  0.22361  -0.5 -0.67082
4  0.67082   0.5  0.22361
```

For the `poly` contrasts it can be seen that the corresponding parameters α^* can be interpreted as the coefficients in an orthogonal polynomial model of degree $r - 1$, *provided* the ordered levels are equally spaced (which is not the case for the example) *and* the class sizes are equal. The α^* parameters corresponding to the Helmert contrasts also have an easy interpretation, as we see in the following. Since both the Helmert and polynomial contrast matrices satisfy $1^T C = 0$, the implied constraint on α will be $1^T \alpha = 0$ in both cases.

     The default contrast matrices can be changed by resetting the `contrasts` option. This is a character vector of length two giving the names of the functions that generate the contrast matrices for unordered and ordered factors respectively. For example,

```
options(contrasts = c("contr.treatment", "contr.poly"))
```

sets the default contrast matrix function for factors to `contr.treatment` and for
R   ordered factors to `contr.poly` (the original default). (This *is* the default in R.) Four supplied contrast functions are as follows:

`contr.helmert` for the Helmert contrasts.

`contr.treatment` for contrasts such that each coefficient represents a comparison of
that level with level 1 (omitting level 1 itself). This corresponds to the constraint
$\alpha_1 = 0$. Note that in this parametrization the coefficients are *not* contrasts in the
usual sense.

`contr.sum` where the coefficients are constrained to add to zero; that is, in this case the
components of $\alpha^\star$ are the same as the first $r - 1$ components of α, with the latter
constrained to add to zero.

`contr.poly` for the equally spaced, equally replicated orthogonal polynomial contrasts.

Others can be written using these as templates (as we do with our function
`contr.sdif`, used on pages 293 and 294). We recommend the use of the treatment contrasts for unbalanced layouts, including generalized linear models and
survival models, because the unconstrained coefficients obtained directly from
the fit are then easy to interpret.

Notice that the `helmert`, `sum` and `poly` contrasts ensure the rank condition
on C is met by choosing C so that the columns of $[\mathbf{1}\,C]$ are mutually orthogonal, whereas the `treatment` contrasts choose C so that $[\mathbf{1}\,C]$ is in echelon
form.

Contrast matrices for particular factors may also be set as an attribute of the
factor itself. This can be done either by the `contrasts` replacement function
or by using the function `C` which takes three arguments: the factor, the matrix
from which contrasts are to be taken (or the abbreviated name of a function that
will generate such a matrix) and the number of contrasts. On some occasions a
p-level factor may be given a contrast matrix with fewer than $p - 1$ columns, in
which case it contributes fewer than $p-1$ degrees of freedom to the model, or the
unreduced parameters α have additional constraints placed on them apart from
the one needed for identification. An alternative method is to use the replacement
form with a specific number of contrasts as the second argument. For example,
suppose we wish to create a factor N2 that would generate orthogonal linear and
quadratic polynomial terms, only. Two equivalent ways of doing this would be

```
> N2 <- N
> contrasts(N2, 2) <- poly(Nlevs, 2)
> N2 <- C(N, poly(Nlevs, 2), 2)        # alternative
> contrasts(N2)
          1        2
0 -0.591608  0.56408
1 -0.253546 -0.32233
2  0.084515 -0.64466
4  0.760639  0.40291
```

In this case the constraints imposed on the α parameters are not merely for identification but actually change the model subspace.

Parameter interpretation

The `poly` contrast matrices lead to $\alpha^\star$ parameters that are sometimes interpretable as coefficients in an orthogonal polynomial regression. The `treatment`

contrasts set $\alpha_1 = 0$ and choose the remaining αs as the α^*s. Other cases are often not so direct, but an interpretation is possible.

To interpret the $\boldsymbol{\alpha}^*$ parameters in general, consider the relationship (6.1). Since the contrast matrix C is of full column rank it has a unique left inverse C^+, so we can reverse this relationship to give

$$\boldsymbol{\alpha}^* = C^+ \boldsymbol{\alpha} \quad \text{where} \quad C^+ = (C^T C)^{-1} C^T \tag{6.2}$$

The pattern in the matrix C^+ then provides an interpretation of each unconstrained parameter as a linear function of the (usually) readily appreciated constrained parameters. For example, consider the Helmert contrasts for $r = 4$. To exhibit the pattern in C^+ more clearly we use the function `fractions` from MASS for rational approximation and display.

```
> fractions(ginv(contr.helmert(n = 4)))
      [,1]    [,2]   [,3]  [,4]
[1,]  -1/2    1/2     0     0
[2,]  -1/6   -1/6    1/3    0
[3,] -1/12  -1/12  -1/12  1/4
```

Hence $\alpha_1^\star = \frac{1}{2}(\alpha_2 - \alpha_1)$, $\alpha_2^\star = \frac{1}{3}\{\alpha_3 - \frac{1}{2}(\alpha_1 + \alpha_2)\}$ and in general $\alpha_j^\star$ is a comparison of α_{j+1} with the average of all preceding αs, divided by $j + 1$. This is a comparison of the (unweighted) mean of class $j + 1$ with that of the preceding classes.

It can sometimes be important to use contrast matrices that give a simple interpretation to the fitted coefficients. This can be done by noting that $(C^+)^+ = C$. For example, suppose we wished to choose contrasts so that the $\alpha_j^\star = \alpha_{j+1} - \alpha_j$, that is, the successive differences of class effects. For $r = 5$, say, the C^+ matrix is then given by

```
> Cp <- diag(-1, 4, 5);  Cp[row(Cp) == col(Cp) - 1] <- 1
> Cp
     [,1] [,2] [,3] [,4] [,5]
[1,]  -1    1    0    0    0
[2,]   0   -1    1    0    0
[3,]   0    0   -1    1    0
[4,]   0    0    0   -1    1
```

Hence the contrast matrix to obtain these linear functions as the estimated coefficients is

```
> fractions(ginv(Cp))
      [,1] [,2] [,3] [,4]
[1,] -4/5 -3/5 -2/5 -1/5
[2,]  1/5 -3/5 -2/5 -1/5
[3,]  1/5  2/5 -2/5 -1/5
[4,]  1/5  2/5  3/5 -1/5
[5,]  1/5  2/5  3/5  4/5
```

Note that again the columns have zero sums, so the implied constraint is that the effects add to zero. (If it were not obvious we could find the induced constraint using our function `Null` (page 63) to find a basis for the null space of the contrast matrix.)

The pattern is obvious from this example and a contrast matrix function for the general case can now be written. To be usable as a component of the `contrasts` option such a function has to conform with a fairly strict convention, but the key computational steps are

```
. . . .
    contr <- col(matrix(nrow = n, ncol = n - 1))
    upper.tri <- !lower.tri(contr)
    contr[upper.tri] <- contr[upper.tri] - n
    contr/n
. . . .
```

The complete function is supplied as `contr.sdif` in MASS. We make use of it on pages 293 and 294.

Higher-way layouts

Two- and higher-way layouts may be specified by two or more factors and formula operators. The way the model matrix is generated is then an extension of the conventions for a one-way layout.

If a and b are r- and s-level factors, respectively, the model formula y ~ a+b specifies an additive model for the two-way layout. Using the redundant specification the algebraic formulation would be

$$y_{ijk} = \mu + \alpha_i + \beta_j + \epsilon_{ijk}$$

and the model matrix would be

$$X = \begin{bmatrix} \mathbf{1} \ X_a \ X_b \end{bmatrix}$$

The reduced model matrix then has the form

$$X^\star = \begin{bmatrix} \mathbf{1} \ X_a C_a \ X_b C_b \end{bmatrix}$$

However, if the intercept term is explicitly removed using, say, y ~ a + b - 1, the reduced form is

$$X^\star = \begin{bmatrix} X_a \ X_b C_b \end{bmatrix}$$

Note that this is asymmetric in a and b and order-dependent.

A two-way non-additive model has a redundant specification of the form

$$y_{ijk} = \mu + \alpha_i + \beta_j + \gamma_{ij} + \epsilon_{ijk}$$

The model matrix can be written as

$$X = \begin{bmatrix} \mathbf{1} \ X_a \ X_b \ X_a{:}X_b \end{bmatrix}$$

where we use the notation $A{:}B$ to denote the matrix obtained by taking each column of A and multiplying it element-wise by each column of B. In the example $X_a{:}X_b$ generates an incidence matrix for the sub-classes defined jointly by a and b. Such a model may be specified by the formula y ~ a + b + a:b

or equivalently by y ~ a*b. The reduced form of the model matrix is then

$$X^\star = \begin{bmatrix} 1 & X_a C_a & X_b C_b & (X_a C_a):(X_b C_b) \end{bmatrix}$$

It may be seen that $(X_a C_a):(X_b C_b) = (X_a:X_b)(C_b \otimes C_a)$, where $\otimes$ denotes the Kronecker product, so the relationship between the γ parameters and the corresponding $\gamma^\star$s is

$$\gamma = (C_b \otimes C_a)\gamma^\star$$

The identification constraints can be most easily specified by writing γ as an $r \times s$ matrix. If this is done, the relationship has the form $\gamma = C_a \gamma^\star C_b^T$ and the constraints have the form $c_a^T \gamma = 0^T$ and $\gamma c_b = 0$, separately.

If the intercept term is removed, however, such as by using y ~ a*b - 1, the form is different, namely,

$$X^\star = \begin{bmatrix} X_a & X_b C_b & X_a^{(-r)}:(X_b C_b) \end{bmatrix}$$

where (somewhat confusingly) $X_a^{(-r)}$ is a matrix obtained by removing the *last* column of X_a. Furthermore, if a model is specified as y ~ - 1 + a + a:b the model matrix generated is $\begin{bmatrix} X_a & X_a:(X_b C_b) \end{bmatrix}$. In general addition of a term a:b extends the previously constructed design matrix to a complete non-additive model in some non-redundant way (unless the design is deficient, of course).

Even though a*b expands to a + b + a:b, it should be noted that a + a:b is not always the same[4] as a*b - b or even a + b - b + a:b. When used in model-fitting functions the last two formulae construct the design matrix for a*b and only then remove any columns corresponding to the b term. (The result is not a statistically meaningful model.) Model matrices are constructed within the fitting functions by arranging the positive terms in order of their complexity, sequentially adding columns to the model matrix according to the redundancy resolution rules and then removing any generated columns corresponding to negative terms. The exception to this rule is the intercept term which is always removed initially. (With update, however, the formula is expanded and all negative terms are removed *before* the model matrix is constructed.)

The model a + a:b generates the same matrix as a/b, which expands in S-PLUS to a + b %in% a. There is no compelling reason for the additional operator,[5] %in%, but it does serve to emphasize that the slash operator should be thought of as specifying separate submodels of the form 1 + b for each level of a. The operator behaves like the colon formula operator when the second main effect term is not given, but is conventionally reserved for nested models.

Star products of more than two terms, such as a*b*c, may be thought of as expanding (1 + a):(1 + b):(1 + c) according to ordinary algebraic rules and may be used to define higher-way non-additive layouts. There is also a power operator, ^, for generating models up to a specified degree of interaction term.

---

[4]In S-PLUS: it is always the same in R.
[5]Which R does not have.

For example, `(a+b+c)^3` generates the same model as `a*b*c` but `(a+b+c)^2` has the highest order interaction absent.

Combinations of factors and non-factors with formula operators are useful in an obvious way. We have seen already that `a/x - 1` generates separate simple linear regressions on `x` within the levels of `a`. The same model may be specified as `a + a:x - 1`, whereas `a*x` generates an equivalent model using a different resolution of the redundancy. It should be noted that `(x + y + z)^3` does *not* generate a general third-degree polynomial regression in the three variables, as might be expected. This is because terms of the form `x:x` are regarded as the same as `x`, not as `I(x^2)`. However, in **S-PLUS** a single term such as `x^2` is   S+ silently promoted to `I(x^2)` and interpreted as a power.

6.3   Regression Diagnostics

The message in the Whiteside example is relatively easy to discover and we did not have to work hard to find an adequate linear model. There is an extensive literature (for example Atkinson, 1985) on examining the fit of linear models to consider whether one or more points are not fitted as well as they should be or have undue influence on the fitting of the model. This can be contrasted with the robust regression methods we discuss in Section 6.5, which automatically take account of anomalous points.

The basic tool for examining the fit is the residuals, and we have already looked for patterns in residuals and assessed the normality of their distribution. The residuals are not independent (they sum to zero if an intercept is present) and they do not have the same variance. Indeed, their variance-covariance matrix is

$$\mathrm{var}\left(e\right) = \sigma^2[I - H] \tag{6.3}$$

where $H = X(X^T X)^{-1} X^T$ is the orthogonal projector matrix onto the model space, or *hat* matrix. If a diagonal entry h_{ii} of H is large, changing y_i will move the fitted surface appreciably towards the altered value. For this reason h_{ii} is said to measure the *leverage* of the observation y_i. The trace of H is p, the dimension of the model space, so 'large' is taken to be greater than two or three times the average, p/n.

Having large leverage has two consequences for the corresponding residual. First, its variance will be lower than average from (6.3). We can compensate for this by rescaling the residuals to have unit variance. The *standardized residuals* are

$$e'_i = \frac{e_i}{s\sqrt{1 - h_{ii}}}$$

where as usual we have estimated σ^2 by s^2, the residual mean square. Second, if one error is very large, the variance estimate s^2 will be too large, and this deflates all the standardized residuals. Let us consider fitting the model omitting observation i. We then get a prediction for the omitted observation, $\hat{y}_{(i)}$, and an estimate of the error variance, $s^2_{(i)}$, from the reduced sample. The *studentized residuals* are

$$e_i^* = \frac{y_i - \hat{y}_{(i)}}{\sqrt{\text{var}\left(y_i - \hat{y}_{(i)}\right)}}$$

but with σ replaced by $s_{(i)}$. Fortunately, it is not necessary to re-fit the model each time an observation is omitted, since it can be shown that

$$e_i^* = e_i' \left/ \left[\frac{n - p - e_i'^2}{n - p - 1}\right]^{1/2}\right.$$

Notice that this implies that the standardized residuals, e_i, must be bounded by $\pm\sqrt{n-p}$.

The terminology used here is not universally adopted; in particular studentized residuals are sometimes called *jackknifed* residuals.

It is usually better to compare studentized residuals rather than residuals; in particular we recommend that they be used for normal probability plots.

We have provided functions `studres` and `stdres` to compute studentized and standardized residuals. There is a function `hat`, but this expects the model matrix as its argument. (There is a useful function, `lm.influence`, for most of the fundamental calculations. The diagonal of the hat matrix can be obtained by `lm.influence(lmobject)$hat`.)

Scottish hill races

As an example of regression diagnostics, let us return to the data on 35 Scottish hill races in our data frame `hills` considered in Chapter 1. The data come from Atkinson (1986) and are discussed further in Atkinson (1988) and Staudte and Sheather (1990). The columns are the overall race distance, the total height climbed and the record time. In Chapter 1 we considered a regression of `time` on `dist`. We can now include `climb`:

```
> (hills.lm <- lm(time ~ dist + climb, data = hills))
Coefficients:
 (Intercept)  dist    climb
      -8.992 6.218 0.011048

Degrees of freedom: 35 total; 32 residual
Residual standard error: 14.676
> frame(); par(fig = c(0, 0.6, 0, 0.55))
> plot(fitted(hills.lm), studres(hills.lm))
> abline(h = 0, lty = 2)
> identify(fitted(hills.lm), studres(hills.lm),
           row.names(hills))
> par(fig = c(0.6, 1, 0, 0.55), pty = "s")
> qqnorm(studres(hills.lm))
> qqline(studres(hills.lm))
> hills.hat <- lm.influence(hills.lm)$hat
> cbind(hills, lev = hills.hat)[hills.hat > 3/35, ]
```

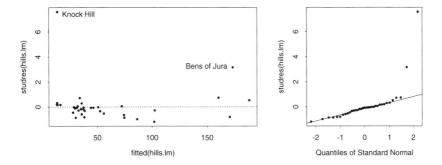

Figure 6.2: Diagnostic plots for Scottish hills data, unweighted model.

```
                  dist climb     time      lev
  Bens of Jura     16  7500 204.617 0.42043
  Lairig Ghru      28  2100 192.667 0.68982
    Ben Nevis      10  4400  85.583 0.12158
Two Breweries      18  5200 170.250 0.17158
Moffat Chase       20  5000 159.833 0.19099
```

so two points have very high leverage, two points have large residuals, and Bens of Jura is in both sets. (See Figure 6.2.)

If we look at Knock Hill we see that the prediction is over an hour less than the reported record:

```
> cbind(hills, pred = predict(hills.lm))["Knock Hill", ]
           dist climb  time    pred
Knock Hill    3   350 78.65 13.529
```

and Atkinson (1988) suggests that the record is one hour out. We drop this observation to be safe:

```
> (hills1.lm <- update(hills.lm, subset = -18))
Coefficients:
 (Intercept)   dist    climb
     -13.53 6.3646 0.011855

Degrees of freedom: 34 total; 31 residual
Residual standard error: 8.8035
```

Since Knock Hill did not have a high leverage, deleting it did not change the fitted model greatly. On the other hand, Bens of Jura had both a high leverage and a large residual and so does affect the fit:

```
> update(hills.lm, subset = -c(7, 18))
Coefficients:
 (Intercept)   dist    climb
    -10.362 6.6921 0.0080468

Degrees of freedom: 33 total; 30 residual
Residual standard error: 6.0538
```

If we consider this example carefully we find a number of unsatisfactory features. First, the prediction is negative for short races. Extrapolation is often unsafe, but on physical grounds we would expect the model to be a good fit with a zero intercept; indeed hill-walkers use a prediction of this sort (3 miles/hour plus 20 minutes per 1 000 feet). We can see from the summary that the intercept is significantly negative:

```
> summary(hills1.lm)
 ....
Coefficients:
            Value Std. Error t value Pr(>|t|)
(Intercept) -13.530    2.649   -5.108    0.000
       dist   6.365    0.361   17.624    0.000
      climb   0.012    0.001    9.600    0.000
 ....
```

Furthermore, we would not expect the predictions of times that range from 15 minutes to over 3 hours to be equally accurate, but rather that the accuracy be roughly proportional to the time. This suggests a log transform, but that would be hard to interpret. Rather we weight the fit using distance as a surrogate for time. We want weights inversely proportional to the variance:

```
> summary(update(hills1.lm, weights = 1/dist^2))
 ....
Coefficients:
            Value Std. Error t value Pr(>|t|)
(Intercept) -5.809    2.034   -2.855    0.008
       dist  5.821    0.536   10.858    0.000
      climb  0.009    0.002    5.873    0.000

Residual standard error: 1.16 on 31 degrees of freedom
```

The intercept is still significantly non-zero. If we are prepared to set it to zero on physical grounds, we can achieve the same effect by dividing the prediction equation by distance, and regressing inverse speed (time/distance) on gradient (climb/distance):

```
> lm(time ~ -1 + dist + climb, hills[-18, ], weights = 1/dist^2)
Coefficients:
 dist    climb
 4.9 0.0084718

Degrees of freedom: 34 total; 32 residual
Residual standard error (on weighted scale): 1.2786
> hills <- hills    # make a local copy (needed in S-PLUS)
> hills$ispeed <- hills$time/hills$dist
> hills$grad <- hills$climb/hills$dist
> (hills2.lm <- lm(ispeed ~ grad, data = hills[-18, ]))
Coefficients:
 (Intercept)      grad
         4.9 0.0084718
```

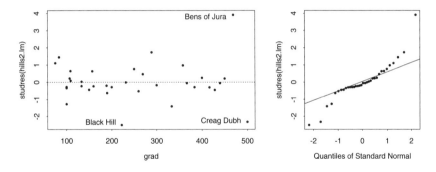

Figure 6.3: Diagnostic plots for Scottish hills data, weighted model.

```
Degrees of freedom: 34 total; 32 residual
Residual standard error: 1.2786
> frame(); par(fig = c(0, 0.6, 0, 0.55))
> plot(hills$grad[-18], studres(hills2.lm), xlab = "grad")
> abline(h = 0, lty = 2)
> identify(hills$grad[-18], studres(hills2.lm),
          row.names(hills)[-18])
> par(fig = c(0.6, 1, 0, 0.55), pty = "s")     # Figure 6.3
> qqnorm(studres(hills2.lm))
> qqline(studres(hills2.lm))
> hills2.hat <- lm.influence(hills2.lm)$hat
> cbind(hills[-18,], lev = hills2.hat)[hills2.hat > 1.8*2/34, ]
              dist climb    time ispeed    grad    lev
Bens of Jura   16  7500 204.617 12.7886 468.75 0.11354
  Creag Dubh    4  2000  26.217  6.5542 500.00 0.13915
```

The two highest-leverage cases are now the steepest two races, and are outliers pulling in opposite directions. We could consider elaborating the model, but this would be to fit only one or two exceptional points; for most of the data we have the formula of 5 minutes/mile plus 8 minutes per 1 000 feet. We return to this example on page 162 where robust fits do support a zero intercept.

6.4   Safe Prediction

A warning is needed on the use of the `predict` method function when polynomials are used (and also splines, see Section 8.8). We illustrate this by the dataset `wtloss`, for which a more appropriate analysis is given in Chapter 8. This has a weight loss `Weight` against `Days`. Consider a quadratic polynomial regression model of `Weight` on `Days`. This may be fitted by either of

```
quad1 <- lm(Weight ~ Days + I(Days^2), data = wtloss)
quad2 <- lm(Weight ~ poly(Days, 2), data = wtloss)
```

The second uses orthogonal polynomials and is the preferred form on grounds of numerical stability.

Suppose we wished to predict future weight loss. The first step is to create a new data frame with a variable x containing the new values, for example,

```
new.x <- data.frame(Days = seq(250, 300, 10),
                     row.names = seq(250, 300, 10))
```

The predict method may now be used:

```
> predict(quad1, newdata = new.x)
    250    260    270    280    290    300
 112.51 111.47 110.58 109.83 109.21 108.74
> predict(quad2, newdata = new.x) # from S-PLUS 6.0
    250    260    270    280    290    300
 244.56 192.78 149.14 113.64 86.29 67.081
```

The first form gives correct answers but the second does not in S-PLUS!

The reason for this is as follows. The predict method for lm objects works by attaching the estimated coefficients to a new model matrix that it constructs using the formula and the new data. In the first case the procedure will work, but in the second case the columns of the model matrix are for a *different* orthogonal polynomial basis, and so the old coefficients do not apply. The same will hold for any function used to define the model that generates mathematically different bases for old and new data, such as spline bases using bs or ns. R retains enough information to predict from the old data.

R

S+   The remedy in S-PLUS is to use the method function predict.gam:

```
> predict.gam(quad2, newdata = new.x) # S-PLUS only
    250    260    270    280    290    300
 112.51 111.47 110.58 109.83 109.21 108.74
```

This constructs a new model matrix by putting old and new data together, re-estimates the regression using the old data only and predicts using these estimates of regression coefficients. This can involve appreciable extra computation, but the results will be correct for polynomials, but not exactly so for splines since the knot positions will change. As a check, predict.gam compares the predictions with the old fitted values for the original data. If these are seriously different, a warning is issued that the process has probably failed.

In our view this is a serious flaw in predict.lm. It would have been better to use the safe method as the default and provide an unsafe argument for the faster method as an option.

6.5 Robust and Resistant Regression

There are a number of ways to perform robust regression in S-PLUS, but many have drawbacks and are not mentioned here. First consider an example. Rousseeuw and Leroy (1987) give data on annual numbers of Belgian telephone calls, given in our dataset phones.

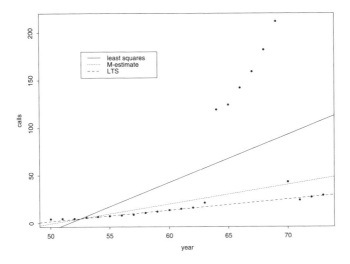

Figure 6.4: Millions of phone calls in Belgium, 1950–73, from Rousseeuw and Leroy (1987), with three fitted lines.

```
# R: library(lqs)
phones.lm <- lm(calls ~ year, data = phones)
attach(phones); plot(year, calls); detach()
abline(phones.lm$coef)
abline(rlm(calls ~ year, phones, maxit=50), lty = 2, col = 2)
abline(lqs(calls ~ year, phones), lty = 3, col = 3)
legend(locator(1), lty = 1:3, col = 1:3,
        legend = c("least squares", "M-estimate", "LTS"))
```

Figure 6.4 shows the least squares line, an M-estimated regression and the least trimmed squares regression (Section 6.5). The `lqs` line is $-56.16 + 1.16\,\mathtt{year}$. Rousseeuw & Leroy's investigations showed that for 1964–9 the total length of calls (in minutes) had been recorded rather than the number, with each system being used during parts of 1963 and 1970.

Next some theory. In a regression problem there are two possible sources of errors, the observations y_i and the corresponding row vector of p regressors x_i. Most robust methods in regression only consider the first, and in some cases (designed experiments?) errors in the regressors can be ignored. This is the case for M-estimators, the only ones we consider in this section.

Consider a regression problem with n cases (y_i, x_i) from the model

$$y = x\beta + \epsilon$$

for a p-variate row vector x.

M-estimators

If we assume a scaled pdf $f(e/s)/s$ for ϵ and set $\rho = -\log f$, the maximum
likelihood estimator minimizes

$$\sum_{i=1}^{n} \rho \left(\frac{y_i - x_i b}{s} \right) + n \log s \tag{6.4}$$

Suppose for now that s is known. Let $\psi = \rho'$. Then the MLE b of β solves

$$\sum_{i=1}^{n} x_i \psi \left(\frac{y_i - x_i b}{s} \right) = 0 \tag{6.5}$$

Let $r_i = y_i - x_i b$ denote the residuals.

The solution to equation (6.5) or to minimizing over (6.4) can be used to define
an M-estimator of β.

A common way to solve (6.5) is by iterated re-weighted least squares, with
weights

$$w_i = \psi \left(\frac{y_i - x_i b}{s} \right) \bigg/ \left(\frac{y_i - x_i b}{s} \right) \tag{6.6}$$

The iteration is guaranteed to converge only for *convex* ρ functions, and for re-
descending functions (such as those of Tukey and Hampel; page 123) equation
(6.5) may have multiple roots. In such cases it is usual to choose a good starting
point and iterate carefully.

Of course, in practice the scale s is not known. A simple and very resistant
scale estimator is the MAD about some centre. This is applied to the residuals
about zero, either to the current residuals within the loop or to the residuals from
a very resistant fit (see the next subsection).

Alternatively, we can estimate s in an MLE-like way. Finding a stationary
point of (6.4) with respect to s gives

$$\sum_{i} \psi \left(\frac{y_i - x_i b}{s} \right) \left(\frac{y_i - x_i b}{s} \right) = n$$

which is not resistant (and is biased at the normal). As in the univariate case we
modify this to

$$\sum_{i} \chi \left(\frac{y_i - x_i b}{s} \right) = (n - p)\gamma \tag{6.7}$$

Our function `rlm`

Our `MASS` library section introduces a new class `rlm` and model-fitting function
`rlm`, building on `lm`. The syntax in general follows `lm`. By default Huber's M-
estimator is used with tuning parameter $c = 1.345$. By default the scale s is
estimated by iterated MAD, but Huber's proposal 2 can also be used.

```
> summary(lm(calls ~ year, data = phones), cor = F)
            Value Std. Error  t value Pr(>|t|)
(Intercept) -260.059  102.607    -2.535    0.019
       year    5.041    1.658     3.041    0.006
Residual standard error: 56.2 on 22 degrees of freedom
> summary(rlm(calls ~ year, maxit = 50, data = phones), cor = F)
            Value Std. Error  t value
(Intercept) -102.622   26.608    -3.857
       year    2.041    0.430     4.748
Residual standard error: 9.03 on 22 degrees of freedom
> summary(rlm(calls ~ year, scale.est = "proposal 2",
              data = phones), cor = F)
Coefficients:
            Value Std. Error  t value
(Intercept) -227.925  101.874    -2.237
       year    4.453    1.646     2.705
Residual standard error: 57.3 on 22 degrees of freedom
```

As Figure 6.4 shows, in this example there is a batch of outliers from a different population in the late 1960s, and these should be rejected completely, which the Huber M-estimators do not. Let us try a re-descending estimator.

```
> summary(rlm(calls ~ year, data = phones, psi = psi.bisquare),
          cor = F)
Coefficients:
            Value Std. Error t value
(Intercept) -52.302    2.753   -18.999
       year   1.098    0.044    24.685
Residual standard error: 1.65 on 22 degrees of freedom
```

This happened to work well for the default least-squares start, but we might want to consider a better starting point, such as that given by init = "lts".

Resistant regression

M-estimators are not very resistant to outliers unless they have redescending ψ functions, in which case they need a good starting point. A succession of more resistant regression estimators was defined in the 1980s. The first to become popular was

$$\min_b \operatorname{median}_i |y_i - x_i b|^2$$

called the *least median of squares* (LMS) estimator. The square is necessary if n is even, when the central median is taken. This fit is very resistant, and needs no scale estimate. It is, however, very inefficient, converging at rate $1/\sqrt[3]{n}$. Furthermore, it displays marked sensitivity to central data values; see Hettmansperger and Sheather (1992) and Davies (1993, §2.3).

Rousseeuw suggested least trimmed squares (LTS) regression:

$$\min_b \sum_{i=1}^{q} |y_i - x_i b|^2_{(i)}$$

as this is more efficient, but shares the same extreme resistance. The recommended sum is over the smallest $q = \lfloor (n+p+1)/2 \rfloor$ squared residuals. (Earlier accounts differed.)

This was followed by *S-estimation*, in which the coefficients are chosen to find the solution to

$$\sum_{i=1}^{n} \chi\left(\frac{y_i - \boldsymbol{x}_i b}{c_0\, s}\right) = (n-p)\beta$$

with smallest scale s. Here χ is usually chosen to be the integral of Tukey's bisquare function

$$\chi(u) = u^6 - 3u^4 + 3u^2,\ |u| \leqslant 1, \qquad 1,\ |u| \geqslant 1$$

$c_0 = 1.548$ and $\beta = 0.5$ is chosen for consistency at the normal distribution of errors. This gives efficiency 28.7% at the normal, which is low but better than LMS and LTS.

In only a few special cases (such as LMS for univariate regression with intercept) can these optimization problems be solved exactly, and approximate search methods are used.

S *implementation*

Various versions of S-PLUS have (different) implementations of LMS and LTS regression in functions `lmsreg` and `ltsreg`[6], but as these are not fully documented and give different results in different releases, we prefer our function `lqs`.[7] The default method is LTS.

```
> lqs(calls ~ year, data = phones)
Coefficients:
 (Intercept)    year
 -56.2          1.16
Scale estimates 1.25 1.13

> lqs(calls ~ year, data = phones, method = "lms")
Coefficients:
 (Intercept)    year
 -55.9          1.15
Scale estimates 0.938 0.909

> lqs(calls ~ year, data = phones, method = "S")
Coefficients:
 (Intercept)  year
 -52.5        1.1
Scale estimates 2.13
```

Two scale estimates are given for LMS and LTS: the first comes from the fit criterion, the second from the variance of the residuals of magnitude no more than 2.5 times the first scale estimate. All the scale estimates are set up to be

---

[6]In S-PLUS this now uses 10% trimming.
[7]Adopted by R in its package `lqs`.

consistent at the normal, but measure different things for highly non-normal data (as here).

MM-estimation

It is possible to combine the resistance of these methods with the efficiency of M-estimation. The MM-estimator proposed by Yohai, Stahel and Zamar (1991) (see also Marazzi, 1993, §9.1.3) is an M-estimator starting at the coefficients given by the S-estimator and with fixed scale given by the S-estimator. This retains (for $c > c_0$) the high-breakdown point of the S-estimator and the high efficiency at the normal. At considerable computational expense, this gives the best of both worlds.

Our function rlm has an option to implement MM-estimation.

```
> summary(rlm(calls ~ year, data=phones, method="MM"), cor = F)
Coefficients:
            Value Std. Error t value
(Intercept) -52.423  2.916   -17.978
      year    1.101  0.047     23.367

Residual standard error: 2.13 on 22 degrees of freedom
```

S-PLUS has a function lmRob in library section robust that implements a   S+
slightly different MM-estimator with similar properties, and comes with a full set of method functions, so it can be used routinely as a replacement for lm. Let us try it on the phones data.

```
> library(robust, first = T) # S-PLUS only
> phones.lmr <- lmRob(calls ~ year, data = phones)
> summary(phones.lmr, cor = F)
Coefficients:
            Value Std. Error t value Pr(>|t|)
(Intercept) -52.541  3.625   -14.493  0.000
      year    1.104  0.061     18.148  0.000

Residual scale estimate: 2.03 on 22 degrees of freedom
Proportion of variation in response explained by model: 0.494

Test for Bias:
            Statistics P-value
 M-estimate      1.401   0.496
LS-estimate      0.243   0.886
> plot(phones.lmr)
```

This works well, rejecting all the spurious observations. The 'test for bias' is of the M-estimator against the initial S-estimator; if the M-estimator appears biased the initial S-estimator is returned.

Library section robust provides a wide range of robust techniques.

Scottish hill races revisited

We return to the data on Scottish hill races studied in the introduction and Section 6.3. There we saw one gross outlier and a number of other extreme observations.

```
> hills.lm
Coefficients:
 (Intercept)   dist     climb
     -8.992   6.218   0.011048
Residual standard error: 14.676

> hills1.lm # omitting Knock Hill
Coefficients:
 (Intercept)   dist     climb
     -13.53   6.3646  0.011855
Residual standard error: 8.8035

> rlm(time ~ dist + climb, data = hills)
Coefficients:
 (Intercept)   dist     climb
    -9.6067   6.5507  0.0082959
Scale estimate: 5.21

> summary(rlm(time ~ dist + climb, data = hills,
            weights = 1/dist^2, method = "MM"), cor = F)
Coefficients:
               Value Std. Error t value
(Intercept)  -1.802    1.664     -1.083
       dist   5.244    0.233     22.549
      climb   0.007    0.001      9.391
Residual standard error: 4.84 on 32 degrees of freedom

> lqs(time ~ dist + climb, data = hills, nsamp = "exact")
Coefficients:
 (Intercept)     dist     climb
    -1.26        4.86    0.00851
Scale estimates 2.94 3.01
```

Notice that the intercept is no longer significant in the robust weighted fits. By default `lqs` uses a random search, but here exhaustive enumeration is possible, so we use it.

If we move to the model for inverse speed:

```
> summary(hills2.lm) # omitting Knock Hill
Coefficients:
               Value Std. Error t value Pr(>|t|)
(Intercept)  4.900    0.474     10.344   0.000
       grad  0.008    0.002      5.022   0.000

Residual standard error: 1.28 on 32 degrees of freedom
```

```
> summary(rlm(ispeed ~ grad, data = hills), cor = F)
Coefficients:
            Value Std. Error t value
(Intercept) 5.176  0.381     13.585
       grad 0.007  0.001      5.428

Residual standard error: 0.869 on 33 degrees of freedom
# method="MM" results are very similar.
> # S: summary(lmRob(ispeed ~ grad, data = hills))
            Value Std. Error t value Pr(>|t|)
(Intercept) 5.082  0.403     12.612   0.000
       grad 0.008  0.002      5.055   0.000

Residual scale estimate: 0.819 on 33 degrees of freedom

> lqs(ispeed ~ grad, data = hills)
Coefficients:
 (Intercept)     grad
 4.75           0.00805
Scale estimates 0.608 0.643
```

The results are in close agreement with the least-squares results after removing
Knock Hill.

6.6 Bootstrapping Linear Models

In frequentist inference we have to consider what might have happened but did
not. Linear models can arise exactly or approximately in a number of ways. The
most commonly considered form is

$$Y = X\beta + \epsilon$$

in which only ϵ is considered to be random. This supposes that in all (hypothet-
ical) repetitions the same x points would have been chosen, but the responses
would vary. This is a plausible assumption for a designed experiment such as the
N, P, K experiment on page 165 and for an observational study such as Quine's
with prespecified factors. It is less clearly suitable for the Scottish hill races, and
clearly not correct for Whiteside's gas consumption data.

Another form of regression is sometimes referred to as the *random regressor*
case in which the pairs (x_i, y_i) are thought of as a random sample from a population and we are interested in the regression function $f(x) = E(Y \mid X = x)$
which is assumed to be linear. This seems appropriate for the gas consumption
data. However, it is common to perform conditional inference in this case and
condition on the observed xs, converting this to a fixed-design problem. For example, in the hill races the inferences drawn depend on whether certain races,
notably Bens of Jura, were included in the sample. As they were included, conclusions conditional on the set of races seems most pertinent. (There are other

ways that linear models can arise, including calibration problems and where both x and y are measured with error about a true linear relationship.)

These considerations are particularly relevant when we consider bootstrap re-sampling. The most obvious form of bootstrapping is to randomly sample pairs (x_i, y_i) with replacement,[8] which corresponds to randomly weighted regressions. However, this may not be appropriate in not mimicking the assumed random variation and in some examples in producing singular fits with high probability. The main alternative, *model-based resampling*, is to resample the residuals. After fitting the linear model we have

$$y_i = x_i \widehat{\beta} + e_i$$

and we create a new dataset by $y_i = x_i \widehat{\beta} + e_i^*$ where the (e_i^*) are resampled with replacement from the residuals (e_i). There are a number of possible objections to this procedure. First, the residuals need not have mean zero if there is no intercept in the model, and it is usual to subtract their mean. Second, they do not have the correct variance or even the same variance. Thus we can adjust their variance by resampling the *modified residuals* $r_i = e_1 / \sqrt{1 - h_{ii}}$, which have variance σ^2 from (6.3).

We see bootstrapping as having little place in least-squares regression. If the errors are close to normal, the standard theory suffices. If not, there are better methods of fitting than least-squares, or perhaps the data should be transformed as in the quine dataset on page 171.

The distribution theory for the estimated coefficients in robust regression is based on asymptotic theory, so we could use bootstrap estimates of variability as an alternative. Resampling the residuals seems most appropriate for the phones data.

```
library(boot)
fit <- lm(calls ~ year, data = phones)
ph <- data.frame(phones, res = resid(fit), fitted = fitted(fit))
ph.fun <- function(data, i) {
  d <- data
  d$calls <- d$fitted + d$res[i]
  coef(update(fit, data=d))
}

(ph.lm.boot <- boot(ph, ph.fun, R = 999))
      ....
      original      bias    std. error
t1* -260.0592  0.210500     95.3262
t2*    5.0415 -0.011469      1.5385

fit <- rlm(calls ~ year, method = "MM", data = phones)
ph <- data.frame(phones, res = resid(fit), fitted = fitted(fit))
(ph.rlm.boot <- boot(ph, ph.fun, R = 999))
      ....
```

---

[8]Davison and Hinkley (1997) call this *case-based resampling*.

Table 6.1: Layout of a classic N, P, K fractional factorial design. The response is yield (in lbs/(1/70)acre-plot).

pk	np	—	nk		n	npk	k	p
49.5	62.8	46.8	57.0		62.0	48.8	45.5	44.2
n	npk	k	p		np	—	nk	pk
59.8	58.5	55.5	56.0		52.0	51.5	49.8	48.8
p	npk	n	k		nk	np	pk	—
62.8	55.8	69.5	55.0		57.2	59.0	53.2	56.0

```
        original      bias  std. error
t1* -52.4230  2.354793    26.98130
t2*   1.1009 -0.014189     0.37449
```

(The `rlm` bootstrap runs took about fifteen minutes,[9] and readers might like to start with a smaller number of resamples.) These results suggest that the asymptotic theory for `rlm` is optimistic for this example, but as the residuals are clearly serially correlated the validity of the bootstrap is equally in doubt. Statistical inference really does depend on what one considers might have happened but did not.

The bootstrap results can be investigated further by using `plot`, and `boot.ci` will give confidence intervals for the coefficients. The robust results have very long tails.

6.7 Factorial Designs and Designed Experiments

Factorial designs are powerful tools in the design of experiments. Experimenters often cannot afford to perform all the runs needed for a complete factorial experiment, or they may not all be fitted into one experimental block. To see what can be achieved, consider the following N, P, K (*nitrogen, phosphate, potassium*) factorial experiment on the growth of peas which was conducted on six blocks shown in Table 6.1.

Half of the design (technically a fractional factorial design) is performed in each of six blocks, so each half occurs three times. (If we consider the variables to take values ± 1, the halves are defined by even or odd parity, equivalently product equal to $+1$ or -1.) Note that the NPK interaction cannot be estimated as it is confounded with block differences, specifically with $(b_2 + b_3 + b_4 - b_1 - b_5 - b_6)$. An ANOVA table may be computed by

```
> (npk.aov <- aov(yield ~ block + N*P*K, data = npk))
    ....
Terms:
```

---

[9] Using S-PLUS under Linux; R took 90 seconds.

```
                     block       N      P      K    N:P    N:K
Sum of Squares 343.29 189.28   8.40  95.20  21.28  33.14
Deg. of Freedom      5       1      1      1      1      1
                     P:K Residuals
Sum of Squares  0.48      185.29
Deg. of Freedom    1          12
```

```
Residual standard error: 3.9294
1 out of 13 effects not estimable
Estimated effects are balanced
> summary(npk.aov)
             Df Sum of Sq Mean Sq F Value   Pr(F)
     block    5    343.29   68.66   4.447 0.01594
         N    1    189.28  189.28  12.259 0.00437
         P    1      8.40    8.40   0.544 0.47490
         K    1     95.20   95.20   6.166 0.02880
       N:P    1     21.28   21.28   1.378 0.26317
       N:K    1     33.14   33.14   2.146 0.16865
       P:K    1      0.48    0.48   0.031 0.86275
Residuals 12    185.29   15.44
```

```
> alias(npk.aov)
   ....
Complete
        (Intercept) block1 block2 block3 block4 block5 N P K
N:P:K             1   0.33   0.17   -0.3   -0.2
        N:P N:K P:K
N:P:K
> coef(npk.aov)
  (Intercept) block1 block2  block3  block4 block5
       54.875 1.7125 1.6792 -1.8229 -1.0137  0.295
            N      P      K     N:P    N:K     P:K
       2.8083 -0.59167 -1.9917 -0.94167 -1.175 0.14167
```

Note how the N:P:K interaction is silently omitted in the summary, although its absence is mentioned in printing npk.aov. The alias command shows which effect is missing (the particular combinations corresponding to the use of Helmert contrasts for the factor block).

Only the N and K main effects are significant (we ignore blocks whose terms are there precisely because we expect them to be important and so we must allow for them). For two-level factors the Helmert contrast is the same as the sum contrast (up to sign) giving -1 to the first level and $+1$ to the second level. Thus the effects of adding nitrogen and potassium are 5.62 and -3.98, respectively. This interpretation is easier to see with treatment contrasts:

```
> options(contrasts = c("contr.treatment", "contr.poly"))
> npk.aov1 <- aov(yield ~ block + N + K, data = npk)
> summary.lm(npk.aov1)
   ....
Coefficients:
```

```
                    Value Std. Error t value Pr(>|t|)
      ....
       N    5.617    1.609       3.490    0.003
       K   -3.983    1.609      -2.475    0.025

Residual standard error: 3.94 on 16 degrees of freedom
```

Note the use of `summary.lm` to give the standard errors. Standard errors of contrasts can also be found from the function `se.contrast`. The full form is quite complex, but a simple use is:

```
> se.contrast(npk.aov1, list(N == "0", N == "1"), data = npk)
Refitting model to allow projection
[1] 1.6092
```

For highly regular designs such as this standard errors may also be found along with estimates of means, effects and other quantities using `model.tables`.

```
> model.tables(npk.aov1, type = "means", se = T)
    ....
 N
       0      1
 52.067 57.683
    ....
Standard errors for differences of means
            block       N       K
           2.7872  1.6092  1.6092
    replic. 4.0000 12.0000 12.0000
```

Generating designs

The three functions[10] `expand.grid`, `fac.design` and `oa.design` can each be used to construct designs such as our example.

Of these, `expand.grid` is the simplest. It is used in a similar way to `data.frame`; the arguments may be named and the result is a data frame with those names. The columns contain all combinations of values for each argument. If the argument values are numeric the column is numeric; if they are anything else, for example, character, the column is a factor. Consider an example:

```
> mp <- c("-", "+")
> (NPK <- expand.grid(N = mp, P = mp, K = mp))
   N P K
 1 - - -
 2 + - -
 3 - + -
 4 + + -
 5 - - +
 6 + - +
 7 - + +
 8 + + +
```

---

[10]Only `expand.grid` is in R.

Note that the first column changes fastest and the last slowest. This is a single complete replicate.

Our example used three replicates, each split into two blocks so that the block comparison is confounded with the highest-order interaction. We can construct such a design in stages. First we find a half-replicate to be repeated three times and form the contents of three of the blocks. The simplest way to do this is to use `fac.design`:

```
blocks13 <- fac.design(levels = c(2, 2, 2),
    factor= list(N=mp, P=mp, K=mp), rep = 3, fraction = 1/2)
```

The first two arguments give the numbers of levels and the factor names and level labels. The third argument gives the number of replications (default 1). The `fraction` argument may only be used for 2^p factorials. It may be given either as a small negative power of 2, as here, or as a *defining contrast formula*. When `fraction` is numerical the function chooses a defining contrast that becomes the `fraction` attribute of the result. For half-replicates the highest-order interaction is chosen to be aliased with the mean. To find the complementary fraction for the remaining three blocks we need to use the defining contrast formula form for `fraction`:

```
blocks46 <- fac.design(levels = c(2, 2, 2),
    factor = list(N=mp, P=mp, K=mp), rep = 3, fraction = ~ -N:P:K)
```

(This is explained in the following.) To complete our design we put the blocks together, add in the `block` factor and randomize:

```
NPK <- design(block = factor(rep(1:6, each = 4)),
            rbind(blocks13, blocks46))
i <- order(runif(6)[NPK$block], runif(24))
NPK <- NPK[i,]  # Randomized
```

Using `design` instead of `data.frame` creates an object of class `design` that inherits from `data.frame`. For most purposes designs and data frames are equivalent, but some generic functions such as `plot`, `formula` and `alias` have useful `design` methods.

Defining contrast formulae resemble model formulae in syntax only; the meaning is quite distinct. There is no left-hand side. The right-hand side consists of colon products of factors only, separated by + or - signs. A plus (or leading blank) specifies that the treatments with *positive* signs for that contrast are to be selected and a minus those with *negative* signs. A formula such as `~A:B:C-A:D:E` specifies a quarter-replicate consisting of the treatments that have a positive sign in the ABC interaction and a negative sign in ADE.

Box, Hunter and Hunter (1978, §12.5) consider a 2^{7-4} design used for an experiment in riding up a hill on a bicycle. The seven factors are Seat (up or down), Dynamo (off or on), Handlebars (up or down), Gears (low or medium), Raincoat (on or off), Breakfast (yes or no) and Tyre pressure (hard or soft). A resolution III design was used, so the main effects are not aliased with each other.

Such a design cannot be constructed using a numerical fraction in `fac.design`, so the defining contrasts have to be known. Box *et al.* use the design relations:

$$D = AB, \quad E = AC, \quad F = BC, \quad G = ABC$$

which mean that ABD, ACE, BCF and $ABCG$ are all aliased with the mean, and form the defining contrasts of the fraction. Whether we choose the positive or negative halves is immaterial here.

```
> lev <- rep(2, 7)
> factors <- list(S=mp, D=mp, H=mp, G=mp, R=mp, B=mp, P=mp)
> (Bike <- fac.design(lev, factors,
      fraction = ~ S:D:G + S:H:R + D:H:B + S:D:H:P))
  S D H G R B P
1 - - - - - - -
2 - + + + + - -
3 + - + + - + -
4 + + - - + + -
5 + + + - - - +
6 + - - + + - +
7 - + - + - + +
8 - - + - + + +

Fraction:   ~ S:D:G + S:H:R + D:H:B + S:D:H:P
```

(We chose P for pressure rather than T for tyres since T and F are reserved identifiers.)

We may check the symmetry of the design using `replications`:

```
> replications(~.^2, data = Bike)
  S D H G R B P S:D S:H S:G S:R S:B S:P D:H D:G D:R D:B D:P H:G
  4 4 4 4 4 4 4   2   2   2   2   2   2   2   2   2   2   2   2
H:R H:B H:P G:R G:B G:P R:B R:P B:P
  2   2   2   2   2   2   2   2   2
```

Fractions may be specified either in a call to `fac.design` or subsequently using the `fractionate` function.

The third function, `oa.design`, provides some resolution III designs (also known as *main effect plans* or *orthogonal arrays*) for factors at two or three levels. Only low-order cases are provided, but these are the most useful in practice.

6.8   An Unbalanced Four-Way Layout

Aitkin (1978) discussed an observational study of S. Quine. The response is the number of days absent from school in a year by children from a large town in rural New South Wales, Australia. The children were classified by four factors, namely,

Age	4 levels: primary, first, second or third form
Eth	2 levels: aboriginal or non-aboriginal
Lrn	2 levels: slow or average learner
Sex	2 levels: male or female.

The dataset is included in the paper of Aitkin (1978) and is available as data frame quine in MASS. This has been explored several times already, but we now consider a more formal statistical analysis.

There were 146 children in the study. The frequencies of the combinations of factors are

```
> attach(quine)
> table(Lrn, Age, Sex, Eth)
, , F, A                     , , F, N
   F0 F1 F2 F3                  F0 F1 F2 F3
AL  4  5  1  9               AL  4  6  1 10
SL  1 10  8  0               SL  1 11  9  0

, , M, A                     , , M, N
   F0 F1 F2 F3                  F0 F1 F2 F3
AL  5  2  7  7               AL  6  2  7  7
SL  3  3  4  0               SL  3  7  3  0
```

(The output has been slightly rearranged to save space.) The classification is unavoidably very unbalanced. There are no slow learners in form F3, but all 28 other cells are non-empty. In his paper Aitkin considers a normal analysis on the untransformed response, but in the reply to the discussion he chooses a transformed response, $\log(\text{Days} + 1)$.

A casual inspection of the data shows that homoscedasticity is likely to be an unrealistic assumption on the original scale, so our first step is to plot the cell variances and standard deviations against the cell means.

```
Means <- tapply(Days, list(Eth, Sex, Age, Lrn), mean)
Vars  <- tapply(Days, list(Eth, Sex, Age, Lrn), var)
SD <- sqrt(Vars)
par(mfrow = c(1, 2))
plot(Means, Vars, xlab = "Cell Means", ylab = "Cell Variances")
plot(Means, SD, xlab = "Cell Means", ylab = "Cell Std Devn.")
```

Missing values are silently omitted from the plot. Interpretation of the result in Figure 6.5 requires some caution because of the small and widely different degrees of freedom on which each variance is based. Nevertheless the approximate linearity of the standard deviations against the cell means suggests a logarithmic transformation or something similar is appropriate. (See, for example, Rao, 1973, §6g.)

Some further insight on the transformation needed is provided by considering a model for the transformed observations

$$y^{(\lambda)} = \begin{cases} (y^\lambda - 1)/\lambda & \lambda \neq 0 \\ \log y & \lambda = 0 \end{cases}$$

where here $y = \text{Days} + \alpha$. (The positive constant α is added to avoid problems with zero entries.) Rather than include α as a second parameter we first consider Aitkin's choice of $\alpha = 1$. Box and Cox (1964) show that the profile likelihood function for λ is

$$\widehat{L}(\lambda) = \text{const} - \tfrac{n}{2} \log \text{RSS}(z^{(\lambda)})$$

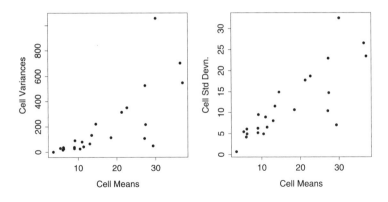

Figure 6.5: Two diagnostic plots for the Quine data.

where $z^{(\lambda)} = y^{(\lambda)}/\dot{y}^{\lambda-1}$, $\dot{y}$ is the geometric mean of the observations and $\mathrm{RSS}(z^{(\lambda)})$ is the residual sum of squares for the regression of $z^{(\lambda)}$.

Box & Cox suggest using the profile likelihood function for the largest linear model to be considered as a guide in choosing a value for λ, which will then remain fixed for any remaining analyses. Ideally other considerations from the context will provide further guidance in the choice of λ, and in any case it is desirable to choose easily interpretable values such as square-root, log or inverse.

Our MASS function boxcox calculates and (optionally) displays the Box–Cox profile likelihood function, together with a horizontal line showing what would be an approximate 95% likelihood ratio confidence interval for λ. The function is generic and several calling protocols are allowed but a convenient one to use here is with the same arguments as lm together with an additional (named) argument, lambda, to provide the sequence at which the marginal likelihood is to be evaluated. (By default the result is extended using a spline interpolation.)

Since the dataset has four empty cells the full model Eth*Sex*Age*Lrn has a rank-deficient model matrix. Hence in **S-PLUS** we must use singular.ok = T   S+
to fit the model.

```
boxcox(Days+1 ~ Eth*Sex*Age*Lrn, data = quine, singular.ok = T,
       lambda = seq(-0.05, 0.45, len = 20))
```

(Alternatively the first argument may be a fitted model object that supplies all needed information apart from lambda.) The result is shown on the left-hand panel of Figure 6.6 which suggests strongly that a log transformation is not optimal when $\alpha = 1$ is chosen. An alternative one-parameter family of transformations that could be considered in this case is

$$t(y, \alpha) = \log(y + \alpha)$$

Using the same analysis as presented in Box and Cox (1964) the profile log likelihood for α is easily seen to be

$$\widehat{L}(\alpha) = \mathrm{const} - \tfrac{n}{2} \log \mathrm{RSS}\{\log(y + \alpha)\} - \sum \log(y + \alpha)$$

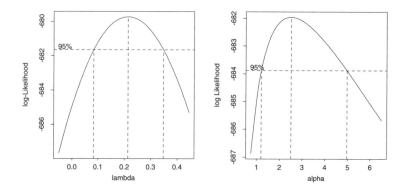

Figure 6.6: Profile likelihood for a Box–Cox transformation model with displacement $\alpha = 1$, left, and a displaced log transformation model, right.

It is interesting to see how this may be calculated directly using low-level tools, in particular the functions `qr` for the QR-decomposition and `qr.resid` for orthogonal projection onto the residual space. Readers are invited to look at our functions `logtrans.default` and `boxcox.default`.

```
logtrans(Days ~ Age*Sex*Eth*Lrn, data = quine,
      alpha = seq(0.75, 6.5, len = 20), singular.ok = T)
```

The result is shown in the right-hand panel of Figure 6.6. If a displaced log transformation is chosen a value $\alpha = 2.5$ is suggested, and we adopt this in our analysis. Note that $\alpha = 1$ is outside the notional 95% confidence interval. It can also be checked that with $\alpha = 2.5$ the log transform is well within the range of reasonable Box–Cox transformations to choose.

Model selection

The complete model, `Eth*Sex*Age*Lrn`, has a different parameter for each identified group and hence contains all possible simpler models for the mean as special cases, but has little predictive or explanatory power. For a better insight into the mean structure we need to find more parsimonious models. Before considering tools to prune or extend regression models it is useful to make a few general points on the process itself.

Marginality restrictions

In regression models it is usually the case that not all terms are on an equal footing as far as inclusion or removal is concerned. For example, in a quadratic regression on a single variable x one would normally consider removing only the highest-degree term, x^2, first. Removing the first-degree term while the second-degree one is still present amounts to forcing the fitted curve to be flat at $x = 0$, and unless there were some good reason from the context to do this it would be an arbitrary imposition on the model. Another way to view this is to note that if we write a polynomial regression in terms of a new variable $x^\star = x - \alpha$ the model remains in predictive terms the same, but only the highest-order coefficient remains invariant. If, as is usually the case, we would like our model selection procedure not to depend on the arbitrary choice of origin we must work only with the highest-degree terms at each stage.

The linear term in x is said to be *marginal* to the quadratic term, and the intercept term is marginal to both. In a similar way if a second-degree term in two variables, $x_1 x_2$, is present, any linear terms in either variable or an intercept term are marginal to it.

There are circumstances where a regression through the origin does make sense, but in cases where the origin is arbitrary one would normally only consider regression models where for each term present all terms marginal to it are also present.

In the case of factors the situation is even more clear-cut. A two-factor interaction `a:b` is marginal to any higher-order interaction that contains a and b. Fitting a model such as `a + a:b` leads to a model matrix where the columns corresponding to `a:b` are extended to compensate for the absent marginal term, b, and the fitted values are the same as if it were present. Fitting models with marginal terms removed such as with `a*b - b` generates a model with no readily understood statistical meaning[11] but updating models specified in this way using `update` changes the model matrix so that the absent marginal term again has no effect on the fitted model. In other words, removing marginal factor terms from a fitted model is either statistically meaningless or futile in the sense that the model simply changes its parametrization to something equivalent.

Variable selection for the Quine data

The `anova` function when given a single fitted-model object argument constructs a *sequential* analysis of variance table. That is, a sequence of models is fitted by expanding the formula, arranging the terms in increasing order of marginality and including one additional term for each row of the table. The process is order-dependent for non-orthogonal designs and several different orders may be needed to appreciate the analysis fully if the non-orthogonality is severe. For an orthogonal design the process is not order-dependent provided marginality restrictions are obeyed.

To explore the effect of adding or dropping terms from a model our two functions `addterm` and `dropterm` are usually more convenient. These allow the effect of, respectively, adding or removing individual terms from a model to be assessed, where the model is defined by a fitted-model object given as the first argument. For `addterm` a second argument is required to specify the scope of the terms considered for inclusion. This may be a formula or an object defining a formula for a larger model. Terms are included or removed in such a way that the marginality principle for factor terms is obeyed; for purely quantitative regressors this has to be managed by the user.

---

[11] Marginal terms are sometimes removed in this way in order to calculate what are known as 'Type III sums of squares' but we have yet to see a situation where this makes compelling statistical sense. If they are needed, they can be computed by `summary.aov` in S-PLUS.

Both functions are generic and compute the change in AIC (Akaike, 1974)

$$\text{AIC} = -2 \text{ maximized log-likelihood} + 2 \,\#\, \text{parameters}$$

Since the log-likelihood is defined only up to a constant depending on the data, this is also true of AIC. For a regression model with n observations, p parameters and normally-distributed errors the log-likelihood is

$$L(\beta, \sigma^2; \boldsymbol{y}) = \text{const} - \tfrac{n}{2} \log \sigma^2 - \tfrac{1}{2\sigma^2} \|\boldsymbol{y} - X\beta\|^2$$

and on maximizing over β we have

$$L(\widehat{\beta}, \sigma^2; \boldsymbol{y}) = \text{const} - \tfrac{n}{2} \log \sigma^2 - \tfrac{1}{2\sigma^2} \text{RSS}$$

Thus if σ^2 is *known*, we can take

$$\text{AIC} = \frac{\text{RSS}}{\sigma^2} + 2p + \text{const}$$

but if σ^2 is *unknown*,

$$L(\widehat{\beta}, \widehat{\sigma}^2; \boldsymbol{y}) = \text{const} - \tfrac{n}{2} \log \widehat{\sigma}^2 - \tfrac{n}{2}, \qquad \widehat{\sigma}^2 = \text{RSS}/n$$

and so

$$\text{AIC} = n \log(\text{RSS}/n) + 2p + \text{const}$$

For *known* σ^2 it is conventional to use Mallows' C_p,

$$C_p = \text{RSS}/\sigma^2 + 2p - n$$

(Mallows, 1973) and in this case `addterm` and `dropterm` label their output as C_p.

For an example consider removing the four-way interaction from the complete model and assessing which three-way terms might be dropped next.

```
> quine.hi <- aov(log(Days + 2.5) ~ .^4, quine)
> quine.nxt <- update(quine.hi, . ~ . - Eth:Sex:Age:Lrn)
> dropterm(quine.nxt, test = "F")
Single term deletions
....
```

	Df	Sum of Sq	RSS	AIC	F Value	Pr(F)
<none>			64.099	-68.184		
Eth:Sex:Age	3	0.9739	65.073	-71.982	0.6077	0.61125
Eth:Sex:Lrn	1	1.5788	65.678	-66.631	2.9557	0.08816
Eth:Age:Lrn	2	2.1284	66.227	-67.415	1.9923	0.14087
Sex:Age:Lrn	2	1.4662	65.565	-68.882	1.3725	0.25743

Clearly dropping `Eth:Sex:Age` most reduces AIC but dropping `Eth:Sex:Lrn` would increase it. Note that only non-marginal terms are included; none are significant in a conventional F-test.

Alternatively we could start from the simplest model and consider adding terms to reduce C_p; in this case the choice of scale parameter is important, since the simple-minded choice is inflated and may over-penalize complex models.

```
> quine.lo <- aov(log(Days+2.5) ~ 1, quine)
> addterm(quine.lo, quine.hi, test = "F")
Single term additions
....
        Df Sum of Sq    RSS     AIC F Value   Pr(F)
<none>                106.79 -43.664
    Eth  1    10.682  96.11 -57.052  16.006 0.00010
    Sex  1     0.597 106.19 -42.483   0.809 0.36981
    Age  3     4.747 102.04 -44.303   2.202 0.09048
    Lrn  1     0.004 106.78 -41.670   0.006 0.93921
```

It appears that only Eth and Age might be useful, although in fact all factors are needed since some higher-way interactions lead to large decreases in the residual sum of squares.

Automated model selection

Our function stepAIC may be used to automate the process of stepwise selection. It requires a fitted model to define the starting process (one somewhere near the final model is probably advantageous), a list of two formulae defining the upper (most complex) and lower (most simple) models for the process to consider and a scale estimate. If a large model is selected as the starting point, the scope and scale arguments have generally reasonable defaults, but for a small model where the process is probably to be one of adding terms, they will usually need both to be supplied. (A further argument, direction, may be used to specify whether the process should only add terms, only remove terms, or do either as needed.)

By default the function produces a verbose account of the steps it takes which we turn off here for reasons of space, but which the user will often care to note. The anova component of the result shows the sequence of steps taken and the reduction in AIC or C_p achieved.

```
> quine.stp <- stepAIC(quine.nxt,
      scope = list(upper = ~Eth*Sex*Age*Lrn, lower = ~1),
      trace = F)
> quine.stp$anova
....
                Step Df Deviance Resid. Df Resid. Dev     AIC
1                                      120     64.099 -68.184
2 - Eth:Sex:Age  3   0.9739         123     65.073 -71.982
3 - Sex:Age:Lrn  2   1.5268         125     66.600 -72.597
```

At this stage we might want to look further at the final model from a significance point of view. The result of stepAIC has the same class as its starting point argument, so in this case dropterm may be used to check each remaining non-marginal term for significance.

```
> dropterm(quine.stp, test = "F")
              Df Sum of Sq    RSS     AIC F Value   Pr(F)
    <none>                  66.600 -72.597
   Sex:Age  3    10.796 77.396 -56.663  6.7542 0.00029
 Eth:Sex:Lrn  1     3.032 69.632 -68.096  5.6916 0.01855
 Eth:Age:Lrn  2     2.096 68.696 -72.072  1.9670 0.14418
```

The term `Eth:Age:Lrn` is not significant at the conventional 5% significance level. This suggests, correctly, that selecting terms on the basis of AIC can be somewhat permissive in its choice of terms, being roughly equivalent to choosing an F-cutoff of 2. We can proceed manually

```
> quine.3 <- update(quine.stp, . ~ . - Eth:Age:Lrn)
> dropterm(quine.3, test = "F")
            Df Sum of Sq    RSS     AIC F Value   Pr(F)
  <none>                 68.696 -72.072
  Eth:Age    3    3.031 71.727 -71.768  1.8679 0.13833
  Sex:Age    3   11.427 80.123 -55.607  7.0419 0.00020
  Age:Lrn    2    2.815 71.511 -70.209  2.6020 0.07807
Eth:Sex:Lrn  1    4.696 73.391 -64.419  8.6809 0.00383
> quine.4 <- update(quine.3, . ~ . - Eth:Age)
> dropterm(quine.4, test = "F")
            Df Sum of Sq    RSS     AIC F Value    Pr(F)
  <none>                 71.727 -71.768
  Sex:Age    3   11.566 83.292 -55.942   6.987 0.000215
  Age:Lrn    2    2.912 74.639 -69.959   2.639 0.075279
Eth:Sex:Lrn  1    6.818 78.545 -60.511  12.357 0.000605
> quine.5 <- update(quine.4, . ~ . - Age:Lrn)
> dropterm(quine.5, test = "F")

Model:
log(Days + 2.5) ~ Eth + Sex + Age + Lrn + Eth:Sex + Eth:Lrn
                + Sex:Age + Sex:Lrn + Eth:Sex:Lrn
            Df Sum of Sq    RSS     AIC F Value    Pr(F)
  <none>                 74.639 -69.959
  Sex:Age    3   9.9002 84.539 -57.774   5.836 0.0008944
Eth:Sex:Lrn  1   6.2988 80.937 -60.130  11.140 0.0010982
```

or by setting `k = 4` in `stepAIC`. We obtain a model equivalent to `Sex/(Age + Eth*Lrn)` which is the same as that found by Aitkin (1978), apart from his choice of $\alpha = 1$ for the displacement constant. (However, when we consider a negative binomial model for the same data in Section 7.4 a more extensive model seems to be needed.)

Standard diagnostic checks on the residuals from our final fitted model show no strong evidence of any failure of the assumptions, as the reader may wish to verify.

It can also be verified that had we started from a very simple model and worked forwards we would have stopped much sooner with a much simpler model, even using the same scale estimate. This is because the major reductions in the residual sum of squares only occur when the third-order interaction `Eth:Sex:Lrn` is included.

There are other tools in S-PLUS for model selection called `stepwise` and `leaps`,[12] but these only apply for quantitative regressors. There is also no possibility of ensuring that marginality restrictions are obeyed.

---

[12]There are equivalent functions in the R package `leaps` on CRAN.

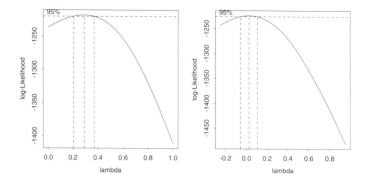

Figure 6.7: Box–Cox plots for the cpus data. Left: original regressors. Right: discretized regressors.

6.9   Predicting Computer Performance

Ein-Dor and Feldmesser (1987) studied data on the performance on a benchmark of a mix of minicomputers and mainframes. The measure was normalized relative to an IBM 370/158-3. There were six machine characteristics: the cycle time (nanoseconds), the cache size (Kb), the main memory size (Kb) and number of channels. (For the latter two there are minimum and maximum possible values; what the actual machine tested had is unspecified.) The original paper gave a linear regression for the square root of performance, but log scale looks more intuitive.

We can consider the Box–Cox family of transformations, Figure 6.7.

```
boxcox(perf ~ syct + mmin + mmax + cach + chmin + chmax,
          data = cpus, lambda = seq(0, 1, 0.1))
```

which tends to suggest a power of around 0.3 (and excludes both 0 and 0.5 from its 95% confidence interval). However, this does not allow for the regressors to be transformed, and many of them would be most naturally expressed on log scale. One way to allow the variables to be transformed is to discretize them; we show a more sophisticated approach in Section 8.8.

```
cpus1 <- cpus
attach(cpus)
for(v in names(cpus)[2:7])
    cpus1[[v]] <- cut(cpus[[v]], unique(quantile(cpus[[v]])),
                      include.lowest = T)
detach()
boxcox(perf ~ syct + mmin + mmax + cach + chmin + chmax,
          data = cpus1, lambda = seq(-0.25, 1, 0.1))
```

which does give a confidence interval including zero.

The purpose of this study is to predict computer performance. We randomly select 100 examples for fitting the models and test the performance on the remaining 109 examples.

```
> set.seed(123)
> cpus2 <- cpus[, 2:8]  # excludes names, authors' predictions
> cpus.samp <- sample(1:209, 100)
> cpus.lm <- lm(log10(perf) ~ ., data = cpus2[cpus.samp, ])
> test.cpus <- function(fit)
    sqrt(sum((log10(cpus2[-cpus.samp, "perf"]) -
            predict(fit, cpus2[-cpus.samp,]))^2)/109)
> test.cpus(cpus.lm)
[1] 0.21295
> cpus.lm2 <- stepAIC(cpus.lm, trace = F)
> cpus.lm2$anova
    Step Df Deviance Resid. Df Resid. Dev      AIC
1                         93     3.2108 -329.86
2 - syct  1 0.013177       94     3.2240 -331.45
> test.cpus(cpus.lm2)
[1] 0.21711
```

So selecting a smaller model does not improve the performance on this random split. We consider a variety of non-linear models for this example in later chapters.

6.10 Multiple Comparisons

As we all know, the theory of p-values of hypothesis tests and of the coverage of confidence intervals applies to pre-planned analyses. However, the only circumstances in which an adjustment is routinely made for testing after looking at the data is in multiple comparisons of contrasts in designed experiments. This is sometimes known as *post hoc* adjustment.

Consider the experiment on yields of barley in our dataset immer.[13] This has the yields of five varieties of barley at six experimental farms in both 1931 and 1932; we average the results for the two years. An analysis of variance gives

```
> immer.aov <- aov((Y1 + Y2)/2 ~ Var + Loc, data = immer)
> summary(immer.aov)
            Df Sum of Sq Mean Sq F Value      Pr(F)
     Var  4      2655    663.7   5.989 0.0024526
     Loc  5     10610   2122.1  19.148 0.0000005
Residuals 20      2217    110.8
```

The interest is in the difference in yield between varieties, and there is a statistically significant difference. We can see the mean yields by a call to model.tables.

```
> model.tables(immer.aov, type = "means", se = T, cterms = "Var")
    ....
   Var
        M      P      S      T      V
    94.39 102.54  91.13 118.20  99.18
```

---

[13]The Trellis dataset barley discussed in Cleveland (1993) is a more extensive version of the same dataset.

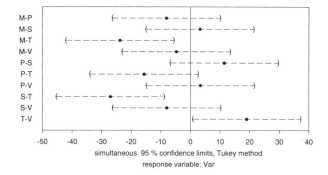

Figure 6.8: Simultaneous 95% confidence intervals for variety comparisons in the `immer` dataset.

```
Standard errors for differences of means
          Var
        6.078
replic. 6.000
```

This suggests that variety T is different from all the others, as a pairwise significant difference at 5% would exceed $6.078 \times t_{20}(0.975) \approx 12.6$; however the comparisons to be made have been selected after looking at the fit.

Function `multicomp`[14] allows us to compute *simultaneous* confidence intervals in this problem, that is, confidence intervals such that the probability that they cover the true values for all of the comparisons considered is bounded above by 5% for 95% confidence intervals. We can also plot the confidence intervals (Figure 6.8) by

```
> multicomp(immer.aov, plot = T)   # S-PLUS only
95 % simultaneous confidence intervals for specified
linear combinations, by the Tukey method

critical point: 2.9925
response variable: (Y1 + Y2)/2

intervals excluding 0 are flagged by '****'
```

	Estimate	Std.Error	Lower Bound	Upper Bound	
M-P	-8.15	6.08	-26.300	10.00	
M-S	3.26	6.08	-14.900	21.40	
M-T	-23.80	6.08	-42.000	-5.62	****
M-V	-4.79	6.08	-23.000	13.40	
P-S	11.40	6.08	-6.780	29.60	
P-T	-15.70	6.08	-33.800	2.53	
P-V	3.36	6.08	-14.800	21.50	
S-T	-27.10	6.08	-45.300	-8.88	****

---

[14] Available in S-PLUS, but not in R.

```
S-V    -8.05    6.08    -26.200    10.10
T-V    19.00    6.08      0.828    37.20 ****
```

This does not allow us to conclude that variety T has a significantly different yield than variety P.

We can do the Tukey multiple comparison test in R by

```
> (tk <- TukeyHSD(immer.aov, which = "Var"))
  Tukey multiple comparisons of means
    95% family-wise confidence level

Fit: aov(formula = (Y1 + Y2)/2 ~ Var + Loc, data = immer)
$Var
         diff      lwr       upr
P-M    8.1500 -10.0376  26.33759
S-M   -3.2583 -21.4459  14.92925
T-M   23.8083   5.6207  41.99592
V-M    4.7917 -13.3959  22.97925
S-P  -11.4083 -29.5959   6.77925
T-P   15.6583  -2.5293  33.84592
V-P   -3.3583 -21.5459  14.82925
T-S   27.0667   8.8791  45.25425
V-S    8.0500 -10.1376  26.23759
V-T  -19.0167 -37.2043  -0.82908

> plot(tk)
```

We may want to restrict the set of comparisons, for example to comparisons with a control treatment. The dataset oats is discussed on page 282; here we ignore the split-plot structure.

```
> oats1 <- aov(Y ~ N + V + B, data = oats)
> summary(oats1)
             Df Sum of Sq Mean Sq F Value     Pr(F)
          N   3     20020  6673.5  28.460 0.000000
          V   2      1786   893.2   3.809 0.027617
          B   5     15875  3175.1  13.540 0.000000
Residuals 61     14304   234.5

> multicomp(oats1, focus = "V")  # S-PLUS only

95 % simultaneous confidence intervals for specified
linear combinations, by the Tukey method

critical point: 2.4022
response variable: Y

intervals excluding 0 are flagged by '****'

                         Estimate Std.Error Lower Bound
Golden.rain-Marvellous      -5.29      4.42      -15.90
```

```
          Golden.rain-Victory      6.88      4.42       -3.74
          Marvellous-Victory      12.20      4.42        1.55
                             Upper Bound
  Golden.rain-Marvellous         5.33
      Golden.rain-Victory       17.50
       Marvellous-Victory       22.80 ****

> # R: (tk <- TukeyHSD(oats1, which = "V"))
  Tukey multiple comparisons of means
    95% family-wise confidence level

Fit: aov(formula = Y ~ N + V + B, data = oats)
$V
                            diff      lwr      upr
Marvellous-Golden.rain    5.2917  -5.3273 15.9107
Victory-Golden.rain      -6.8750 -17.4940  3.7440
Victory-Marvellous      -12.1667 -22.7857 -1.5477

> plot(tk)

> multicomp(oats1, focus = "N", comparisons = "mcc", control = 1)
    ....
             Estimate Std.Error Lower Bound Upper Bound
0.2cwt-0.0cwt    19.5       5.1        7.24        31.8 ****
0.4cwt-0.0cwt    34.8       5.1       22.60        47.1 ****
0.6cwt-0.0cwt    44.0       5.1       31.70        56.3 ****
```

Note that we need to specify the control level; perversely by default the last level
is chosen. We might also want to know if all the increases in nitrogen give significant increases in yield, which we can examine by

```
> lmat <- matrix(c(0,-1,1,rep(0, 11), 0,0,-1,1, rep(0,10),
                  0,0,0,-1,1,rep(0,9)), , 3,
       dimnames = list(NULL,
         c("0.2cwt-0.0cwt", "0.4cwt-0.2cwt", "0.6cwt-0.4cwt")))
> multicomp(oats1, lmat = lmat, bounds = "lower",
           comparisons = "none")
    ....
             Estimate Std.Error Lower Bound
0.2cwt-0.0cwt    19.50       5.1        8.43 ****
0.4cwt-0.2cwt    15.30       5.1        4.27 ****
0.6cwt-0.4cwt     9.17       5.1       -1.90
```

There is a bewildering variety of methods for multiple comparisons reflected
in the options for `multicomp`. Miller (1981), Hsu (1996) and Yandell (1997,
Chapter 6) give fuller details. Do remember that this tackles only part of the
problem; the analyses here have been done after selecting a model and specific
factors on which to focus. The allowance for multiple comparisons is only over
contrasts of one selected factor in one selected model.

Chapter 7

Generalized Linear Models

Generalized linear models (GLMs) extend linear models to accommodate both non-normal response distributions and transformations to linearity. (We assume that Chapter 6 has been read before this chapter.) The essay by Firth (1991) gives a good introduction to GLMs; the comprehensive reference is McCullagh and Nelder (1989).

A generalized linear model may be described by the following assumptions.

- There is a response y observed independently at fixed values of stimulus variables $x_1, \ldots, x_p$.

- The stimulus variables may only influence the distribution of y through a single linear function called the *linear predictor* $\eta = \beta_1 x_1 + \cdots + \beta_p x_p$.

- The distribution of y has density of the form

$$f(y_i; \theta_i, \varphi) = \exp\left[A_i \{y_i \theta_i - \gamma(\theta_i)\} / \varphi + \tau(y_i, \varphi/A_i)\right] \qquad (7.1)$$

  where φ is a *scale parameter* (possibly known), A_i is a *known* prior weight and parameter θ_i depends upon the linear predictor.

- The mean μ is a smooth invertible function of the linear predictor:

$$\mu = m(\eta), \qquad \eta = m^{-1}(\mu) = \ell(\mu)$$

The inverse function $\ell(\cdot)$ is called the *link function*.

Note that θ is also an invertible function of μ, in fact $\theta = (\gamma')^{-1}(\mu)$ as we show in the following. If φ were known the distribution of y would be a one-parameter canonical exponential family. An unknown φ is handled as a nuisance parameter by moment methods.

GLMs allow a unified treatment of statistical methodology for several important classes of models. We consider a few examples.

Gaussian For a normal distribution $\varphi = \sigma^2$ and we can write

$$\log f(y) = \frac{1}{\varphi} \left\{ y\mu - \tfrac{1}{2}\mu^2 - \tfrac{1}{2}y^2 \right\} - \tfrac{1}{2} \log(2\pi\varphi)$$

so $\theta = \mu$ and $\gamma(\theta) = \theta^2/2$.

Table 7.1: Families and link functions. The default link is denoted by D.

Link	binomial	Gamma	gaussian	inverse.-gaussian	poisson
logit	D				
probit	•				
cloglog	•				
identity		•	D		•
inverse		D			
log		•			D
1/mu^2				D	
sqrt					•

Poisson For a Poisson distribution with mean μ we have

$$\log f(y) = y \log \mu - \mu - \log(y!)$$

so $\theta = \log \mu$, $\varphi = 1$ and $\gamma(\theta) = \mu = e^\theta$.

Binomial For a binomial distribution with fixed number of trials a and parameter p we take the response to be $y = s/a$ where s is the number of 'successes'. The density is

$$\log f(y) = a \left[y \log \frac{p}{1-p} + \log(1-p) \right] + \log \binom{a}{ay}$$

so we take $A_i = a_i$, $\varphi = 1$, θ to be the logit transform of p and $\gamma(\theta) = -\log(1-p) = \log(1+e^\theta)$.

The functions supplied with S for handling generalized linear modelling distributions include gaussian, binomial, poisson, inverse.gaussian and Gamma.

Each response distribution allows a variety of link functions to connect the mean with the linear predictor. Those automatically available are given in Table 7.1. The combination of response distribution and link function is called the *family* of the generalized linear model.

For n observations from a GLM the log-likelihood is

$$l(\theta, \varphi; Y) = \sum_i \left[A_i \{ y_i \theta_i - \gamma(\theta_i) \} / \varphi + \tau(y_i, \varphi/A_i) \right] \quad (7.2)$$

and this has score function for θ of

$$U(\theta) = A_i \{ y_i - \gamma'(\theta_i) \} / \varphi \quad (7.3)$$

From this it is easy to show that

$$\mathrm{E}(y_i) = \mu_i = \gamma'(\theta_i) \qquad \mathrm{var}(y_i) = \frac{\varphi}{A_i} \gamma''(\theta_i)$$

Table 7.2: Canonical (default) links and variance functions.

Family	Canonical link	Name	Variance	Name
binomial	$\log(\mu/(1-\mu))$	logit	$\mu(1-\mu)$	mu(1-mu)
Gamma	$-1/\mu$	inverse	μ^2	mu^2
gaussian	μ	identity	1	constant
inverse.gaussian	$-2/\mu^2$	1/mu^2	μ^3	mu^3
poisson	$\log\mu$	log	μ	mu

(See, for example, McCullagh and Nelder, 1989, §2.2.) It follows immediately that

$$\mathrm{E}\left(\frac{\partial^2 l(\theta, \varphi; y)}{\partial\theta\,\partial\varphi}\right) = 0$$

Hence θ and φ, or more generally β and φ, are *orthogonal parameters*.

The function defined by $V(\mu) = \gamma''(\theta(\mu))$ is called the *variance function*.

For each response distribution the link function $\ell = (\gamma')^{-1}$ for which $\theta \equiv \eta$ is called the *canonical link*. If X is the model matrix, so $\eta = X\beta$, it is easy to see that with φ known, $A = \mathrm{diag}\,A_i$ and the canonical link that $X^T A y$ is a minimal sufficient statistic for β. Also using (7.3) the score equations for the regression parameters β reduce to

$$X^T A y = X^T A \widehat{\mu} \tag{7.4}$$

This relation is sometimes described by saying that the "observed and fitted values have the same (weighted) marginal totals." Equation (7.4) is the basis for certain simple fitting procedures, for example, Stevens' algorithm for fitting an additive model to a non-orthogonal two-way layout by alternate row and column sweeps (Stevens, 1948), and the Deming and Stephan (1940) or *iterative proportional scaling* algorithm for fitting log-linear models to frequency tables (see Darroch and Ratcliff, 1972).

Table 7.2 shows the canonical links and variance functions. The canonical link function is the default link for the families catered for by the S software (except for the inverse Gaussian, where the factor of two is dropped).

Iterative estimation procedures

Since explicit expressions for the maximum likelihood estimators are not usually available, estimates must be calculated iteratively. It is convenient to give an outline of the iterative procedure here; for a more complete description the reader is referred to McCullagh and Nelder (1989, §2.5, pp. 40ff) or Firth (1991, §3.4). The scheme is sometimes known by the acronym IWLS, for *iterative weighted least squares*.

An initial estimate to the linear predictor is found by some guarded version of $\widehat{\eta}_0 = \ell(y)$. (Guarding is necessary to prevent problems such as taking logarithms of zero.) Define *working weights* W and *working values* z by

$$W = \frac{A}{V}\left(\frac{\mathrm{d}\mu}{\mathrm{d}\eta}\right)^2, \qquad z = \eta + \frac{y-\mu}{\mathrm{d}\mu/\mathrm{d}\eta}$$

Initial values for W_0 and z_0 can be calculated from the initial linear predictor.

At iteration k a new approximation to the estimate of β is found by a weighted regression of the working values z_k on X with weights W_k. This provides a new linear predictor and hence new working weights and values which allow the iterative process to continue. The difference between this process and a Newton–Raphson scheme is that the Hessian matrix is replaced by its expectation. This statistical simplification was apparently first used in Fisher (1925), and is often called *Fisher scoring*. The estimate of the large sample variance of $\widehat{\beta}$ is $\widehat{\varphi}(X^T \widehat{W} X)^{-1}$, which is available as a by-product of the iterative process. It is easy to see that the iteration scheme for $\widehat{\beta}$ does not depend on the scale parameter φ.

This iterative scheme depends on the response distribution only through its mean and variance functions. This has led to ideas of *quasi-likelihood*, implemented in the family `quasi`.

The analysis of deviance

A *saturated model* is one in which the parameters θ_i, or almost equivalently the linear predictors η_i, are free parameters. It is then clear from (7.3) that the maximum likelihood estimator of $\theta_i = \theta(y_i)$ is obtained by $y_i = \gamma'(\widehat{\theta}_i) = \widehat{\mu}_i$ which is also a special case of (7.4). Denote the saturated model by S.

Assume temporarily that the scale parameter φ is known and has value 1. Let M be a model involving $p < n$ regression parameters, and let $\hat{\theta}_i$ be the maximum likelihood estimate of θ_i under M. Twice the log-likelihood ratio statistic for testing M within S is given by

$$D_M = 2\sum_{i=1}^{n} A_i \left[\left\{ y_i\theta(y_i) - \gamma\left(\theta(y_i)\right) \right\} - \left\{ y_i\widehat{\theta}_i - \gamma(\widehat{\theta}_i) \right\} \right] \qquad (7.5)$$

The quantity D_M is called the *deviance* of model M, even when the scale parameter is unknown or is known to have a value other than one. In the latter case D_M/φ, the difference in twice the log-likelihood, is known as the *scaled deviance*. (Confusingly, sometimes either is called the *residual deviance*, for example, McCullagh and Nelder, 1989, p. 119.)

For a Gaussian family with identity link the scale parameter φ is the variance and D_M is the residual sum of squares. Hence in this case the scaled deviance has distribution

$$D_M/\varphi \sim \chi^2_{n-p} \qquad (7.6)$$

leading to the customary unbiased estimator

$$\widehat{\varphi} = \frac{D_M}{n-p} \qquad (7.7)$$

An alternative estimator is the sum of squares of the standardized residuals divided by the residual degrees of freedom

$$\widetilde{\varphi} = \frac{1}{n-p} \sum_i \frac{(y_i - \hat{\mu}_i)^2}{V(\hat{\mu}_i)/A_i} \tag{7.8}$$

where $V(\mu)$ is the variance function. Note that $\widetilde{\varphi} = \widehat{\varphi}$ for the Gaussian family, but in general they differ.

In other cases the distribution (7.6) for the deviance under M may be approximately correct suggesting $\widehat{\varphi}$ as an approximately unbiased estimator of φ. It should be noted that sufficient (if not always necessary) conditions under which (7.6) becomes approximately true are that the individual distributions for the components y_i should become closer to normal form and the link effectively closer to an identity link. The approximation will often *not* improve as the sample size n increases since the number of parameters under S also increases and the usual likelihood ratio approximation argument does not apply. Nevertheless, (7.6) may sometimes be a good approximation, for example, in a binomial GLM with large values of a_i. Firth (1991, p. 69) discusses this approximation, including the extreme case of a binomial GLM with only one trial per case, that is, with $a_i = 1$.

Let $M_0 \subset M$ be a submodel with $q < p$ regression parameters and consider testing M_0 within M. If φ is known, by the usual likelihood ratio argument under M_0 we have a test given by

$$\frac{D_{M_0} - D_M}{\varphi} \overset{.}{\sim} \chi^2_{p-q} \tag{7.9}$$

where $\overset{.}{\sim}$ denotes "is approximately distributed as." The distribution is exact only in the Gaussian family with identity link. If φ is not known, by analogy with the Gaussian case it is customary to use the approximate result

$$\frac{(D_{M_0} - D_M)}{\widehat{\varphi}(p-q)} \overset{.}{\sim} F_{p-q,n-p} \tag{7.10}$$

although this must be used with some caution in non-Gaussian cases.

7.1   Functions for Generalized Linear Modelling

The linear predictor part of a generalized linear model may be specified by a model formula using the same notation and conventions as linear models. Generalized linear models also require the family to be specified, that is, the response distribution, the link function and perhaps the variance function for `quasi` models.

The fitting function is `glm` for which the main arguments are

```
glm(formula, family, data, weights, control)
```

The `family` argument is usually given as the name of one of the standard family functions listed under "Family Name" in Table 7.1. Where there is a choice of links, the name of the link may also be supplied in parentheses as a parameter, for example `binomial(link=probit)`. (The variance function for the `quasi` family may also be specified in this way.) For user-defined families (such as our `negative.binomial` discussed in Section 7.4) other arguments to the family function may be allowed or even required.

Prior weights A_i may be specified using the `weights` argument.

The iterative process can be controlled by many parameters. The only ones that are at all likely to need altering are `maxit`, which controls the maximum number of iterations and whose default value of `10` is occasionally too small, `trace` which will often be set as `trace=T` to trace the iterative process, and `epsilon` which controls the stopping rule for convergence. The convergence criterion is to stop if

$$|\text{deviance}^{(i)} - \text{deviance}^{(i-1)}| < \epsilon(\text{deviance}^{(i-1)} + \epsilon)$$

(This comes from reading the S-PLUS code[1] and is not as described in Chambers and Hastie, 1992, p. 243.) It is quite often necessary to reduce the tolerance `epsilon` whose default value is 10^{-4}. Under some circumstances the convergence of the IWLS algorithm can be extremely slow, when the change in deviance at each step can be small enough for premature stopping to occur with the default ϵ.

Generic functions with methods for `glm` objects include `coef`, `resid`, `print`, `summary` and `deviance`. It is useful to have a `glm` method function to extract the variance-covariance matrix of the estimates. This can be done using part of the result of `summary`:

```
vcov.glm <- function(obj) {
    so <- summary(obj, corr = F)
    so$dispersion * so$cov.unscaled
}
```

Our library section `MASS` contains the generic function, `vcov`, and methods for objects of classes `lm` and `nls` as well as `glm`.

For `glm` fitted-model objects the `anova` function allows an additional argument `test` to specify which test is to be used. Two possible choices are `test = "Chisq"` for chi-squared tests using (7.9) and `test = "F"` for F-tests using (7.10). The default is `test = "Chisq"` for the binomial and Poisson families, otherwise `test = "F"`.

The scale parameter ϕ is used only within the `summary` and `anova` methods and for standard errors for prediction. It can be supplied via the `dispersion` argument, defaulting to 1 for binomial and Poisson fits, and to (7.8) otherwise. (See Section 7.5 for estimating ϕ in the binomial and Poisson families.)

---

[1] R adds 0.1 rather than ϵ.

Prediction and residuals

The predict method function for glm has a type argument to specify what is to be predicted. The default is type = "link" which produces predictions of the linear predictor η. Predictions on the scale of the mean μ (for example, the fitted values $\widehat{\mu}_i$) are specified by type = "response". There is a se.fit argument that if true asks for standard errors to be returned.

For glm models there are four types of residual that may be requested, known as *deviance, working, Pearson* and *response* residuals. The response residuals are simply $y_i - \widehat{\mu}_i$. The Pearson residuals are a standardized version of the response residuals, $(y_i - \widehat{\mu}_i)/\sqrt{\widehat{V}_i}$. The working residuals come from the last stage of the iterative process, $(y_i - \widehat{\mu}_i) \ / \ \mathrm{d}\mu_i/\mathrm{d}\eta_i$. The deviance residuals d_i are defined as the signed square roots of the summands of the deviance (7.5) taking the same sign as $y_i - \widehat{\mu}_i$.

For Gaussian families all four types of residual are identical. For binomial and Poisson GLMs the sum of the squared Pearson residuals is the Pearson chi-squared statistic, which often approximates the deviance, and the deviance and Pearson residuals are usually then very similar.

Method functions for the resid function have an argument type that defaults to type = "deviance" for objects of class glm. Other values are "response", "pearson" or "working"; these may be abbreviated to the initial letter. Deviance residuals are the most useful for diagnostic purposes.

Concepts of leverage and its effect on the fit are as important for GLMs as they are in linear regression, and are discussed in some detail by Davison and Snell (1991) and extensively for binomial GLMs by Collett (1991). On the other hand, they seem less often used, as GLMs are most often used either for simple regressions or for contingency tables where, as in designed experiments, high leverage cannot occur.

Model formulae

The model formula language for GLMs is slightly more general than that described in Section 6.2 for linear models in that the function offset may be used. Its effect is to evaluate its argument and to add it to the linear predictor, equivalent to enclosing the argument in I( ) and forcing the coefficient to be one.

Note that offset can be used with the formulae of lm or aov, but it is completely ignored in S-PLUS, without any warning. Of course, with ordinary linear models the same effect can be achieved by subtracting the offset from the dependent variable, but if the more intuitive formula with offset is desired, it can be used with glm and the gaussian family. For example, library section MASS contains a dataset anorexia that contains the pre- and post-treatment weights for a clinical trial of three treatment methods for patients with anorexia. A natural model to consider would be

```
ax.1 <- glm(Postwt ~ Prewt + Treat + offset(Prewt),
            family = gaussian, data = anorexia)
```

Sometimes as here a variable is included both as a free regressor and as an offset to allow a test of the hypothesis that the regression coefficient is one or, by extension, any specific value.

If we had fitted this model omitting the offset but using `Postwt - Prewt` on the left side of the formula, predictions from the model would have been of weight gains, not the actual post-treatment weights. Hence another reason to use an offset rather than an adjusted dependent variable is to allow direct predictions from the fitted model object.

The default Gaussian family

A call to `glm` with the default `gaussian` family achieves the same purpose as a call to `lm` but less efficiently. In S-PLUS the `gaussian` family only allows the
R identity link; in R the identity, log and inverse links can be used. (If a problem requires a Gaussian family with a non-identity link, this can usually be handled using the `quasi` family. Indeed yet another, even more inefficient, way to emulate `lm` with `glm` is to use the family `quasi(link = identity, variance = constant)`.) Although the `gaussian` family is the default, it is virtually never used in practice other than when an offset is needed.

7.2 Binomial Data

Consider first a small example. Collett (1991, p. 75) reports an experiment on the toxicity to the tobacco budworm *Heliothis virescens* of doses of the pyrethroid *trans*-cypermethrin to which the moths were beginning to show resistance. Batches of 20 moths of each sex were exposed for three days to the pyrethroid and the number in each batch that were dead or knocked down was recorded. The results were

	Dose					
Sex	1	2	4	8	16	32
Male	1	4	9	13	18	20
Female	0	2	6	10	12	16

The doses were in μg. We fit a logistic regression model using $\log_2$ (dose) since the doses are powers of two. To do so we must specify the numbers of trials of a_i. This is done using `glm` with the `binomial` family in one of three ways.

1. If the response is a numeric vector it is assumed to hold the data in ratio form, $y_i = s_i/a_i$, in which case the a_is must be given as a vector of weights using the `weights` argument. (If the a_i are all one the default `weights` suffices.)

2. If the response is a logical vector or a two-level factor it is treated as a 0/1 numeric vector and handled as previously.
   If the response is a multi-level factor, the first level is treated as 0 (failure) and all others as 1 (success).

3. If the response is a two-column matrix it is assumed that the first column holds the number of successes, s_i, and the second holds the number of failures, $a_i - s_i$, for each trial. In this case no weights argument is required.

The less-intuitive third form allows the fitting function in S-PLUS to select a better starting value, so we tend to favour it.

In all cases the response is the relative frequency $y_i = s_i/a_i$, so the means μ_i are the probabilities p_i. Hence fitted yields probabilities, not binomial means.

Since we have binomial data we use the third possibility:

```
> options(contrasts = c("contr.treatment", "contr.poly"))
> ldose <- rep(0:5, 2)
> numdead <- c(1, 4, 9, 13, 18, 20, 0, 2, 6, 10, 12, 16)
> sex <- factor(rep(c("M", "F"), c(6, 6)))
> SF <- cbind(numdead, numalive = 20 - numdead)
> budworm.lg <- glm(SF ~ sex*ldose, family = binomial)
> summary(budworm.lg, cor = F)
    ....
Coefficients:
              Value Std. Error  t value
(Intercept) -2.99354    0.55253 -5.41789
        sex  0.17499    0.77816  0.22487
      ldose  0.90604    0.16706  5.42349
  sex:ldose  0.35291    0.26994  1.30735
    ....
    Null Deviance: 124.88 on 11 degrees of freedom
Residual Deviance: 4.9937 on 8 degrees of freedom
    ....
```

This shows slight evidence of a difference in slope between the sexes. Note that we use treatment contrasts to make interpretation easier. Since female is the first level of sex (they are in alphabetical order) the parameter for sex:ldose represents the increase in slope for males just as the parameter for sex measures the increase in intercept. We can plot the data and the fitted curves by

```
plot(c(1,32), c(0,1), type = "n", xlab = "dose",
     ylab = "prob", log = "x")
text(2^ldose, numdead/20, labels = as.character(sex))
ld <- seq(0, 5, 0.1)
lines(2^ld, predict(budworm.lg, data.frame(ldose = ld,
    sex = factor(rep("M", length(ld)), levels = levels(sex))),
    type = "response"), col = 3)
lines(2^ld, predict(budworm.lg, data.frame(ldose = ld,
    sex = factor(rep("F", length(ld)), levels = levels(sex))),
    type = "response"), lty = 2, col = 2)
```

see Figure 7.1. Note that when we set up a factor for the new data, we must specify all the levels or both lines would refer to level one of sex. (Of course, here we could have predicted for both sexes and plotted separately, but it helps to understand the general mechanism needed.)

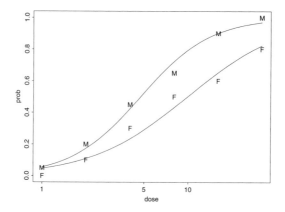

Figure 7.1: Tobacco budworm destruction versus dosage of *trans*-cypermethrin by sex.

The apparently non-significant sex effect in this analysis has to be interpreted carefully; it is marginal to the `sex:ldose` term. Since we are fitting separate lines for each sex, it tests the (uninteresting) hypothesis that the lines do not differ at zero log-dose. If we re-parametrize to locate the intercepts at dose 8 we find

```
> budworm.lgA <- update(budworm.lg, . ~ sex * I(ldose - 3))
> summary(budworm.lgA, cor = F)$coefficients
                    Value Std. Error t value
    (Intercept) -0.27543    0.23049 -1.1950
            sex  1.23373    0.37694  3.2730
  I(ldose - 3)   0.90604    0.16706  5.4235
sex:I(ldose - 3)  0.35291    0.26994  1.3074
```

which shows a significant difference between the sexes at dose 8. The model fits very well as judged by the residual deviance (4.9937 is a small value for a χ^2_8 variate, and the estimated probabilities are based on a reasonable number of trials), so there is no suspicion of curvature. We can confirm this by an analysis of deviance:

```
> anova(update(budworm.lg, . ~ . + sex * I(ldose^2)),
        test = "Chisq")
 ....
Terms added sequentially (first to last)
              Df Deviance Resid. Df Resid. Dev Pr(Chi)
        NULL                     11     124.88
         sex  1     6.08         10     118.80 0.01370
       ldose  1   112.04          9       6.76 0.00000
   I(ldose^2)  1     0.91          8       5.85 0.34104
   sex:ldose  1     1.24          7       4.61 0.26552
sex:I(ldose^2)  1     1.44          6       3.17 0.23028
```

This isolates a further two degrees of freedom that if curvature were appreciable would most likely be significant, but are not. Note how `anova` when given a single fitted-model object produces a sequential analysis of deviance, which will nearly always be order-dependent for glm objects. The additional argument test = "Chisq" to the anova method may be used to specify tests using equation (7.9). The default corresponds to test = "none". The other possible choices are "F" and "Cp", neither of which is appropriate here.

Our analysis so far suggests a model with parallel lines (on logit scale) for each sex. We estimate doses corresponding to a given probability of death. The first step is to re-parametrize the model to give separate intercepts for each sex and a common slope.

```
> budworm.lg0 <- glm(SF ~ sex + ldose - 1, family = binomial)
> summary(budworm.lg0, cor = F)$coefficients
          Value Std. Error t value
sexF  -3.4732    0.46829  -7.4166
sexM  -2.3724    0.38539  -6.1559
ldose  1.0642    0.13101   8.1230
```

Let ξ_p be the log-dose for which the probability of response is p. (Historically $2^{\xi_{0.5}}$ was called the "50% lethal dose" or LD50.) Clearly

$$\xi_p = \frac{\ell(p) - \beta_0}{\beta_1}, \qquad \frac{\partial \xi_p}{\partial \beta_0} = -\frac{1}{\beta_1}, \qquad \frac{\partial \xi_p}{\partial \beta_1} = -\frac{\ell(p) - \beta_0}{\beta_1^2} = -\frac{\xi_p}{\beta_1}$$

where β_0 and β_1 are the slope and intercept. Our library section MASS contains functions to calculate and print $\hat{\xi}_p$ and its asymptotic standard error, namely,

```
dose.p <- function(obj, cf = 1:2, p = 0.5) {
  eta <- family(obj)$link(p)
  b <- coef(obj)[cf]
  x.p <- (eta - b[1])/b[2]
  names(x.p) <- paste("p = ", format(p), ":", sep = "")
  pd <-  - cbind(1, x.p)/b[2]
  SE <- sqrt(((pd %*% vcov(obj)[cf, cf]) * pd) %*% c(1, 1))
  structure(x.p, SE = SE, p = p, class = "glm.dose")
}
print.glm.dose <- function(x, ...) {
  M <- cbind(x, attr(x, "SE"))
  dimnames(M) <- list(names(x), c("Dose", "SE"))
  x <- M
  NextMethod("print")
}
```

Notice how the family function can be used to extract the link function, which can then be used to obtain values of the linear predictor. For females the values of ξ_p at the quartiles may be calculated as

```
> dose.p(budworm.lg0, cf = c(1,3), p = 1:3/4)
            Dose      SE
p = 0.25: 2.2313 0.24983
p = 0.50: 3.2636 0.22971
p = 0.75: 4.2959 0.27462
```

For males the corresponding log-doses are lower.

In biological assays the *probit* link used to be more conventional and the technique was called *probit analysis*. Unless the estimated probabilities are concentrated in the tails, probit and logit links tend to give similar results with the values of ξ_p near the centre almost the same. We can demonstrate this for the budworm assay by fitting a probit model and comparing the estimates of ξ_p for females.

```
> dose.p(update(budworm.lg0, family = binomial(link = probit)),
        cf = c(1, 3), p = 1:3/4)
            Dose      SE
p = 0.25: 2.1912 0.23841
p = 0.50: 3.2577 0.22405
p = 0.75: 4.3242 0.26685
```

The differences are insignificant. This occurs because the logistic and standard normal distributions can approximate each other very well, at least between the 10th and 90th percentiles, by a simple scale change in the abscissa. (See, for example, Cox and Snell, 1989, p. 21.) The `menarche` data frame in `MASS` has data with a substantial proportion of the responses at very high probabilities and provides an example where probit and logit models appear noticeably different.

A binary data example: Low birth weight in infants

Hosmer and Lemeshow (1989) give a dataset on 189 births at a US hospital, with the main interest being in low birth weight. The following variables are available in our data frame `birthwt`:

low	birth weight less than 2.5 kg (0/1),
age	age of mother in years,
lwt	weight of mother (lbs) at last menstrual period,
race	white / black / other,
smoke	smoking status during pregnancy (0/1),
ptl	number of previous premature labours,
ht	history of hypertension (0/1),
ui	has uterine irritability (0/1),
ftv	number of physician visits in the first trimester,
bwt	actual birth weight (grams).

Although the actual birth weights are available, we concentrate here on predicting if the birth weight is low from the remaining variables. The dataset contains a small number of pairs of rows that are identical apart from the ID; it is possible that these refer to twins but identical birth weights seem unlikely.

We use a logistic regression with a binomial (in fact 0/1) response. It is worth considering carefully how to use the variables. It is unreasonable to expect a linear response with `ptl`. Since the numbers with values greater than one are so small we reduce it to an indicator of past history. Similarly, `ftv` can be reduced to three levels. With non-Gaussian GLMs it is usual to use treatment contrasts.

```
> options(contrasts = c("contr.treatment", "contr.poly"))
> attach(birthwt)
> race <- factor(race, labels = c("white", "black", "other"))
> table(pt1)
   0  1 2 3
 159 24 5 1
> ptd <- factor(pt1 > 0)
> table(ftv)
   0  1  2 3 4 6
 100 47 30 7 4 1
> ftv <- factor(ftv)
> levels(ftv)[-(1:2)] <- "2+"
> table(ftv)   # as a check
   0  1 2+
 100 47 42
> bwt <- data.frame(low = factor(low), age, lwt, race,
     smoke = (smoke > 0), ptd, ht = (ht > 0), ui = (ui > 0), ftv)
> detach(); rm(race, ptd, ftv)
```

We can then fit a full logistic regression, and omit the rather large correlation matrix from the summary.

```
> birthwt.glm <- glm(low ~ ., family = binomial, data = bwt)
> summary(birthwt.glm, cor = F)
   ....
Coefficients:
                 Value Std. Error  t value
(Intercept)  0.823013 1.2440732  0.66155
        age -0.037234 0.0386777 -0.96267
        lwt -0.015653 0.0070759 -2.21214
  raceblack  1.192409 0.5357458  2.22570
  raceother  0.740681 0.4614609  1.60508
      smoke  0.755525 0.4247645  1.77869
        ptd  1.343761 0.4804445  2.79691
         ht  1.913162 0.7204344  2.65557
         ui  0.680195 0.4642156  1.46526
       ftv1 -0.436379 0.4791611 -0.91071
      ftv2+  0.179007 0.4562090  0.39238

   Null Deviance: 234.67 on 188 degrees of freedom
Residual Deviance: 195.48 on 178 degrees of freedom
```

Since the responses are binary, even if the model is correct there is no guarantee that the deviance will have even an approximately chi-squared distribution, but since the value is about in line with its degrees of freedom there seems no serious reason to question the fit. Rather than select a series of sub-models by hand, we make use of the `stepAIC` function. By default argument `trace` is true and produces voluminous output.

```
> birthwt.step <- stepAIC(birthwt.glm, trace = F)
> birthwt.step$anova
```

```
Initial Model:
low ~ age + lwt + race + smoke + ptd + ht + ui + ftv
Final Model:
low ~ lwt + race + smoke + ptd + ht + ui

    Step Df Deviance Resid. Df Resid. Dev    AIC
1                           178     195.48 217.48
2 - ftv  2   1.3582         180     196.83 214.83
3 - age  1   1.0179         181     197.85 213.85
> birthwt.step2 <- stepAIC(birthwt.glm, ~ .^2 + I(scale(age)^2)
                            + I(scale(lwt)^2), trace = F)
> birthwt.step2$anova
Initial Model:
low ~ age + lwt + race + smoke + ptd + ht + ui + ftv
Final Model:
low ~ age + lwt + smoke + ptd + ht + ui + ftv + age:ftv
      + smoke:ui

         Step Df Deviance Resid. Df Resid. Dev    AIC
1                              178     195.48 217.48
2  + age:ftv  2   12.475       176     183.00 209.00
3 + smoke:ui  1    3.057       175     179.94 207.94
4     - race  2    3.130       177     183.07 207.07

> summary(birthwt.step2, cor = F)$coef
                 Value Std. Error  t value
(Intercept) -0.582520   1.418729 -0.41059
        age  0.075535   0.053880  1.40190
        lwt -0.020370   0.007465 -2.72878
      smoke  0.780057   0.419249  1.86061
        ptd  1.560205   0.495741  3.14722
         ht  2.065549   0.747204  2.76437
         ui  1.818252   0.664906  2.73460
       ftv1  2.920800   2.278843  1.28170
      ftv2+  9.241693   2.631548  3.51188
     ageftv1 -0.161809   0.096472 -1.67726
    ageftv2+ -0.410873   0.117548 -3.49537
    smoke:ui -1.916401   0.970786 -1.97407
> table(bwt$low, predict(birthwt.step2) > 0)
   FALSE TRUE
0    116   14
1     28   31
```

Note that although both `age` and `ftv` were previously dropped, their interaction is now included, the slopes on `age` differing considerably within the three `ftv` groups. The AIC criterion penalizes terms less severely than a likelihood ratio or Wald's test would, and so although adding the term `smoke:ui` reduces the AIC, its t-statistic is only just significant at the 5% level. We also considered three-way interactions, but none were chosen.

Residuals are not always very informative with binary responses but at least none are particularly large here.

An alternative approach is to predict the actual live birth weight and later threshold at 2.5 kilograms. This is left as an exercise for the reader; surprisingly it produces somewhat worse predictions with around 52 errors.

We can examine the linearity in age and mother's weight more flexibly using generalized additive models. These stand in the same relationship to additive models (Section 8.8) as generalized linear models do to regression models; replace the linear predictor in a GLM by an additive model, the sum of linear and smooth terms in the explanatory variables. We use function `gam` from S-PLUS. (R has a somewhat different function `gam` in package `mgcv` by Simon Wood.)

```
> attach(bwt)
> age1 <- age*(ftv=="1"); age2 <- age*(ftv=="2+")
> birthwt.gam <- gam(low ~ s(age) + s(lwt) + smoke + ptd +
      ht + ui + ftv + s(age1) + s(age2) + smoke:ui, binomial,
      bwt, bf.maxit=25)
> summary(birthwt.gam)

Residual Deviance: 170.35 on 165.18 degrees of freedom

DF for Terms and Chi-squares for Nonparametric Effects

          Df Npar Df Npar Chisq  P(Chi)
  s(age)   1    3.0     3.1089  0.37230
  s(lwt)   1    2.9     2.3392  0.48532
 s(age1)   1    3.0     3.2504  0.34655
 s(age2)   1    3.0     3.1472  0.36829

> table(low, predict(birthwt.gam) > 0)
    FALSE TRUE
0   115   15
1    28   31
> plot(birthwt.gam, ask = T, se = T)
```

Creating the variables `age1` and `age2` allows us to fit smooth terms for the *difference* in having one or more visits in the first trimester. Both the summary and the plots show no evidence of non-linearity. The convergence of the fitting algorithm is slow in this example, so we increased the control parameter `bf.maxit` from 10 to 25. The parameter `ask = T` allows us to choose plots from a menu. Our choice of plots is shown in Figure 7.2.

See Chambers and Hastie (1992) for more details on `gam`.

Problems with binomial GLMs

There is a little-known phenomenon for binomial GLMs that was pointed out by Hauck and Donner (1977). The standard errors and t values derive from the Wald approximation to the log-likelihood, obtained by expanding the log-likelihood in a second-order Taylor expansion at the maximum likelihood estimates. If there

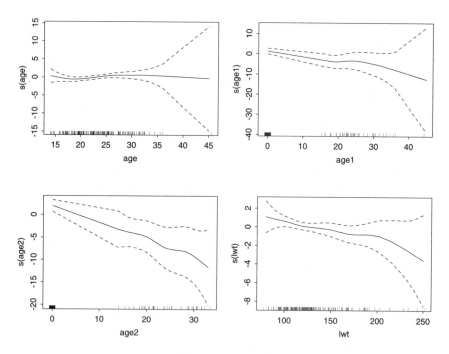

Figure 7.2: Plots of smooth terms in a generalized additive model for the data on low birth weight. The dashed lines indicate plus and minus two pointwise standard deviations.

are some $\widehat{\beta}_i$ that are large, the curvature of the log-likelihood at $\widehat{\beta}$ can be much less than near $\beta_i = 0$, and so the Wald approximation underestimates the change in log-likelihood on setting $\beta_i = 0$. This happens in such a way that as $|\widehat{\beta}_i| \to \infty$, the t statistic tends to zero. Thus highly significant coefficients according to the likelihood ratio test may have non-significant t ratios. (Curvature of the log-likelihood surface may to some extent be explored *post hoc* using the `glm` method for the `profile` generic function and the associated `plot` and `pairs` methods supplied with our `MASS` library section. The `profile` generic function is discussed in Chapter 8 on page 220.)

There is one fairly common circumstance in which both convergence problems and the Hauck–Donner phenomenon can occur. This is when the fitted probabilities are extremely close to zero or one. Consider a medical diagnosis problem with thousands of cases and around 50 binary explanatory variables (which may arise from coding fewer categorical factors); one of these indicators is rarely true but always indicates that the disease is present. Then the fitted probabilities of cases with that indicator should be one, which can only be achieved by taking $\widehat{\beta}_i = \infty$. The result from `glm` will be warnings and an estimated coefficient of around ± 10. (Such cases are not hard to spot, but might occur during a series of stepwise fits.) There has been fairly extensive discussion of this in the statistical literature, usually claiming the non-existence of maximum likelihood estimates; see Santner and Duffy (1989, p. 234). However, the phenomenon was discussed much earlier (as a desirable outcome) in the pattern recognition literature (Duda and Hart, 1973; Ripley, 1996).

It is straightforward to fit binomial GLMs by direct maximization (see page 445) and this provides a route to study several extensions of logistic regression.

7.3 Poisson and Multinomial Models

The canonical link for the Poisson family is `log`, and the major use of this family is to fit surrogate Poisson log-linear models to what are actually multinomial frequency data. Such log-linear models have a large literature. (For example, Plackett, 1974; Bishop, Fienberg and Holland, 1975; Haberman, 1978, 1979; Goodman, 1978; Whittaker, 1990.) Poisson log-linear models are often applied directly to rates, for example of accidents with the log-exposure time given by an `offset` term.

It is convenient to divide the factors classifying a multi-way frequency table into *response* and *stimulus* factors. Stimulus factors have their marginal totals fixed in advance (or for the purposes of inference). The main interest lies in the conditional probabilities of the response factor given the stimulus factors.

It is well known that the conditional distribution of a set of independent Poisson random variables given their sum is multinomial with probabilities given by the ratios of the Poisson means to their total. This result is applied to the counts for the multi-way response within each combination of stimulus factor levels. This allows models for multinomial data with a multiplicative probability specification to be fitted and tested using Poisson log-linear models.

Identifying the multinomial model corresponding to any surrogate Poisson model is straightforward. Suppose A, B, ... are factors classifying a frequency table. The *minimum model* is the interaction of all stimulus factors, and must be included for the analysis to respect the fixed totals over response factors. This model is usually of no interest, corresponding to a uniform distribution over response factors independent of the stimulus factors. Interactions between response and stimulus factors indicate interesting structure. For large models it is often helpful to use a graphical structure to represent the dependency relations (Whittaker, 1990; Lauritzen, 1996; Edwards, 2000).

A four-way frequency table example

As an example of a surrogate Poisson model analysis, consider the data in Table 7.3. This shows a four-way classification of 1 681 householders in Copenhagen who were surveyed on the *type* of rental accommodation they occupied, the degree of *contact* they had with other residents, their feeling of *influence* on apartment management and their level of *satisfaction* with their housing conditions. The data were originally given by Madsen (1976) and later analysed by Cox and Snell (1981). The data frame `housing` in MASS gives the same dataset with four factors and a numeric column of frequencies. A visualization of these data is shown in Figure 11.14 on page 327.

Table 7.3: A four-way frequency table of 1 681 householders from a study of satisfaction with housing conditions in Copenhagen.

Contact		Low			High		
Satisfaction		Low	Med.	High	Low	Med.	High
Housing	**Influence**						
Tower blocks	Low	21	21	28	14	19	37
	Medium	34	22	36	17	23	40
	High	10	11	36	3	5	23
Apartments	Low	61	23	17	78	46	43
	Medium	43	35	40	48	45	86
	High	26	18	54	15	25	62
Atrium houses	Low	13	9	10	20	23	20
	Medium	8	8	12	10	22	24
	High	6	7	9	7	10	21
Terraced houses	Low	18	6	7	57	23	13
	Medium	15	13	13	31	21	13
	High	7	5	11	5	6	13

Cox and Snell (1981, pp. 155ff) present an analysis with type of housing as the only explanatory factor and contact, influence and satisfaction treated symmetrically as response factors and mention an alternative analysis with satisfaction as the only response variable, and hence influence and contact as (conditional) explanatory variables. We consider such an alternative analysis.

Our initial model may be described as having the conditional probabilities for each of the three satisfaction classes the same for all type × contact × influence groups. In other words, satisfaction is independent of the other explanatory factors.

```
> names(housing)
[1] "Sat"  "Infl" "Type" "Cont" "Freq"
> house.glm0 <- glm(Freq ~ Infl*Type*Cont + Sat,
                    family = poisson, data = housing)
> summary(house.glm0, cor = F)
    ....
    Null Deviance: 833.66 on 71 degrees of freedom
Residual Deviance: 217.46 on 46 degrees of freedom
```

The high residual deviance clearly indicates that this simple model is inadequate, so the probabilities do appear to vary with the explanatory factors. We now consider adding the simplest terms to the model that allow for some variation of this kind.

```
> addterm(house.glm0, ~. + Sat:(Infl+Type+Cont), test = "Chisq")
    ....
          Df Deviance    AIC   LRT Pr(Chi)
   <none>      217.46 269.46
Sat:Infl  4   111.08 171.08 106.37 0.00000
Sat:Type  6   156.79 220.79  60.67 0.00000
Sat:Cont  2   212.33 268.33   5.13 0.07708
```

The 'influence' factor achieves the largest single term reduction in the AIC[2] and our next step would be to incorporate the term Sat:Infl in our model and re-assess. It turns out that all three terms are necessary, so we now update our initial model by including all three at once.

```
> house.glm1 <- update(house.glm0, . ~ . + Sat:(Infl+Type+Cont))
> summary(house.glm1, cor = F)
    ....
    Null Deviance: 833.66 on 71 degrees of freedom
 Residual Deviance: 38.662 on 34 degrees of freedom
> 1 - pchisq(deviance(house.glm1), house.glm1$df.resid)
[1] 0.26714
```

The deviance indicates a satisfactorily fitting model, but we might look to see if some adjustments to the model might be warranted.

```
> dropterm(house.glm1, test = "Chisq")
    ....
               Df Deviance    AIC   LRT Pr(Chi)
        <none>      38.66 114.66
      Sat:Infl  4  147.78 215.78 109.12 0.00000
      Sat:Type  6  100.89 164.89  62.23 0.00000
      Sat:Cont  2   54.72 126.72  16.06 0.00033
 Infl:Type:Cont 6   43.95 107.95   5.29 0.50725
```

Note that the final term here is part of the minimum model and hence may *not* be removed. Only terms that contain the response factor, Sat, are of any interest to us for this analysis. Now consider adding possible interaction terms.

```
> addterm(house.glm1, ~. + Sat:(Infl+Type+Cont)^2, test = "Chisq")
    ....
              Df Deviance    AIC    LRT Pr(Chi)
       <none>      38.662 114.66
Sat:Infl:Type 12  16.107 116.11 22.555 0.03175
Sat:Infl:Cont  4  37.472 121.47  1.190 0.87973
Sat:Type:Cont  6  28.256 116.26 10.406 0.10855
```

The first term, a type $\times$ influence interaction, appears to be mildly significant, but as it increases the AIC we choose not include it on the grounds of simplicity, although in some circumstances we might view this decision differently.

We have now shown (subject to checking assumptions) that the three explanatory factors, type, influence and contact do affect the probabilities of each of the

---

[2]The S-PLUS and R versions of stepAIC use different additive constants for Poisson GLMs.

Table 7.4: Estimated probabilities from a main effects model for the Copenhagen housing conditions study.

Contact		Low			High		
Satisfaction		Low	Med.	High	Low	Med.	High
Housing	**Influence**						
Tower blocks	Low	0.40	0.26	0.34	0.30	0.28	0.42
	Medium	0.26	0.27	0.47	0.18	0.27	0.54
	High	0.15	0.19	0.66	0.10	0.19	0.71
Apartments	Low	0.54	0.23	0.23	0.44	0.27	0.30
	Medium	0.39	0.26	0.34	0.30	0.28	0.42
	High	0.26	0.21	0.53	0.18	0.21	0.61
Atrium houses	Low	0.43	0.32	0.25	0.33	0.36	0.31
	Medium	0.30	0.35	0.36	0.22	0.36	0.42
	High	0.19	0.27	0.54	0.13	0.27	0.60
Terraced houses	Low	0.65	0.22	0.14	0.55	0.27	0.19
	Medium	0.51	0.27	0.22	0.40	0.31	0.29
	High	0.37	0.24	0.39	0.27	0.26	0.47

satisfaction classes in a simple, 'main effect'–only way. Our next step is to look at these estimated probabilities under the model and assess what effect these factors are having. The picture becomes clear if we normalize the means to probabilities.

```
hnames <- lapply(housing[, -5], levels) # omit Freq
house.pm <- predict(house.glm1, expand.grid(hnames),
                    type = "response")  # poisson means
house.pm <- matrix(house.pm, ncol = 3, byrow = T,
                   dimnames = list(NULL, hnames[[1]]))
house.pr <- house.pm/drop(house.pm %*% rep(1, 3))
cbind(expand.grid(hnames[-1]), prob = round(house.pr, 2))
```

The result is shown in conventional typeset form in Table 7.4. The message of the fitted model is now clear. The factor having most effect on the probabilities is influence, with an increase in influence reducing the probability of low satisfaction and increasing that of high. The next most important factor is the type of housing itself. Finally, as contact with other residents rises, the probability of low satisfaction tends to fall and that of high to rise, but the effect is relatively small.

The reader should compare the model-based probability estimates with the relative frequencies from the original data. In a few cases the smoothing effect of the model is perhaps a little larger than might have been anticipated, but there are no very surprising differences. A conventional residual analysis also shows up no serious problem with the assumptions underlying the model, and the details are left as an informal exercise.

Fitting by iterative proportional scaling

The function `loglin` fits log-linear models by iterative proportional scaling. This starts with an array of fitted values that has the correct multiplicative structure, (for example with all values equal to 1) and makes multiplicative adjustments so that the observed and fitted values come to have the same marginal totals, in accordance with equation (7.4). (See Darroch and Ratcliff, 1972.) This is usually very much faster than GLM fitting but is less flexible.

Function `loglin`, is rather awkward to use as it requires the frequencies to be supplied as a multi-way array. Our front-end to `loglin` called `loglm` (in library section MASS) accepts the frequencies in a variety of forms and can be used with essentially the same convenience as any model fitting function. Methods for standard generic functions to deal with fitted-model objects are also provided.

```
> loglm(Freq ~ Infl*Type*Cont + Sat*(Infl+Type+Cont),
        data = housing)
Statistics:
                    X^2 df P(> X^2)
Likelihood Ratio 38.662 34  0.26714
        Pearson 38.908 34  0.25823
```

The output shows that the deviance is the same as for the Poisson log-linear model and the (Pearson) chi-squared statistic is very close, as would be expected.

Fitting as a multinomial model

We can fit a multinomial model directly rather than use a surrogate Poisson model by using our function `multinom` from library section nnet. No interactions are needed:

```
> library(nnet)
> house.mult <- multinom(Sat ~ Infl + Type + Cont,
                         weights = Freq, data = housing)
> house.mult
Coefficients:
        (Intercept) InflMedium InflHigh TypeApartment TypeAtrium
Medium    -0.41923    0.44644  0.66497      -0.43564    0.13134
  High    -0.13871    0.73488  1.61264      -0.73566   -0.40794
        TypeTerrace    Cont
Medium    -0.66658 0.36085
  High    -1.41234 0.48188

Residual Deviance: 3470.10
AIC: 3498.10
```

Here the deviance is comparing with the model that correctly predicts each person, not the multinomial response for each cell of the minimum model: we can compare with the usual saturated model by

```
> house.mult2 <- multinom(Sat ~ Infl*Type*Cont,
                          weights = Freq, data = housing)
> anova(house.mult, house.mult2, test = "none")
   ....
```

	Model	Resid. df	Resid. Dev	Test	Df	LR stat.
1	Infl + Type + Cont	130	3470.1			
2	Infl * Type * Cont	96	3431.4	1 vs 2	34	38.662

A table of fitted probabilities can be found by

```
house.pm <- predict(house.mult, expand.grid(hnames[-1]),
                    type = "probs")
cbind(expand.grid(hnames[-1]), round(house.pm, 2))
```

A proportional-odds model

Since the response in this example is ordinal, a natural model to consider would be a proportional-odds model. Under such a model the odds ratio for cumulative probabilities for low and medium satisfaction does not depend on the cell to which the three probabilities belong. For a definition and good discussion see the seminal paper of McCullagh (1980) and McCullagh and Nelder (1989, §5.2.2). A proportional-odds logistic regression for a response factor with K levels has

$$\operatorname{logit} P(Y \leqslant k \mid \boldsymbol{x}) = \zeta_k - \eta$$

for $\zeta_0 = -\infty < \zeta_1 < \cdots < \zeta_K = \infty$ and η the usual linear predictor.

One check to see if the proportional-odds assumption is reasonable is to look at the differences of logits of the cumulative probabilities either using the simple relative frequencies or the model-based estimates. Since the multinomial model appears to fit well, we use the latter here.

```
> house.cpr <- apply(house.pr, 1, cumsum)
> logit <- function(x) log(x/(1-x))
> house.ld <- logit(house.cpr[2, ]) - logit(house.cpr[1, ])
> sort(drop(house.ld))
 [1] 0.93573 0.98544 1.05732 1.06805 1.07726 1.08036 1.08249
 [8] 1.09988 1.12000 1.15542 1.17681 1.18664 1.20915 1.24350
[15] 1.27241 1.27502 1.28499 1.30626 1.31240 1.39047 1.45401
[22] 1.49478 1.49676 1.60688
> mean(.Last.value)
[1] 1.2238
```

The average log-odds ratio is about 1.2 and variations from it are not great, so such a model may prove a useful simplification.

Our MASS library section contains a function, polr, for proportional-odds logistic regression that behaves in a very similar way to the standard fitting functions.

```
> (house.plr <- polr(Sat ~ Infl + Type + Cont,
                     data = housing, weights = Freq))
   ....
```

```
Coefficients:
 InflMedium InflHigh TypeApartment TypeAtrium TypeTerrace
    0.56639   1.2888      -0.57234    -0.36619      -1.091
    Cont
 0.36029

Intercepts:
 Low|Medium Medium|High
   -0.49612     0.69072


Residual Deviance: 3479.10
AIC: 3495.10
```

The residual deviance is comparable with that of the multinomial model fitted before, showing that the increase is only 9.0 for a reduction from 14 to 8 parameters. The AIC criterion is consequently much reduced. Note that the difference in intercepts, 1.19, agrees fairly well with the average logit difference of 1.22 found previously.

We may calculate a matrix of estimated probabilities comparable with the one we obtained from the surrogate Poisson model by

```
house.pr1 <- predict(house.plr, expand.grid(hnames[-1]),
                     type = "probs")
cbind(expand.grid(hnames[-1]), round(house.pr1, 2))
```

These fitted probabilities are close to those of the multinomial model given in Table 7.4.

Finally we can calculate directly the likelihood ratio statistic for testing the proportional-odds model within the multinomial:

```
> Fr <- matrix(housing$Freq, ncol = 3, byrow = T)
> 2 * sum(Fr * log(house.pr/house.pr1))
[1] 9.0654
```

which agrees with the difference in deviance noted previously. For six degrees of freedom this is clearly not significant.

The advantage of the proportional-odds model is not just that it is so much more parsimonious than the multinomial, but with the smaller number of parameters the action of the covariates on the probabilities is much easier to interpret and to describe. However, as it is more parsimonious, stepAIC will select a more complex model linear predictor:

```
> house.plr2 <- stepAIC(house.plr, ~.^2)
> house.plr2$anova
    ....
           Step Df Deviance Resid. Df Resid. Dev    AIC
1                                1673     3479.1 3495.1
2 + Infl:Type  6   22.509        1667     3456.6 3484.6
3 + Type:Cont  3    7.945        1664     3448.7 3482.7
```

7.4   A Negative Binomial Family

Once the link function and its derivative, the variance function, the deviance function and some method for obtaining starting values are known, the fitting procedure is the same for all generalized linear models. This particular information is all taken from the *family object*, which in turn makes it fairly easy to handle a new GLM family by writing a *family object generator* function. We illustrate the procedure with a family for negative binomial models with known shape parameter, a restriction that is relaxed later.

Using the negative binomial distribution in modelling is important in its own right and has a long history. An excellent reference is Lawless (1987). The variance is greater than the mean, suggesting that it might be useful for describing frequency data where this is a prominent feature. (The next section takes an alternative approach.)

The negative binomial can arise from a two-stage model for the distribution of a discrete variable Y. We suppose there is an unobserved random variable E having a gamma distribution gamma$(\theta)/\theta$, that is with mean 1 and variance $1/\theta$. Then the model postulates that conditionally on E, Y is Poisson with mean μE. Thus:

$$Y \mid E \sim \text{Poisson}(\mu E), \qquad \theta E \sim \text{gamma}(\theta)$$

The marginal distribution of Y is then negative binomial with mean, variance and probability function given by

$$\text{E}(Y) = \mu, \quad \text{var}(Y) = \mu + \mu^2/\theta, \quad f_Y(y; \theta, \mu) = \frac{\Gamma(\theta + y)}{\Gamma(\theta)\, y!} \frac{\mu^y\, \theta^\theta}{(\mu + \theta)^{\theta + y}}$$

If θ is known, this distribution has the general form (7.1); we provide a family `negative.binomial` to fit it.

A Poisson model for the Quine data has an excessively large deviance:

```
> glm(Days ~ .^4, family = poisson, data = quine)
    ....
Degrees of Freedom: 146 Total; 118 Residual
Residual Deviance: 1173.9
```

Inspection of the mean–variance relationship in Figure 6.5 on page 171 suggests a negative binomial model with $\theta \approx 2$ might be appropriate. We assume at first that $\theta = 2$ is known. A negative binomial model may be fitted by

```
quine.nb <- glm(Days ~ .^4, family = negative.binomial(2),
                data = quine)
```

The standard generic functions may now be used to fit sub-models, produce analysis of variance tables, and so on. For example, let us check the final model found in Chapter 6. (The output has been edited.)

```
> quine.nb0 <- update(quine.nb, . ~ Sex/(Age + Eth*Lrn))
> anova(quine.nb0, quine.nb, test = "Chisq")
  Resid. Df Resid. Dev   Test  Df  Deviance  Pr(Chi)
1       132     198.51
2       118     171.98 1 vs. 2  14    26.527 0.022166
```

which suggests that model to be an over-simplification in the fixed–θ negative
binomial setting.

Consider now what happens when θ is estimated rather than held fixed. We
supply a function glm.nb, a modification of glm that incorporates maximum
likelihood estimation of θ. This has summary and anova methods; the latter
produces likelihood ratio tests for the sequence of fitted models. (For deviance
tests to be applicable, the θ parameter has to be held constant for all fitted models.)

The following models summarize the results of a selection by stepAIC.

```
> quine.nb <- glm.nb(Days ~ .^4, data = quine)
> quine.nb2 <- stepAIC(quine.nb)
> quine.nb2$anova
Initial Model:
Days ~ (Eth + Sex + Age + Lrn)^4

Final Model:
Days ~ Eth + Sex + Age + Lrn + Eth:Sex + Eth:Age + Eth:Lrn +
       Sex:Age + Sex:Lrn + Age:Lrn + Eth:Sex:Lrn +
       Eth:Age:Lrn + Sex:Age:Lrn


                  Step Df Deviance Resid. Df Resid. Dev    AIC
1                                        118    167.45 1095.3
2 - Eth:Sex:Age:Lrn  2  0.09786        120    167.55 1092.7
3     - Eth:Sex:Age  3  0.10951        123    167.44 1089.4
```

This model is Lrn/(Age + Eth + Sex)^2, but AIC tends to overfit, so we test
the terms Sex:Age:Lrn and Eth:Sex:Lrn.

```
> dropterm(quine.nb2, test = "Chisq")
             Df    AIC    LRT  Pr(Chi)
<none>           -10228
Eth:Sex:Lrn   1 -10225 5.0384 0.024791
Eth:Age:Lrn   2 -10226 5.7685 0.055896
Sex:Age:Lrn   2 -10226 5.7021 0.057784
```

This indicates that we might try removing both terms,

```
> quine.nb3 <-
    update(quine.nb2, . ~ . - Eth:Age:Lrn - Sex:Age:Lrn)
> anova(quine.nb2, quine.nb3)
    theta Resid. df  2 x log-lik.   Test    df LR stat.
1 1.7250      127         -1053.4
2 1.8654      123         -1043.4 1 vs 2     4   10.022
    theta Resid. df 2 x log-lik   Test  df LR stat.  Pr(Chi)
1 1.7250      127       1053.4
2 1.8654      123       1043.4 1 vs 2   4   10.022 0.040058
```

which suggests removing either term but not both! Clearly there is a close-run
choice among models with one, two or three third-order interactions.

The estimate of θ and its standard error are available from the summary or as
components of the fitted model object:

```
> c(theta = quine.nb2$theta, SE = quine.nb2$SE)
  theta      SE
 1.8654 0.25801
```

We can perform some diagnostic checks by examining the deviance residuals:

```
rs <- resid(quine.nb2, type = "deviance")
plot(predict(quine.nb2), rs, xlab = "Linear predictors",
     ylab = "Deviance residuals")
abline(h = 0, lty = 2)
qqnorm(rs, ylab = "Deviance residuals")
qqline(rs)
```

Such plots show nothing untoward for any of the candidate models.

7.5 Over-Dispersion in Binomial and Poisson GLMs

The role of the dispersion parameter φ in the theory and practice of GLMs is often misunderstood. For a Gaussian family with identity link and constant variance function, the moment estimator used for φ is the usual unbiased modification of the maximum likelihood estimator (see equations (7.6) and (7.7)).

For binomial and Poisson families the theory specifies that $\varphi = 1$, but in some cases we estimate φ as if it were an unknown parameter and use that value in standard error calculations and as a denominator in approximate F-tests rather than use chi-squared tests (by specifying argument dispersion = 0 in

R the summary and predict methods, or in R by using the quasibinomial and quasipoisson families). This is an *ad hoc* adjustment for over-dispersion[3] but the corresponding likelihood may not correspond to any family of error distributions. (Of course, for the Poisson family the negative binomial family introduced in Section 7.4 provides a parametric alternative way of modelling over-dispersion.)

We begin with a warning. A common way to 'discover' over- or under-dispersion is to notice that the residual deviance is appreciably different from the residual degrees of freedom, since in the usual theory the expected value of the residual deviance should equal the degrees of freedom. *This can be seriously misleading.* The theory is asymptotic, and only applies for large $n_i p_i$ for a binomial and for large μ_i for a Poisson. Figure 7.3 shows the exact expected value. The estimate of φ used by summary.glm (if allowed to estimate the dispersion) is the (weighted) sum of the squared Pearson residuals divided by the residual degrees of freedom (equation (7.8) on page 187). This has much less bias than the other estimator sometimes proposed, namely the deviance (or sum of squared *deviance* residuals) divided by the residual degrees of freedom. (See the example on page 296.)

Many authors (for example Finney, 1971; Collett, 1991; Cox and Snell, 1989; McCullagh and Nelder, 1989) discuss over-dispersion in binomial GLMs, and Aitkin *et al.* (1989) also discuss over-dispersion in Poisson GLMs. For binomial

---

[3]Or 'heterogeneity', apparently originally proposed by Finney (1971).

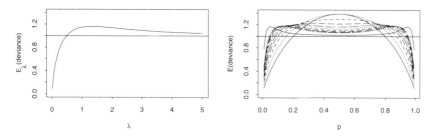

Figure 7.3: Plots of the expected residual deviance against (left) the parameter of a Poisson and (right) the p for a binomial(n, p) for $n = 1, 2, \ldots, 10, 25$. The code for the left panel is given on page 82.

GLMs, the accounts all concentrate on sampling regimes that can give rise to over-dispersion in a binomial (n, p) observation Y for $n > 1$. Suppose that p is in fact a random variable θ with mean p; this might arise if there were random effects in a linear logistic model specifying p. Then if we assume that $\operatorname{var} \theta = \phi p(1 - p)$ we find that

$$EY = np, \qquad \operatorname{var} Y = np(1 - p)[1 + (n - 1)\phi]$$

One example occurs if θ has a beta (α, β) distribution, in which case $p = \mathrm{E}\,\theta = \alpha/(\alpha + \beta)$, and $\operatorname{var} \theta = p(1 - p)/(\alpha + \beta + 1)$. In the special case that the n_i in a binomial GLM are all equal, we have

$$\operatorname{var} Y = np(1 - p)[1 + (n - 1)\phi] = \varphi np(1 - p)$$

say, so this appears to provide an explanation for the *ad hoc* adjustment. However, there are problems with this.

- It is not applicable for $n \equiv 1$, a common circumstance in which to observe over-dispersion.

- There is an upper bound on ϕ and hence φ. The most extreme distribution for θ has $\theta = 1$ with probability p and $\theta = 0$ with probability $1 - p$, hence variance $p(1-p)$. Thus $\phi \leqslant 1$ and $\varphi \leqslant n$. Plausible beta-binomial models will lead to much lower bounds, say $n/5$.

- If this model is accepted, the *ad hoc* adjustment of the GLM fit is not maximum likelihood estimation, even for the regression parameters.

McCullagh and Nelder (1989, pp. 125–6) prefer a variation on this model, in which the n data points are assumed to have been sampled from k clusters, and there is independent binomial sampling within the clusters (whose size now varies with n), but the clusters have probabilities drawn independently from a distribution of the same form as before. Then it is easy to show that

$$EY = np, \qquad \operatorname{var} Y = np(1 - p)[1 + (k - 1)\phi]$$

This does provide an explanation for the *ad hoc* adjustment model for variable n, but the assumption of the same number of (equally-sized) clusters for each observation seems rather artificial to us.

Asymptotic theory for this model suggests (McCullagh and Nelder, 1989) that changes in deviance and residual deviances *scaled by* φ have asymptotic chi-squared distributions with the appropriate degrees of freedom. Since φ must be estimated, this suggests that F tests are used in place of chi-squared tests in, for example, the analysis of deviance and addterm and dropterm. At the level of the asymptotics there is no difference between the use of estimators (7.7) and (7.8), but we have seen that (7.8) has much less bias, and it is this that is used by anova.glm and addterm and dropterm.

Another explanation that leads to the same conclusion is to assume that n trials that make up the binomial observations are exchangeable but not necessarily independent. Then the results for any pair of trials might have correlation δ, and this leads to

$$\operatorname{var} Y = np(1-p)[1 + (n-1)\delta] = \varphi np(1-p)$$

say. In this model there is no constraint that $\delta \geqslant 0$, but only limited negative correlation is possible. (Indeed, $\operatorname{var} Y \geqslant 0$ implies $\delta \geqslant -1/(n-1)$, and assuming that the trials are part of an infinite population does require $\delta \geqslant 0$.)

All these explanations are effectively quasi-likelihood models, in that just the mean and variance of the observations are specified. We believe that they are best handled as quasi models. In R the binomial and Poisson families never allow φ to be estimated, but there are additional families quasibinomial and quasipoisson for which φ is always estimated. Since there are quasi models, there is no likelihood and hence no AIC, and so stepAIC cannot be used for model selection.

Chapter 8

Non-Linear and Smooth Regression

In linear regression the mean surface is a plane in sample space; in non-linear regression it may be an arbitrary curved surface but in all other respects the models are the same. Fortunately the mean surface in most non-linear regression models met in practice will be approximately planar in the region of highest likelihood, allowing some good approximations based on linear regression to be used, but non-linear regression models can still present tricky computational and inferential problems.

A thorough treatment of non-linear regression is given in Bates and Watts (1988). Another encyclopaedic reference is Seber and Wild (1989), and the books of Gallant (1987) and Ross (1990) also offer some practical statistical advice. The S software is described by Bates and Chambers (1992), who state that its methods are based on those described in Bates and Watts (1988). An alternative approach is described by Huet *et al.* (1996).

Another extension of linear regression is *smooth regression*, in which linear terms are extended to smooth functions whose exact form is not pre-specified but chosen from a flexible family by the fitting procedures. The methods are all fairly computer-intensive, and so only became feasible in the era of plentiful computing power. There are few texts covering this material. Although they do not cover all our topics in equal detail, for what they do cover Hastie and Tibshirani (1990), Simonoff (1996), Bowman and Azzalini (1997) and Hastie *et al.* (2001) are good references, as well as Ripley (1996, Chapters 4 and 5).

8.1   An Introductory Example

Obese patients on a weight reduction programme tend to lose adipose tissue at a diminishing rate. Our dataset `wtloss` has been supplied by Dr T. Davies (personal communication). The two variables are `Days`, the time (in days) since start of the programme, and `Weight`, the patient's weight in kilograms measured under standard conditions. The dataset pertains to a male patient, aged 48, height 193 cm ($6'4''$) with a large body frame. The results are illustrated in Figure 8.1, produced by

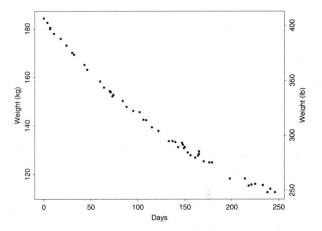

Figure 8.1: Weight loss for an obese patient.

```
attach(wtloss)
# alter margin 4; others are default
oldpar <- par(mar = c(5.1, 4.1, 4.1, 4.1))
plot(Days, Weight, type = "p", ylab = "Weight (kg)")
Wt.lbs <- pretty(range(Weight*2.205))
axis(side = 4, at = Wt.lbs/2.205, lab = Wt.lbs, srt = 90)
mtext("Weight (lb)", side = 4, line = 3)
par(oldpar) # restore settings
```

Although polynomial regression models may describe such data very well within the observed range, they can fail spectacularly outside this range. A more useful model with some theoretical and empirical support is non-linear in the parameters, of the form

$$y = \beta_0 + \beta_1 2^{-t/\theta} + \epsilon \tag{8.1}$$

Notice that all three parameters have a ready interpretation, namely

β_0 is the ultimate lean weight, or asymptote,
β_1 is the total amount to be lost and
θ is the time taken to lose half the amount remaining to be lost,

which allows us to find rough initial estimates directly from the plot of the data.

The parameters β_0 and β_1 are called *linear parameters* since the second partial derivative of the model function with respect to them is identically zero. The parameter, θ, for which this is not the case, is called a *non-linear parameter*.

8.2   Fitting Non-Linear Regression Models

The general form of a non-linear regression model is

$$y = \eta(\boldsymbol{x}, \boldsymbol{\beta}) + \epsilon \tag{8.2}$$

where x is a vector of covariates, β is a p-component vector of unknown parameters and ϵ is a $N(0, \sigma^2)$ error term. In the weight loss example the parameter vector is $\beta = (\beta_0, \beta_1, \theta)^T$. (As x plays little part in the discussion that follows, we often omit it from the notation.)

Suppose y is a sample vector of size n and $\eta(\beta)$ is its mean vector. It is easy to show that the maximum likelihood estimate of β is a least-squares estimate, that is, a minimizer of $\|y - \eta(\beta)\|^2$. The variance parameter σ^2 is then estimated by the residual mean square as in linear regression.

For varying β the vector $\eta(\beta)$ traces out a p-dimensional surface in $\mathbb{R}^n$ that we refer to as the *solution locus*. The parameters β define a coordinate system within the solution locus. From this point of view a linear regression model is one for which the solution locus is a plane through the origin and the coordinate system within it defined by the parameters is affine; that is, it has no curvature. The computational problem in both cases is then to find the coordinates of the point on the solution locus closest to the sample vector y in the sense of Euclidean distance.

The process of fitting non-linear regression models in S is similar to that for fitting linear models, with two important differences:

1. there is no explicit formula for the estimates, so iterative procedures are required, for which initial values must be supplied;

2. linear model formulae that define only the model matrix are not adequate to specify non-linear regression models. A more flexible protocol is needed.

The main S function for fitting a non-linear regression model is nls.[1] We can fit the weight loss model by

```
> wtloss.st <- c(b0 = 90, b1 = 95, th = 120)
> wtloss.fm <- nls(Weight ~ b0 + b1*2^(-Days/th),
      data = wtloss, start = wtloss.st, trace = T)
67.5435 : 90 95 120
40.1808 : 82.7263 101.305 138.714
39.2449 : 81.3987 102.658 141.859
39.2447 : 81.3737 102.684 141.911
> wtloss.fm
Residual sum of squares : 39.245
parameters:
      b0     b1      th
  81.374 102.68 141.91
formula: Weight ~ b0 + b1 * 2^( - Days/th)
52 observations
```

The arguments to nls are the following.

formula   A non-linear model formula. The form is response ~ mean, where the right-hand side can have either of two forms. The standard form is an ordinary algebraic expression containing both parameters and determining variables. Note that the operators now have their usual arithmetical

---

[1]In package nls in R.

meaning. (The second form is used with the `plinear` fitting algorithm, discussed in Section 8.3 on page 218.)

`data` An optional data frame for the variables (and sometimes parameters).

`start` A list or numeric vector specifying the starting values for the parameters in the model.

The `names` of the components of `start` are also used to specify which of the variables occurring on the right-hand side of the model formula are parameters. All other variables are then assumed to be determining variables.[2]

`control` An optional argument allowing some features of the default iterative procedure to be changed.

`algorithm` An optional character string argument allowing a particular fitting algorithm to be specified. The default procedure is simply `"default"`.

`trace` An argument allowing tracing information from the iterative procedure to be printed. By default none is printed.

In our example the names of the parameters were specified as b0, b1 and th. The initial values of 90, 95 and 120 were found by inspection of Figure 8.1. From the trace output the procedure is seen to converge in three iterations.

Weighted data

The `nls` function has no `weights` argument, but non-linear regressions with *known* weights may be handled by writing the formula as ~ sqrt(W)*(y - M) rather than y ~ M. (The algorithm minimizes the sum of squared differences between left- and right-hand sides and an empty left-hand side counts as zero.) If W contains unknown parameters to be estimated the log-likelihood function has an extra term and the problem must be handled by the more general optimization methods such as those discussed in Chapter 16.

Using function derivative information

Most non-linear regression fitting algorithms operate in outline as follows. The first-order Taylor-series approximation to η_k at an initial value $\beta^{(0)}$ is

$$\eta_k(\beta) \approx \eta_k(\beta^{(0)}) + \sum_{j=1}^{p}(\beta_j - \beta_j^{(0)}) \frac{\partial \eta_k}{\partial \beta_j}\bigg|_{\beta=\beta^{(0)}}$$

In vector terms these may be written

$$\eta(\beta) \approx \omega^{(0)} + Z^{(0)}\beta \qquad (8.3)$$

where

$$Z_{kj}^{(0)} = \frac{\partial \eta_k}{\partial \beta_j}\bigg|_{\beta=\beta^{(0)}} \qquad \text{and} \qquad \omega_k^{(0)} = \eta_k(\beta^{(0)}) - \sum_{j=1}^{p}\beta_j^{(0)} Z_{kj}^{(0)}$$

[2]In S-PLUS there is a bug that may be avoided if the order in which the parameters appear in the start vector is the same as the order in which they first appear in the model. It is as if the order in the names attribute were ignored.

Equation (8.3) defines the tangent plane to the surface at the coordinate point $\beta = \beta^{(0)}$. The process consists of regressing the observation vector y onto the tangent plane defined by $Z^{(0)}$ with *offset* vector $w^{(0)}$ to give a new approximation, $\beta = \beta^{(1)}$, and iterating to convergence. For a linear regression the offset vector is 0 and the matrix $Z^{(0)}$ is the model matrix X, a constant matrix, so the process converges in one step. In the non-linear case the next approximation is

$$\beta^{(1)} = \left(Z^{(0)\,T}Z^{(0)}\right)^{-1}Z^{(0)\,T}\left(y - w^{(0)}\right)$$

With the `default` algorithm the Z matrix is computed approximately by numerical methods unless formulae for the first derivatives are supplied. Providing derivatives often (but not always) improves convergence.

Derivatives can be provided as an attribute of the model. The standard way to do this is to write an S function to calculate the mean vector η and the Z matrix. The result of the function is η with the Z matrix included as a `gradient` attribute.

For our simple example the three derivatives are

$$\frac{\partial \eta}{\partial \beta_0} = 1, \qquad \frac{\partial \eta}{\partial \beta_1} = 2^{-x/\theta}, \qquad \frac{\partial \eta}{\partial \theta} = \frac{\log(2)\,\beta_1 x 2^{-x/\theta}}{\theta^2}$$

so an S function to specify the model including derivatives is

```
expn <- function(b0, b1, th, x) {
    temp <- 2^(-x/th)
    model.func <- b0 + b1 * temp
    Z <- cbind(1, temp, (b1 * x * temp * log(2))/th^2)
    dimnames(Z) <- list(NULL, c("b0", "b1", "th"))
    attr(model.func, "gradient") <- Z
    model.func
}
```

Note that the gradient matrix must have column names matching those of the corresponding parameters.

We can fit our model again using first derivative information:

```
> wtloss.gr <- nls(Weight ~ expn(b0, b1, th, Days),
      data = wtloss, start = wtloss.st, trace = T)
67.5435 : 90 95 120
40.1808 : 82.7263 101.305 138.714
39.2449 : 81.3987 102.658 141.859
39.2447 : 81.3738 102.684 141.911
```

This appears to make no difference to the speed of convergence, but tracing the function `expn` shows that only 6 evaluations are required when derivatives are supplied compared with 21 if they are not supplied.

Functions such as `expn` can often be generated automatically using the symbolic differentiation function `deriv`. It is called with three arguments:

(a) the model formula, with the left-hand side optionally left blank,

(b) a character vector giving the names of the parameters and

(c) an empty function with an argument specification as required for the result.

An example makes the process clearer. For the weight loss data with the exponential model, we can use:

```
expn1 <- deriv(y ~ b0 + b1 * 2^(-x/th), c("b0", "b1", "th"),
               function(b0, b1, th, x) {})
```

The result in S-PLUS is the function (R's result is marginally different)

```
expn1 <- function(b0, b1, th, x)
{
    .expr3 <- 2^(( - x)/th)
    .value <- b0 + (b1 * .expr3)
    .grad <- array(0, c(length(.value), 3),
                   list(NULL, c("b0", "b1", "th")))
    .grad[, "b0"] <- 1
    .grad[, "b1"] <- .expr3
    .grad[, "th"] <- b1 *
        (.expr3 * (0.693147180559945 * (x/(th^2))))
    attr(.value, "gradient") <- .grad
    .value
}
```

Self-starting non-linear regressions

Very often reasonable starting values for a non-linear regression can be calculated by some fairly simple automatic procedure. Setting up such a *self-starting* non-linear model is somewhat technical, but several examples[3] are supplied.

Consider once again a negative exponential decay model such as that used in the weight loss example but this time written in the more usual exponential form:

$$y = \beta_0 + \beta_1 \exp(-x/\theta) + \epsilon$$

One effective initial value procedure follows.

 (i) Fit an initial quadratic regression in x.

 (ii) Find the fitted values, say, y_0, y_1 and y_2 at three equally spaced points x_0, $x_1 = x_0 + \delta$ and $x_2 = x_0 + 2\delta$.

 (iii) Equate the three fitted values to their expectation under the non-linear model to give an initial value for θ as

$$\theta_0 = \delta \,/ \log\left(\frac{y_0 - y_1}{y_1 - y_2}\right)$$

 (iv) Initial values for β_0 and β_1 can then be obtained by linear regression of y on $\exp(-x/\theta_0)$.

---

[3] Search for objects with names starting with SS, in R in package nls.

An S function to implement this procedure (with a few extra checks) called
negexp.SSival is supplied in MASS; interested readers should study it carefully.

We can make a self-starting model with both first derivative information and
this initial value routine by.

```
negexp <- selfStart(model = ~ b0 + b1*exp(-x/th),
      initial = negexp.SSival, parameters = c("b0", "b1", "th"),
      template = function(x, b0, b1, th) {})
```

where the first, third and fourth arguments are the same as for deriv. We may
now fit the model without explicit initial values.

```
> wtloss.ss <- nls(Weight ~ negexp(Days, B0, B1, theta),
                  data = wtloss, trace = T)
     B0    B1   theta
 82.713 101.49 200.16
39.5453 : 82.7131 101.495 200.160
39.2450 : 81.3982 102.659 204.652
39.2447 : 81.3737 102.684 204.734
```

(The first two lines of output come from the initial value procedure and the last
three from the nls trace.)

8.3   Non-Linear Fitted Model Objects and Method Functions

The result of a call to nls is an object of class nls. The standard method functions are available.

For the preceding example the summary function gives:

```
> summary(wtloss.gr)
Formula: Weight ~ expn1(b0, b1, th, Days)
Parameters:
       Value Std. Error t value
b0   81.374     2.2690  35.863
b1  102.684     2.0828  49.302
th  141.911     5.2945  26.803
Residual standard error: 0.894937 on 49 degrees of freedom
Correlation of Parameter Estimates:
        b0      b1
b1 -0.989
th -0.986   0.956
```

Surprisingly, no working deviance method function exists but such a method
function is easy to write and is included in MASS. It merely requires

```
> deviance.nls <- function(object) sum(object$residuals^2)
> deviance(wtloss.gr)
[1] 39.245
```

MASS also has a generic function vcov that will extract the estimated variance
matrix of the mean parameters:

```
> vcov(wtloss.gr)
          b0        b1       th
b0    5.1484   -4.6745  -11.841
b1   -4.6745    4.3379   10.543
th  -11.8414   10.5432   28.032
```

Taking advantage of linear parameters

If all non-linear parameters were known the model would be linear and standard linear regression methods could be used. This simple idea lies behind the "plinear" algorithm. It requires a different form of model specification that combines aspects of linear and non-linear model formula protocols. In this case the right-hand side expression specifies a *matrix* whose columns are functions of the non-linear parameters. The linear parameters are then implied as the regression coefficients for the columns of the matrix. Initial values are only needed for the non-linear parameters. Unlike the linear model case there is no implicit intercept term.

There are several advantages in using the partially linear algorithm. It can be much more stable than methods that do not take advantage of linear parameters, it requires fewer initial values and it can often converge from poor starting positions where other procedures fail.

Asymptotic regressions with different asymptotes

As an example of a case where the partially linear algorithm is very convenient, we consider a dataset first discussed in Linder, Chakravarti and Vuagnat (1964). The object of the experiment was to assess the influence of calcium in solution on the contraction of heart muscle in rats. The left auricle of 21 rat hearts was isolated and on several occasions electrically stimulated and dipped into various concentrations of calcium chloride solution, after which the shortening was measured. The data frame `muscle` in MASS contains the data as variables `Strip`, `Conc` and `Length`.

The particular model posed by the authors is of the form

$$\log y_{ij} = \alpha_j + \beta \rho^{x_{ij}} + \varepsilon_{ij} \tag{8.4}$$

where i refers to the concentration and j to the muscle strip. This model has 1 non-linear and 22 linear parameters. We take the initial estimate for ρ to be 0.1. Our first step is to construct a matrix to select the appropriate α.

```
> A <- model.matrix(~ Strip - 1, data = muscle)
> rats.nls1 <- nls(log(Length) ~ cbind(A, rho^Conc),
    data = muscle, start = c(rho = 0.1), algorithm = "plinear")
> (B <- coef(rats.nls1))
      rho    .lin1   .lin2   .lin3   .lin4   .lin5   .lin6  .lin7
  0.077778 3.0831  3.3014  3.4457  2.8047  2.6084  3.0336  3.523
   .lin8   .lin9  .lin10  .lin11  .lin12  .lin13  .lin14  .lin15 .lin16
  3.3871  3.4671  3.8144  3.7388  3.5133  3.3974  3.4709   3.729 3.3186
  .lin17  .lin18  .lin19  .lin20  .lin21   .lin22
  3.3794  2.9645  3.5847  3.3963    3.37 -2.9601
```

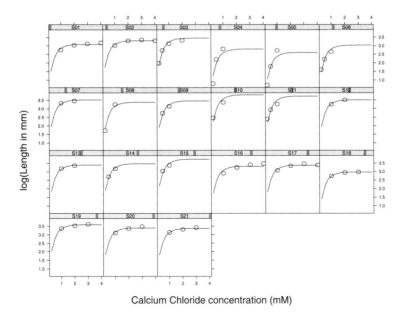

Figure 8.2: The heart contraction data of Linder *et al.* (1964): points and fitted model.

We can now use this coefficient vector as a starting value for a fit using the conventional algorithm.

```
> st <- list(alpha = B[2:22], beta = B[23], rho = B[1])
> rats.nls2 <- nls(log(Length) ~ alpha[Strip] + beta*rho^Conc,
                    data = muscle, start = st)
```

Notice that if a parameter in the non-linear regression is indexed, such as `alpha[Strip]` here, the starting values must be supplied as a list with named separate components.

A trace will show that this converges in one step with unchanged parameter estimates, as expected.

Having the fitted model object in standard rather than `"plinear"` form allows us to predict from it using the standard generic function `predict`. We now show the data and predicted values in a Trellis display.

```
attach(muscle)
Muscle <- expand.grid(Conc = sort(unique(Conc)),
                      Strip = levels(Strip))
Muscle$Yhat <- predict(rats.nls2, Muscle)
Muscle$logLength <- rep(NA, nrow(Muscle))
ind <- match(paste(Strip, Conc),
             paste(Muscle$Strip, Muscle$Conc))
Muscle$logLength[ind] <- log(Length)
detach()
```

```
xyplot(Yhat ~ Conc | Strip, Muscle, as.table = T,
    ylim = range(c(Muscle$Yhat, Muscle$logLength), na.rm = T),
    subscripts = T, xlab = "Calcium Chloride concentration (mM)",
    ylab = "log(Length in mm)", panel =
    function(x, y, subscripts, ...) {
        lines(spline(x, y))    # llines in R
        panel.xyplot(x, Muscle$logLength[subscripts], ...)
    })
```

The result is shown in Figure 8.2. The model appears to describe the situation fairly well, but for some purposes a non-linear mixed effects model might be more suitable (which is left as an exercise on Section 10.3).

8.4 Confidence Intervals for Parameters

For some non-linear regression problems the standard error may be an inadequate summary of the uncertainty of a parameter estimate and an asymmetric confidence interval may be more appropriate. In non-linear regression it is customary to invert the "extra sum of squares" to provide such an interval, although the result is usually almost identical to the likelihood ratio interval.

Suppose the parameter vector is $\theta = (\theta_1, \theta_2)^T$ and without loss of generality we wish to test an hypothesis $H_0 : \theta_1 = \theta_{10}$. Let $\widehat{\theta}$ be the overall least-squares estimate, $\widehat{\theta}_{2|1}$ be the conditional least-squares estimate of θ_2 with $\theta_1 = \theta_{10}$ and put $\widehat{\theta}(\theta_{10}) = (\theta_{10}, \widehat{\theta}_{2|1})^T$. The extra sum of squares principle then leads to the test statistic:

$$F(\theta_{10}) = \frac{\text{RSS}\{\widehat{\theta}(\theta_{10})\} - \text{RSS}(\widehat{\theta})}{s^2}$$

which under the null hypothesis has an approximately $F_{1,\,n-p}$-distribution. (Here RSS denotes the residual sum of squares.) Similarly the *signed square root*

$$\tau(\theta_{10}) = \text{sign}(\theta_{10} - \widehat{\theta}_1)\sqrt{F(\theta_{10})} \tag{8.5}$$

has an approximate t_{n-p}-distribution under H_0. The confidence interval for θ_1 is then the set $\{\theta_1 \mid -t < \tau(\theta_1) < t\}$, where t is the appropriate percentage point from the t_{n-p}-distribution.

The `profile` function is generic. The method for `nls` objects explores the residual sum of squares surface by taking each parameter θ_i in turn and fixing it at a series of values above and below $\widehat{\theta}_i$ so that (if possible) $\tau(\theta_i)$ varies from 0 by at least some pre-set value t in both directions. (If the sum of squares function "flattens out" in either direction, or if the fitting process fails in some way, this may not be achievable, in which case the profile is incomplete.) The value returned is a list of data frames, one for each parameter, and named by the parameter names. Each data frame component consists of

(a) a vector, `tau`, of values of $\tau(\theta_i)$ and

(b) a matrix, par.vals, giving the values of the corresponding parameter estimates $\widehat{\boldsymbol{\theta}}(\theta_i)$, sometimes called the 'parameter traces'.

The argument which of profile may be used to specify only a subset of the
parameters to be so profiled. (Note that profiling is not available for fitted model
objects in which the plinear algorithm is used.)

   Finding a confidence interval then requires the tau values for that parameter
to be interpolated and the set of θ_i values to be found (here always of the form
$\underline{\theta}_i < \theta_i < \overline{\theta}_i$). The function confint in MASS is a generic function to perform
this interpolation. It works for glm or nls fitted models, although the latter must
be fitted with the default fitting method. The critical steps in the method function
are

```
confint.profile.nls <-
   function(object, parm = seq(along = pnames),
            level = 0.95) {
   ....
   for(pm in parm) {
     pro <- object[[pm]]
     if(length(pnames) > 1)
        sp <- spline(x = pro[, "par.vals"][, pm], y = pro$tau)
     else sp <- spline(x = pro[, "par.vals"], y = pro$tau)
     ci[pnames[pm], ] <- approx(sp$y, sp$x, xout = cutoff)$y
   }
   drop(ci)
}
```

The three steps inside the loop are, respectively, extract the appropriate component of the *profile* object, do a spline interpolation in the t-statistic and linearly
interpolate in the digital spline to find the parameter values at the two cutoffs. If
the function is used on the fitted object itself the first step is to make a profile
object from it and this will take most of the time. If a profiled object is already
available it should be used instead.

The weight loss data, continued

For the person on the weight loss programme an important question is how long
it might be before he achieves some goal weight. If δ_0 is the time to achieve a
predicted weight of w_0, then solving the equation $w_0 = \beta_0 + \beta_1 2^{-\delta_0/\theta}$ gives

$$\delta_0 = -\theta \log_2\{(w_0 - \beta_0)/\beta_1\}$$

We now find a confidence interval for δ_0 for various values of w_0 by the outlined
method.

   The first step is to re-write the model function using δ_0 as one of the parameters; the most convenient parameter for it to replace is θ and the new expression
of the model becomes

$$y = \beta_0 + \beta_1 \left(\frac{w_0 - \beta_0}{\beta_1}\right)^{x/\delta_0} + \epsilon \qquad (8.6)$$

In order to find a confidence interval for δ_0 we need to fit the model with this parametrization. First we build a model function:

```
expn2 <- deriv(~b0 + b1*((w0 - b0)/b1)^(x/d0),
               c("b0","b1","d0"), function(b0, b1, d0, x, w0) {})
```

It is also convenient to have a function that builds an initial value vector from the estimates of the present fitted model:

```
wtloss.init <- function(obj, w0) {
  p <- coef(obj)
  d0 <-  - log((w0 - p["b0"])/p["b1"])/log(2) * p["th"]
  c(p[c("b0", "b1")], d0 = as.vector(d0))
}
```

(Note the use of as.vector to remove unwanted names.) The main calculations are done in a short loop:

```
> out <- NULL
> w0s <- c(110, 100, 90)
> for(w0 in w0s) {
    fm <- nls(Weight ~ expn2(b0, b1, d0, Days, w0),
              wtloss, start = wtloss.init(wtloss.gr, w0))
    out <- rbind(out, c(coef(fm)["d0"], confint(fm, "d0")))
  }

Waiting for profiling to be done...
    ....
> dimnames(out)[[1]] <- paste(w0s,"kg:")
> out
              d0    2.5%   97.5%
110 kg: 261.51 256.23 267.50
100 kg: 349.50 334.74 368.02
 90 kg: 507.09 457.56 594.97
```

As the weight decreases the confidence intervals become much wider and more asymmetric. For weights closer to the estimated asymptotic weight the profiling procedure can fail.

A bivariate region: The Stormer viscometer data

The following example comes from Williams (1959). The Stormer viscometer measures the viscosity of a fluid by measuring the time taken for an inner cylinder in the mechanism to perform a fixed number of revolutions in response to an actuating weight. The viscometer is calibrated by measuring the time taken with varying weights while the mechanism is suspended in fluids of accurately known viscosity. The dataset comes from such a calibration, and theoretical considerations suggest a non-linear relationship between time T, weight w and viscosity v of the form

$$T = \frac{\beta_1 v}{w - \beta_2} + \epsilon$$

Table 8.1: The Stormer viscometer calibration data. The body of the table shows the times in seconds.

Weight				Viscosity (poise)				
(grams)	14.7	27.5	42.0	75.7	89.7	146.6	158.3	161.1	298.3
20	35.6	54.3	75.6	121.2	150.8	229.0	270.0		
50	17.6	24.3	31.4	47.2	58.3	85.6	101.1	92.2	187.2
100				24.6	30.0	41.7	50.3	45.1	89.0, 86.5

where β_1 and β_2 are unknown parameters to be estimated. Note that β_1 is a linear parameter and β_2 is non-linear. The dataset is given in Table 8.1.

Williams suggested that a suitable initial value may be obtained by writing the regression model in the form

$$wT = \beta_1 v + \beta_2 T + (w - \beta_2)\epsilon$$

and regressing wT on v and T using ordinary linear regression. With the data available in a data frame `stormer` with variables `Viscosity`, `Wt` and `Time`, we may proceed as follows.

```
> fm0 <- lm(Wt*Time ~ Viscosity + Time - 1,   data = stormer)
> b0 <- coef(fm0);  names(b0) <- c("b1", "b2"); b0
      b1      b2
  28.876 2.8437
> storm.fm <- nls(Time ~ b1*Viscosity/(Wt-b2), data = stormer,
                  start = b0, trace = T)
885.365 : 28.8755 2.84373
825.110 : 29.3935 2.23328
825.051 : 29.4013 2.21823
```

Since there are only two parameters we can display a confidence region for the regression parameters as a contour map. To this end put:

```
bc <- coef(storm.fm)
se <- sqrt(diag(vcov(storm.fm)))
dv <- deviance(storm.fm)
```

Define $d(\beta_1, \beta_2)$ as the sum of squares function:

$$d(\beta_1, \beta_2) = \sum_{i=1}^{23}\left(T_i - \frac{\beta_1 v_i}{w_i - \beta_2}\right)^2$$

Then `dv` contains the minimum value, $d_0 = d(\widehat{\beta}_1, \widehat{\beta}_2)$, the residual sum of squares or model deviance.

If β_1 and β_2 are the true parameter values the "extra sum of squares" statistic

$$F(\beta_1, \beta_2) = \frac{(d(\beta_1, \beta_2) - d_0)/2}{d_0/21}$$

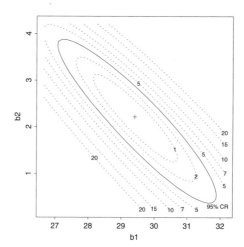

Figure 8.3: The Stormer data. The F-statistic surface and a confidence region for the regression parameters.

is approximately distributed as $F_{2,21}$. An approximate confidence set contains those values in parameter space for which $F(\beta_1, \beta_2)$ is less than the 95% point of the $F_{2,21}$ distribution. We construct a contour plot of the $F(\beta_1, \beta_2)$ function and mark off the confidence region. The result is shown in Figure 8.3.

A suitable region for the contour plot is three standard errors either side of the least-squares estimates in each parameter. Since these ranges are equal in their respective standard error units it is useful to make the plotting region square.

```
par(pty = "s")
b1 <- bc[1] + seq(-3*se[1], 3*se[1], length = 51)
b2 <- bc[2] + seq(-3*se[2], 3*se[2], length = 51)
bv <- expand.grid(b1, b2)
```

The simplest way to calculate the sum of squares function is to use the `apply` function:

```
attach(stormer)
ssq <- function(b)
        sum((Time - b[1] * Viscosity/(Wt-b[2]))^2)
dbetas <- apply(bv, 1, ssq)
```

However, using a function such as `outer` makes better use of the vectorizing facilities of S, and in this case a direct calculation is the most efficient:

```
cc <- matrix(Time - rep(bv[,1],rep(23, 2601)) *
        Viscosity/(Wt - rep(bv[,2], rep(23, 2601))), 23)
dbetas <- matrix(drop(rep(1, 23) %*% cc^2), 51)
```

The F-statistic array is then:

```
fstat <- matrix( ((dbetas - dv)/2) / (dv/21), 51, 51)
```

We can now produce a contour map of the F-statistic, taking care that the contours occur at relatively interesting levels of the surface. Note that the confidence region contour is at about 3.5:

```
> qf(0.95, 2, 21)
[1] 3.4668
```

Our intention is to produce a slightly non-standard contour plot and for this the traditional plotting functions are more flexible than Trellis graphics functions. Rather than use `contour` to set up the plot directly, we begin with a call to `plot`:

```
plot(b1, b2, type = "n")
lev <- c(1, 2, 5, 7, 10, 15, 20)
contour(b1, b2, fstat, levels = lev, labex = 0.75, lty = 2, add = T)
contour(b1, b2, fstat, levels = qf(0.95,2,21), add = T, labex = 0)
text(31.6, 0.3, labels = "95% CR", adj = 0, cex = 0.75)
points(bc[1], bc[2], pch = 3, mkh = 0.1)
par(pty = "m")
```

Since the likelihood function has the same contours as the F-statistic, the near elliptical shape of the contours is an indication that the approximate theory based on normal linear regression is probably accurate, although more than this is needed to be confident. (See the next section.) Given the way the axis scales have been chosen, the elongated shape of the contours shows that the estimates $\widehat{\beta}_1$ and $\widehat{\beta}_2$ are highly (negatively) correlated.

Note that finding a bivariate confidence region for two regression parameters where there are several others present can be a difficult calculation, at least in principle, since each point of the F-statistic surface being contoured must be calculated by optimizing with respect to the other parameters.

Bootstrapping

An alternative way to explore the distribution of the parameter estimates is to use the bootstrap. This was a designed experiment, so we illustrate model-based bootstrapping using the functions from library section `boot`. As this is a nonlinear model, the residuals may have a non-zero mean, which we remove.

```
> library(boot)
> storm.fm <- nls(Time ~ b*Viscosity/(Wt - c), stormer,
                  start = c(b = 29.401, c = 2.2183))
> summary(storm.fm)$parameters
    Value Std. Error t value
b 29.4010    0.91553 32.1135
c  2.2183    0.66553  3.3332
> st <- cbind(stormer, fit = fitted(storm.fm))
> storm.bf <- function(rs, i) {
      st <- st # for S-PLUS
      st$Time <-  st$fit + rs[i]
      coef(nls(Time ~ b * Viscosity/(Wt - c), st,
```

```
                        start = coef(storm.fm)))
     }
> rs <- scale(resid(storm.fm), scale = F) # remove the mean
> (storm.boot <- boot(rs, storm.bf, R = 9999)) ## slow
    ....
Bootstrap Statistics :
      original    bias     std. error
t1*   28.7156   0.69599      0.83153
t2*    2.4799  -0.26790      0.60794
> boot.ci(storm.boot, index = 1,
          type = c("norm", "basic", "perc", "bca"))
Intervals :
Level        Normal                  Basic
95%    (26.39, 29.65 )    (26.46, 29.69 )

Level       Percentile              BCa
95%    (27.74, 30.97 )    (26.18, 29.64 )
Calculations and Intervals on Original Scale
Warning : BCa Intervals used Extreme Quantiles
> boot.ci(storm.boot, index = 2,
          type = c("norm", "basic", "perc", "bca"))
Intervals :
Level        Normal                  Basic
95%   ( 1.556,  3.939 )    ( 1.553,  3.958 )

Level       Percentile              BCa
95%   ( 1.001,  3.406 )    ( 1.571,  4.031 )
Calculations and Intervals on Original Scale
```

The 'original' here is not the original fit as the residuals were adjusted. Figure 8.3 suggests that a likelihood-based analysis would support the percentile intervals. Note that despite the large number of bootstrap samples (which took 2 mins), this is still not really enough for BC_a intervals.

8.5 Profiles

One way of assessing departures from the planar assumption (both of the surface itself and of the coordinate system within the surface) is to look at the low-dimensional profiles of the sum of squares function, as discussed in Section 8.4. Indeed this is the original intended use for the `profile` function.

For coordinate directions along which the approximate linear methods are accurate, a plot of the non-linear t-statistics, $\tau(\beta_i)$ against β_i over several standard deviations on either side of the maximum likelihood estimate should be straight. (The $\tau(\beta_i)$ were defined in equation (8.5) on page 220.) Any deviations from straightness serve as a warning that the linear approximation may be misleading in that direction.

Appropriate plot methods for objects produced by the `profile` function are available (and MASS contains an enhanced version using Trellis). For the weight loss data we can use (Figure 8.4)

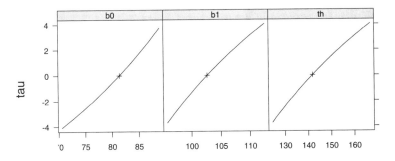

Figure 8.4: Profile plots for the negative exponential weight loss model.

```
plot(profile(wtloss.gr))
```

A limitation of these plots is that they only examine each coordinate direction separately. Profiles can also be defined for two or more parameters simultaneously. These present much greater difficulties of computation and to some extent of visual assessment. An example is given in the previous section with the two parameters of the `stormer` data model. In Figure 8.3 an assessment of the linear approximation requires checking both that the contours are approximately elliptical and that each one-dimensional cross-section through the minimum is approximately quadratic. In one dimension it is much easier visually to check the linearity of the signed square root than the quadratic nature of the sum of squares itself.

The profiling process itself actually accumulates much more information than that shown in the plot. As well as recording the value of the optimized likelihood for a fixed value of a parameter β_j, it also finds and records the values of all other parameters for which the conditional optimum is achieved. The `pairs` method function for `profile` objects supplied in MASS can display this so-called *profile trace* information in graphical form, which can shed considerably more light on the local features of the log-likelihood function near the optimum. This is discussed in more detail in the on-line complements.

8.6   Constrained Non-Linear Regression

All the functions for regression we have seen so far assume that the parameters can take any real value. Constraints on the parameters can be often incorporated by reparametrization (such as $\beta_i = e^\theta$, possibly thereby making a linear model non-linear), but this can be undesirable if, for example, the constraint is $\beta_i \geqslant 0$ and 0 is expected to occur. Two functions for constrained regression are provided in S-PLUS, `nnls.fit` (for linear regression with non-negative coefficients) and `nlregb`. Neither is available in R. The general optimization methods considered in Chapter 16 can also be used.

8.7   One-Dimensional Curve-Fitting

Let us return to the simplest possible case, of one response y and one explanatory variable x. We want to draw a smooth curve through the scatterplot of y vs x. (So these methods are often called *scatterplot smoothers*.) In contrast to the first part of this chapter, we may have little idea of the functional form of the curve.

A wide range of methods is available, including splines, kernel regression, running means and running lines. The classical way to introduce non-linear functions of dependent variables is to add a limited range of transformed variables to the model, for example, quadratic and cubic terms, or to split the range of the variable and use a piecewise constant or piecewise functions. (In S this can be achieved using the functions `poly` and `cut`.) But there are almost always better alternatives.

As an illustrative example, we consider the data on the concentration of a chemical GAG in the urine of 314 children aged from 0 to 17 years in data frame `GAGurine` (Figure 8.5). Forwards selection suggests a degree 6 polynomial:

```
attach(GAGurine)
plot(Age, GAG, main = "Degree 6 polynomial")
GAG.lm <- lm(GAG ~ Age + I(Age^2) + I(Age^3) + I(Age^4) +
   I(Age^5) + I(Age^6) + I(Age^7) + I(Age^8))
anova(GAG.lm)
   ....
```

```
Terms added sequentially (first to last)
          Df Sum of Sq Mean Sq F Value    Pr(F)
     Age   1     12590   12590  593.58 0.00000
I(Age^2)   1      3751    3751  176.84 0.00000
I(Age^3)   1      1492    1492   70.32 0.00000
I(Age^4)   1       449     449   21.18 0.00001
I(Age^5)   1       174     174    8.22 0.00444
I(Age^6)   1       286     286   13.48 0.00028
I(Age^7)   1        57      57    2.70 0.10151
I(Age^8)   1        45      45    2.12 0.14667
```

```
GAG.lm2 <- lm(GAG ~ Age + I(Age^2) + I(Age^3) + I(Age^4) +
                I(Age^5) + I(Age^6))
xx <- seq(0, 17, len = 200)
lines(xx, predict(GAG.lm2, data.frame(Age = xx)))
```

Splines

A modern alternative is to use *spline* functions. (Green and Silverman, 1994, provide a fairly gentle introduction to splines.) We only need cubic splines. Divide the real line by an ordered set of points $\{z_i\}$ known as *knots*. On the interval $[z_i, z_{i+1}]$ the spline is a cubic polynomial, and it is continuous and has continuous first and second derivatives, imposing 3 conditions at each knot. With n knots, $n + 4$ parameters are needed to represent the cubic spline (from $4(n + 1)$ for the cubic polynomials minus $3n$ continuity conditions). Of the many possible parametrizations, that of B-splines has desirable properties. (See Mackenzie

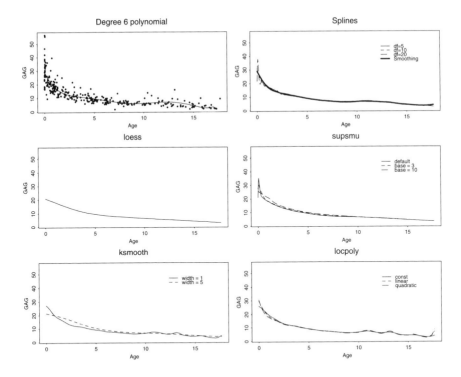

Figure 8.5: Scatter smoothers applied to GAGurine.

and Abrahamowicz, 1996, and Hastie *et al.*, 2001, pp. 160–3, for gentle introductions.) The S function[4] bs generates a matrix of B-splines, and so can be
included in a linear-regression fit; doing so is called using *regression splines*.

A restricted form of B-splines known as *natural splines* (implemented by the
S function ns) is linear on $(-\infty, z_1]$ and $[z_n, \infty)$ and thus would have n parameters. However, ns adds an extra knot at each of the maximum and minimum
of the data points, and so has $n + 2$ parameters, dropping the requirement for the
derivative to be continuous at z_1 and z_n. The functions bs and ns may have
the knots specified or be allowed to choose the knots as quantiles of the empirical
distribution of the variable to be transformed, by specifying the number df of
parameters.

Prediction from models including splines needs care, as the basis for the functions depends on the observed values of the independent variable. If predict.lm
is used, it will form a new set of basis functions and then erroneously apply the
fitted coefficients. The S-PLUS function predict.gam will work more nearly
correctly. (It uses both the old and new data to choose the knots.)

```
plot(Age, GAG, type = "n", main = "Splines")
lines(Age, fitted(lm(GAG ~ ns(Age, df = 5))))
lines(Age, fitted(lm(GAG ~ ns(Age, df = 10))), lty = 3)
```

---

[4]In R bs and ns are in package splines .

```
lines(Age, fitted(lm(GAG ~ ns(Age, df = 20))), lty = 4)
```

Smoothing splines

Suppose we have n pairs (x_i, y_i). A *smoothing spline* minimizes a compromise between the fit and the degree of smoothness of the form

$$\sum w_i[y_i - f(x_i)]^2 + \lambda \int (f''(x))^2 \, dx$$

over all (measurably twice-differentiable) functions f. It is a cubic spline with knots at the x_i, but does not interpolate the data points for $\lambda > 0$ and the degree of fit is controlled by λ. The S function `smooth.spline` allows λ[5] or the (equivalent) degrees of freedom to be specified, otherwise it will choose the degree of smoothness automatically by cross-validation. There are various definitions of equivalent degrees of freedom; see Green and Silverman (1994, pp. 37–8) and Hastie and Tibshirani (1990, Appendix B). That used in `smooth.spline` is the trace of the smoother matrix S; as fitting a smoothing spline is a linear operation, there is an $n \times n$ matrix S such $\hat{y} = Sy$. (In a regression fit S is the hat matrix (page 152); this has trace equal to the number of free parameters.)

```
lines(smooth.spline(Age, GAG), lwd = 3)
legend(12, 50, c("df=5", "df=10", "df=20", "Smoothing"),
       lty = c(1, 3, 4, 1), lwd = c(1,1,1,3), bty = "n")
```

For $\lambda = 0$ the smoothing spline will interpolate the data points if the x_i are distinct. There are simpler methods to fit interpolating cubic splines, implemented in the function `spline`. This can be useful to draw smooth curves through the result of some expensive smoothing algorithm: we could also use linear interpolation implemented by `approx`.

Local regression

There are several smoothers that work by fitting a linear (or polynomial regression) to the data points in the vicinity of x and then using as the smoothed value the predicted value at x.

The algorithm used by `lowess` is quite complex; it uses robust locally linear fits. A window is placed about x; data points that lie inside the window are weighted so that nearby points get the most weight and a robust weighted regression is used to predict the value at x. The parameter f controls the window size and is the proportion of the data that is included. The default, `f = 2/3`, is often too large for scatterplots with appreciable structure. The function[6] `loess` is an extension of the ideas of `lowess` which will work in one, two or more dimensions in a similar way. The function `scatter.smooth` plots a `loess` line on a scatterplot, using `loess.smooth`.

```
plot(Age, GAG, type = "n", main = "loess")
lines(loess.smooth(Age, GAG))
```

---

[5]More precisely λ with the (x_i) scaled to $[0, 1]$ and the weights scaled to average 1.
[6]In R, loess , supsmu and ksmooth are in package modreg .

Friedman's smoother `supsmu` is the one used by the S-PLUS functions `ace`, `avas` and `ppreg`, and by our function `ppr`. It is based on a symmetric k-nearest neighbour linear least squares procedure. (That is, $k/2$ data points on each side of x are used in a linear regression to predict the value at x.) This is run for three values of k, $n/2$, $n/5$ and $n/20$, and cross-validation is used to choose a value of k for each x that is approximated by interpolation between these three. Larger values of the parameter `bass` (up to 10) encourage smooth functions.

```
plot(Age, GAG, type = "n", main = "supsmu")
lines(supsmu(Age, GAG))
lines(supsmu(Age, GAG, bass = 3), lty = 3)
lines(supsmu(Age, GAG, bass = 10), lty = 4)
legend(12, 50, c("default", "base = 3", "base = 10"),
       lty = c(1, 3, 4), bty = "n")
```

A *kernel smoother* is of the form

$$\hat{y}_i = \sum_{j=1}^{n} y_i K\left(\frac{x_i - x_j}{b}\right) \bigg/ \sum_{j=1}^{n} K\left(\frac{x_i - x_j}{b}\right) \tag{8.7}$$

where b is a bandwidth parameter, and K a kernel function, as in density estimation. In our example we use the function `ksmooth` and take K to be a standard normal density. The critical parameter is the bandwidth b. The function `ksmooth` seems very slow in S-PLUS, and the faster alternatives in library section Matt Wand's `KernSmooth` are recommended. These also include ways to select the bandwidth.

```
plot(Age, GAG, type = "n", main = "ksmooth")
lines(ksmooth(Age, GAG, "normal", bandwidth = 1))
lines(ksmooth(Age, GAG, "normal", bandwidth = 5), lty = 3)
legend(12, 50, c("width = 1", "width = 5"), lty = c(1, 3), bty = "n")

library(KernSmooth)
plot(Age, GAG, type = "n", main = "locpoly")
(h <- dpill(Age, GAG))
[1] 0.49592
lines(locpoly(Age, GAG, degree = 0, bandwidth = h))
```

Kernel regression can be seen as local fitting of a constant. Theoretical work (see Wand and Jones, 1995) has shown advantages in local fitting of polynomials, especially those of odd order. Library sections `KernSmooth` and `locfit` contain code for local polynomial fitting and bandwidth selection.

```
lines(locpoly(Age, GAG, degree = 1, bandwidth = h), lty = 3)
lines(locpoly(Age, GAG, degree = 2, bandwidth = h), lty = 4)
legend(12, 50, c("const", "linear", "quadratic"),
       lty = c(1, 3, 4), bty = "n")
detach()
```

Finding derivatives of fitted curves

There are problems in which the main interest is in estimating the first or second derivative of a smoother. One idea is to fit locally a polynomial of high enough order and report its derivative (implemented in `locpoly`). We can differentiate a spline fit, although it may be desirable to fit splines of higher order than cubic; library section `pspline` by Jim Ramsay provides a suitable generalization of `smooth.spline`.

8.8 Additive Models

For linear regression we have a dependent variable Y and a set of predictor variables $X_1, \ldots, X_p$, and model

$$Y = \alpha + \sum_{j=1}^{p} \beta_j X_j + \epsilon$$

Additive models replace the linear function $\beta_j X_j$ by a non-linear function to get

$$Y = \alpha + \sum_{j=1}^{p} f_j(X_j) + \epsilon \tag{8.8}$$

Since the functions f_j are rather general, they can subsume the α. Of course, it is not useful to allow an arbitrary function f_j, and it helps to think of it as a *smooth* function. We may also allow terms to depends on a small number (in practice rarely more than two) terms.

As we have seen, smooth terms parametrized by regression splines can be fitted by `lm`. For smoothing splines it would be possible to set up a penalized least-squares problem and minimize that, but there would be computational difficulties in choosing the smoothing parameters simultaneously (Wahba, 1990; Wahba *et al.*, 1995). Instead an iterative approach is often used. The *backfitting* algorithm fits the smooth functions f_j in (8.8) one at a time by taking the residuals

$$Y - \alpha - \sum_{k \neq j} f_k(X_k)$$

and smoothing them against X_j using one of the scatterplot smoothers of the previous subsection. The process is repeated until it converges. Linear terms in the model (including any linear terms in the smoother) are fitted by least squares.

This procedure is implemented in S-PLUS by the function `gam`. The model formulae are extended to allow the terms `s(x)` and `lo(x)` which, respectively, specify a smoothing spline and a loess smoother. These have similar parameters to the scatterplot smoothers; for `s()` the default degrees of freedom is 4, and for `lo()` the window width is controlled by `span` with default 0.5. There is a plot method, `plot.gam`, which shows the smooth function fitted for each term in the additive model. (R has a somewhat different function `gam` in package `mgcv`.)

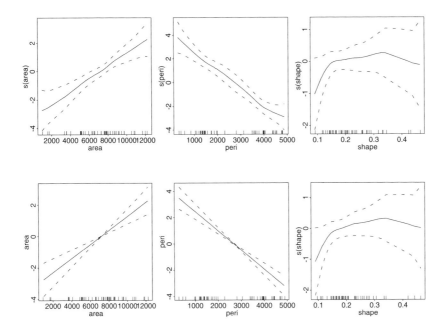

Figure 8.6: The results of fitting additive models to the dataset `rock` with smooth functions of all three predictors (top row) and linear functions of `area` and `perimeter` (bottom row). The dashed lines are approximate 95% pointwise confidence intervals. The tick marks show the locations of the observations on that variable.

Our dataset `rock` contains measurements on four cross-sections of each of 12 oil-bearing rocks; the aim is to predict permeability y (a property of fluid flow) from the other three measurements. As permeabilities vary greatly (6.3–1300), we use a log scale. The measurements are the end product of a complex image-analysis procedure and represent the total area, total perimeter and a measure of 'roundness' of the pores in the rock cross-section.

We first fit a linear model, then a full additive model. In this example convergence of the backfitting algorithm is unusually slow, so the `control` limits must be raised. The plots are shown in Figure 8.6.

```
rock.lm <- lm(log(perm) ~ area + peri + shape, data = rock)
summary(rock.lm)
rock.gam <- gam(log(perm) ~ s(area) + s(peri) + s(shape),
    control = gam.control(maxit = 50, bf.maxit = 50), data = rock)
summary(rock.gam)
anova(rock.lm, rock.gam)
par(mfrow = c(2, 3), pty = "s")
plot(rock.gam, se = T)
rock.gam1 <- gam(log(perm) ~ area + peri + s(shape), data = rock)
plot(rock.gam1, se = T)
anova(rock.lm, rock.gam1, rock.gam)
```

It is worth showing the output from `summary.gam` and the analysis of variance table from anova(rock.lm, rock.gam):

```
Deviance Residuals:
     Min        1Q  Median      3Q     Max
  -1.6855 -0.46962 0.12531 0.54248 1.2963


(Dispersion Parameter ... 0.7446 )

    Null Deviance: 126.93 on 47 degrees of freedom
Residual Deviance: 26.065 on 35.006 degrees of freedom

Number of Local Scoring Iterations: 1

DF for Terms and F-values for Nonparametric Effects

            Df Npar Df Npar F   Pr(F)
(Intercept)  1
   s(area)   1        3 0.3417 0.79523
   s(peri)   1        3 0.9313 0.43583
  s(shape)   1        3 1.4331 0.24966

Analysis of Variance Table

                           Terms Resid. Df    RSS    Test
1           area + peri + shape   44.000 31.949
2 s(area) + s(peri) + s(shape)   35.006 26.065 1 vs. 2
        Df Sum of Sq F Value   Pr(F)
1
2 8.9943    5.8835  0.8785 0.55311
```

This shows that each smooth term has one linear and three non-linear degrees of freedom. The reduction of RSS from 31.95 (for the linear fit) to 26.07 is not significant with an extra nine degrees of freedom, but Figure 8.6 shows that only one of the functions appears non-linear, and even that is not totally convincing. Although suggestive, the non-linear term for peri is not significant. With just a non-linear term in shape the RSS is 29.00.

One of the biggest difficulties with using the S-PLUS version of gam is specifying the degree of smoothing for each term (*via* the degrees of freedom for an s() term or the span for an lo() term). (The R version chooses the amount of smoothing automatically, by default.) This is overcome by using BRUTO (Friedman and Silverman, 1989; Hastie and Tibshirani, 1990) which fits additive models with smooth functions selected by smoothing splines and will choose between a smooth function, a linear term or omitting the variable altogether. (This is implemented in library section mda by Hastie and Tibshirani.)

For the rock data bruto fits an essentially linear model.

```
> library(mda)
> rock.bruto <- bruto(rock[, -4], rock[, 4])
> rock.bruto$type
[1] smooth linear linear
```

```
> rock.bruto$df
   area peri shape
 1.0167    1     1
```

Let us consider again the cpus data:

```
Xin <- as.matrix(cpus2[cpus.samp, 1:6])
test2 <- function(fit) {
  Xp <- as.matrix(cpus2[-cpus.samp, 1:6])
  sqrt(sum((log10(cpus2[-cpus.samp, "perf"]) -
           predict(fit, Xp))^2)/109)
}
cpus.bruto <- bruto(Xin, log10(cpus2[cpus.samp, 7]))
test2(cpus.bruto)
[1] 0.21336

cpus.bruto$type
[1] excluded smooth    linear    smooth    smooth    linear
cpus.bruto$df
 syct    mmin mmax    cach  chmin chmax
    0 1.5191    1 1.0578 1.1698      1

# examine the fitted functions
par(mfrow = c(3, 2))
Xp <- matrix(sapply(cpus2[cpus.samp, 1:6], mean), 100, 6,
             byrow = T)
for(i in 1:6) {
  xr <- sapply(cpus2, range)
  Xp1 <- Xp; Xp1[, i] <- seq(xr[1, i], xr[2, i], len = 100)
  Xf <- predict(cpus.bruto, Xp1)
  plot(Xp1[ ,i], Xf, xlab=names(cpus2)[i], ylab=  "", type = "l")
}
```

The result (not shown) indicates that the non-linear terms have a very slight curvature, as might be expected from the equivalent degrees of freedom that are reported.

Multiple Adaptive Regression Splines

The function mars in library section mda implements the MARS (Multiple Adaptive Regression Splines) method of Friedman (1991). By default this is an additive method, fitting splines of order 1 (piecewise linear functions) to each variable; again the number of pieces is selected by the program so that variables can be entered linearly, non-linearly or not at all.

The library section polymars of Kooperberg and O'Connor implements a restricted form of MARS (for example, allowing only pairwise interactions) suggested by Kooperberg *et al.* (1997).

We can use mars to fit a piecewise linear model with additive terms to the cpus data.

```
cpus.mars <- mars(Xin, log10(cpus2[cpus.samp,7]))
showcuts <- function(obj)
{
  tmp <- obj$cuts[obj$sel, ]
  dimnames(tmp) <- list(NULL, dimnames(Xin)[[2]])
  tmp
}
> showcuts(cpus.mars)
     syct   mmin   mmax cach chmin chmax
[1,]    0 0.0000 0.0000    0     0     0
[2,]    0 0.0000 3.6021    0     0     0
[3,]    0 0.0000 3.6021    0     0     0
[4,]    0 3.1761 0.0000    0     0     0
[5,]    0 0.0000 0.0000    0     8     0
[6,]    0 0.0000 0.0000    0     0     0
> test2(cpus.mars)
[1] 0.21366
# examine the fitted functions
Xp <- matrix(sapply(cpus2[cpus.samp, 1:6], mean), 100, 6,
             byrow = T)
for(i in 1:6) {
  xr <- sapply(cpus2, range)
  Xp1 <- Xp; Xp1[, i] <- seq(xr[1, i], xr[2, i], len = 100)
  Xf <- predict(cpus.mars, Xp1)
  plot(Xp1[ ,i], Xf, xlab = names(cpus2)[i], ylab = "", type = "l")
}
## try degree 2
> cpus.mars2 <- mars(Xin, log10(cpus2[cpus.samp,7]), degree = 2)
> showcuts(cpus.mars2)
     syct   mmin   mmax cach chmin chmax
[1,]    0 0.0000 0.0000    0     0     0
[2,]    0 0.0000 3.6021    0     0     0
[3,]    0 1.9823 3.6021    0     0     0
[4,]    0 0.0000 0.0000   16     8     0
[5,]    0 0.0000 0.0000    0     0     0
> test2(cpus.mars2)
[1] 0.21495
> cpus.mars6 <- mars(Xin, log10(cpus2[cpus.samp,7]), degree = 6)
> showcuts(cpus.mars6)
       syct   mmin   mmax cach chmin chmax
[1,] 0.0000 0.0000 0.0000    0     0     0
[2,] 0.0000 1.9823 3.6021    0     0     0
[3,] 0.0000 0.0000 0.0000   16     8     0
[4,] 0.0000 0.0000 0.0000   16     8     0
[5,] 0.0000 0.0000 3.6990    0     8     0
[6,] 2.3979 0.0000 0.0000   16     8     0
[7,] 2.3979 0.0000 3.6990   16     8     0
[8,] 0.0000 0.0000 0.0000    0     0     0
> test2(cpus.mars6)
[1] 0.20604
```

Allowing pairwise interaction terms (by `degree = 2`) or allowing arbitrary interactions make little difference to the effectiveness of the predictions. The plots (not shown) indicate an additive model with non-linear terms in `mmin` and `chmin` only.

Response transformation models

If we want to predict Y, it may be better to transform Y as well, so we have

$$\theta(Y) = \alpha + \sum_{j=1}^{p} f_j(X_j) + \epsilon \tag{8.9}$$

for an invertible smooth function $\theta()$, for example the log function we used for the `rock` dataset.

The ACE (alternating conditional expectation) algorithm of Breiman and Friedman (1985) chooses the functions θ and $f_1, \ldots, f_j$ to maximize the correlation between the predictor $\alpha + \sum_{j=1}^{p} f_j(X_j)$ and $\theta(Y)$. Tibshirani's (1988) procedure AVAS (additivity and variance stabilising transformation) fits the same model (8.9), but with the aim of achieving constant variance of the residuals for monotone θ.

The functions[7] `ace` and `avas` fit (8.9). Both allow the functions f_j and θ to be constrained to be monotone or linear, and the functions f_j to be chosen for circular or categorical data. Thus these functions provide another way to fit additive models (with linear θ) but they do not provide measures of fit nor standard errors. They do, however, automatically choose the degree of smoothness. Our experience has been that AVAS is much more reliable than ACE.

We can consider the `cpus` data; we have already log-transformed some of the variables.

```
attach(cpus2)
cpus.avas <- avas(cpus2[, 1:6], perf)
plot(log10(perf), cpus.avas$ty)
par(mfrow = c(2, 3))
for(i in 1:6) {
  o <- order(cpus2[, i])
  plot(cpus2[o, i], cpus.avas$tx[o, i], type = "l",
       xlab=names(cpus2[i]), ylab = "")
}
detach()
```

This accepts the log-scale for the response, but has some interesting shapes for the regressors (Figure 8.7). The strange shape of the transformation for `chmin` and `chmax` is probably due to local collinearity as there are five machines without any channels.

For the `rock` dataset we can use the following code. The S function `rug` produces the ticks to indicate the locations of data points, as used by `gam.plot`.

---

[7]In R these are in package `acepack` .

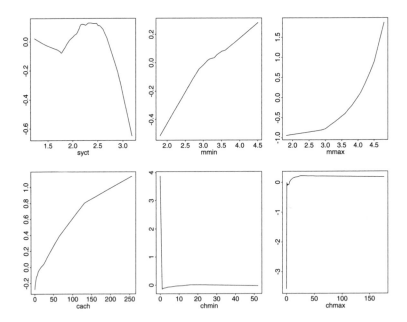

Figure 8.7: AVAS transformations of the regressors in the cpus dataset.

```
attach(rock)
x <- cbind(area, peri, shape)
o1 <- order(area); o2 <- order(peri); o3 <- order(shape)
a <- avas(x, perm)
par(mfrow = c(2, 2))
plot(area[o1], a$tx[o1, 1], type = "l")        see Figure 8.8
rug(area)
plot(peri[o2], a$tx[o2, 2], type = "l")
rug(peri)
plot(shape[o3], a$tx[o3, 3], type = "l")
rug(shape)
plot(perm, a$ty, log = "x")                     note log scale
a <- avas(x, log(perm))                         looks like $\log(y)$
a <- avas(x, log(perm), linear=0)               so force $\theta(y) = \log(y)$
# repeat plots
```

Here AVAS indicates a log transformation of permeabilities (expected on physical grounds) but little transformation of area or perimeter as we might by now expect.

8.9 Projection-Pursuit Regression

Now suppose that the explanatory vector $X = (X_1, \dots, X_p)$ is of high dimension. The additive model (8.8) may be too flexible as it allows a few degrees of

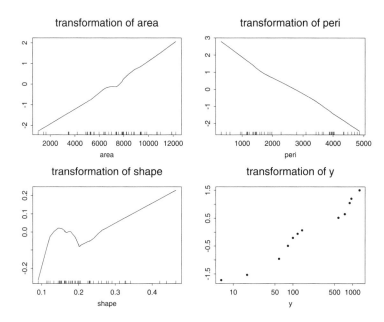

Figure 8.8: AVAS on permeabilities

freedom per X_j, yet it does not cover the effect of interactions between the independent variables. Projection pursuit regression (Friedman and Stuetzle, 1981) applies an additive model to projected variables. That is, it is of the form:

$$Y = \alpha_0 + \sum_{j=1}^{M} f_j(\boldsymbol{\alpha}_j^T \boldsymbol{X}) + \epsilon \tag{8.10}$$

for vectors $\boldsymbol{\alpha}_j$, and a dimension M to be chosen by the user. Thus it uses an additive model on predictor variables that are formed by projecting $\boldsymbol{X}$ in M carefully chosen directions. For large enough M such models can approximate (uniformly on compact sets and in many other senses) arbitrary continuous functions of $\boldsymbol{X}$ (for example, Diaconis and Shahshahani, 1984). The terms of (8.10) are called ridge functions, since they are constant in all but one direction.

The function `ppr`[8] in `MASS` fits (8.10) by least squares, and constrains the vectors $\boldsymbol{\alpha}_k$ to be of unit length. It first fits $M_{\max}$ terms sequentially, then prunes back to M by at each stage dropping the least effective term and re-fitting. The function returns the proportion of the variance explained by all the fits from $M, \dots, M_{\max}$.

For the `rock` example we can use[9]

```
> attach(rock)
```

---

[8]Like the S-PLUS function `ppreg` based on the SMART program described in Friedman (1984), but using double precision internally.

[9]The exact results depend on the machine used.

```
> rock1 <- data.frame(area = area/10000, peri = peri/10000,
                      shape = shape, perm = perm)
> detach()
> (rock.ppr <- ppr(log(perm) ~ area + peri + shape, data = rock1,
                   nterms = 2, max.terms = 5))
Call:
ppr.formula(formula = log(perm) ~ area + peri + shape,
           data = rock1, nterms = 2, max.terms = 5)

Goodness of fit:
 2 terms 3 terms 4 terms 5 terms
 11.2317  7.3547  5.9445  3.1141
```

The summary method gives a little more information.

```
> summary(rock.ppr)
Call:
ppr.formula(formula = log(perm) ~ area + peri + shape,
           data = rock1, nterms = 2, max.terms = 5)

Goodness of fit:
 2 terms 3 terms 4 terms 5 terms
 11.2317  7.3547  5.9445  3.1141

Projection direction vectors:
          term 1     term 2
 area   0.314287   0.428802
 peri  -0.945525  -0.860929
shape   0.084893   0.273732

Coefficients of ridge terms:
 term 1  term 2
0.93549 0.81952
```

The added information is the direction vectors α_k and the coefficients β_{ij} in

$$Y_i = \alpha_{i0} + \sum_{j=1}^{M} \beta_{ij} f_j(\alpha_j^T X) + \epsilon \tag{8.11}$$

Note that this is the extension of (8.10) to multiple responses, and so we separate the scalings from the smooth functions f_j (which are scaled to have zero mean and unit variance over the projections of the dataset).

We can examine the fitted functions f_j by (see Figure 8.9)

```
par(mfrow = c(3, 2))
plot(rock.ppr)
plot(update(rock.ppr, bass = 5))
plot(rock.ppr2 <- update(rock.ppr, sm.method = "gcv",
                         gcvpen = 2))
```

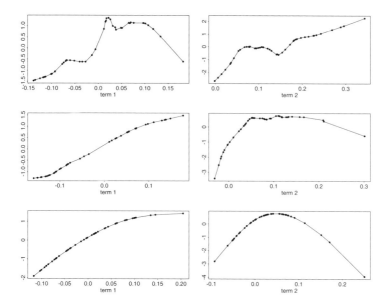

Figure 8.9: Plots of the ridge functions for three two-term projection pursuit regressions fitted to the `rock` dataset. The top two fits used `supsmu`, whereas the bottom fit used smoothing splines.

We first increase the amount of smoothing in the 'super smoother' `supsmu` to fit a smoother function, and then change to using a smoothing spline with smoothness chosen by GCV (generalized cross-validation) with an increased complexity penalty. We can then examine the details of this fit by

```
summary(rock.ppr2)
    ....

Goodness of fit:
 2 terms 3 terms 4 terms 5 terms
 22.523  21.564  21.564   0.000

Projection direction vectors:
          term 1     term 2
  area   0.348531   0.442850
  peri  -0.936987  -0.856179
 shape   0.024124  -0.266162

Coefficients of ridge terms:
  term 1   term 2
 1.46479  0.20077

Equivalent df for ridge terms:
  term 1 term 2
   2.68      2
```

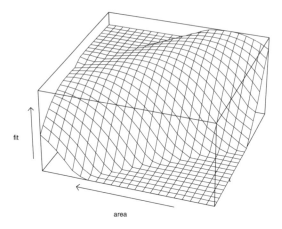

Figure 8.10: A two-dimensional fitted section of a projection pursuit regression surface fitted to the `rock` data. Note that the prediction extends the ridge functions as constant beyond the fitted functions, hence the planar regions shown. For display on paper we set `drape = F`.

This fit is substantially slower since the effort put into choosing the amount of smoothing is much greater. Note that here only two effective terms could be found, and that `area` and `peri` dominate. We can arrange to view the surface for a typical value of `shape` (Figure 8.10). Users of S-PLUS or R under Windows can use interactive rotation: see page 422.

```
summary(rock1) # to find the ranges of the variables
Xp <- expand.grid(area = seq(0.1, 1.2, 0.05),
                  peri = seq(0, 0.5, 0.02), shape = 0.2)
trellis.device()
rock.grid <- cbind(Xp, fit = predict(rock.ppr2, Xp))
## S: Trellis 3D plot
wireframe(fit ~ area + peri, rock.grid,
        screen = list(z = 160, x = -60),
        aspect = c(1, 0.5), drape = T)
## R: use persp
persp(seq(0.1, 1.2, 0.05), seq(0, 0.5, 0.02),
      matrix(rock.grid$fit, 23),
      d = 5, theta = -160, phi = 30, zlim = c(-1, 15))
```

An example: The `cpus` data

We can also consider the `cpus` test problem. Our experience suggests that smoothing the terms rather more than the default for `supsmu` is a good idea.

```
> (cpus.ppr <- ppr(log10(perf) ~ ., data = cpus2[cpus.samp,],
                  nterms = 2, max.terms = 10, bass = 5))
    ....
Goodness of fit:
```

```
 2 terms 3 terms 4 terms 5 terms 6 terms 7 terms 8 terms
 2.1371  2.4223  1.9865  1.8331  1.5806  1.5055  1.3962
 9 terms 10 terms
 1.2723  1.2338
> cpus.ppr <- ppr(log10(perf) ~ ., data = cpus2[cpus.samp,],
                  nterms = 8, max.terms = 10, bass = 5)
> test.cpus(cpus.ppr)
[1] 0.18225
> ppr(log10(perf) ~ ., data = cpus2[cpus.samp,],
       nterms = 2, max.terms = 10, sm.method = "spline")
Goodness of fit:
 2 terms 3 terms 4 terms 5 terms 6 terms 7 terms 8 terms
 2.6752  2.2854  2.0998  2.0562  1.6744  1.4438  1.3948
 9 terms 10 terms
 1.3843  1.3395
> cpus.ppr2 <- ppr(log10(perf) ~ ., data = cpus2[cpus.samp,],
      nterms = 7, max.terms = 10, sm.method = "spline")
> test.cpus(cpus.ppr2)
[1] 0.18739
> res3 <- log10(cpus2[-cpus.samp, "perf"]) -
               predict(cpus.ppr, cpus2[-cpus.samp,])
> wilcox.test(res2^2, res3^2, paired = T, alternative = "greater")
signed-rank normal statistic with correction Z = 0.6712,
  p-value = 0.2511
```

In these experiments projection pursuit regression outperformed all the additive models, but not by much, and is not significantly better than the best linear model with discretized variables.

8.10   Neural Networks

Feed-forward neural networks provide a flexible way to generalize linear regression functions. General references are Bishop (1995); Hertz, Krogh and Palmer (1991) and Ripley (1993, 1996). They are non-linear regression models in the sense of Section 8.2, but with so many parameters that they are extremely flexible, flexible enough to approximate any smooth function.

We start with the simplest but most common form with one hidden layer as shown in Figure 8.11. The input units just provide a 'fan out' and distribute the inputs to the 'hidden' units in the second layer. These units sum their inputs, add a constant (the 'bias') and take a fixed function ϕ_h of the result. The output units are of the same form, but with output function ϕ_o. Thus

$$y_k = \phi_o \left(\alpha_k + \sum_h w_{hk}\, \phi_h \left(\alpha_h + \sum_i w_{ih}\, x_i \right) \right) \qquad (8.12)$$

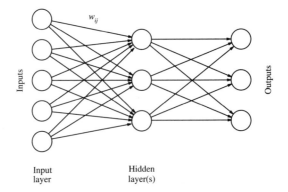

Figure 8.11: A generic feed-forward neural network.

The 'activation function' ϕ_h of the hidden layer units is almost always taken to be the logistic function

$$\ell(z) = \frac{\exp(z)}{1 + \exp(z)}$$

and the output units are linear, logistic or threshold units. (The latter have $\phi_o(x) = I(x > 0)$.) Note the similarity to projection pursuit regression (*cf* (8.10)), which has linear output units but general smooth hidden units. (However, arbitrary smooth functions can be approximated by sums of rescaled logistics.)

The general definition allows more than one hidden layer, and also allows 'skip-layer' connections from input to output when we have

$$y_k = \phi_o\left(\alpha_k + \sum_{i \to k} w_{ik}x_i + \sum_{j \to k} w_{jk}\phi_h\left(\alpha_j + \sum_{i \to j} w_{ij}x_i\right)\right) \qquad (8.13)$$

which allows the non-linear units to perturb a linear functional form.

We can eliminate the biases α_i by introducing an input unit 0 which is permanently at $+1$ and feeds every other unit. The regression function f is then parametrized by the set of weights w_{ij}, one for every link in the network (or zero for links that are absent).

The original biological motivation for such networks stems from McCulloch and Pitts (1943) who published a seminal model of a neuron as a binary thresholding device in discrete time, specifically that

$$n_i(t) = H\left(\sum_{j \to i} w_{ji}n_j(t-1) - \theta_i\right)$$

the sum being over neurons j connected to neuron i. Here H denotes the Heaviside or threshold function $H(x) = I(x > 0)$, $n_i(t)$ is the output of neuron i at time t and $0 < w_{ij} < 1$ are attenuation weights. Thus the effect is to threshold a weighted sum of the inputs at value θ_i. Real neurons are now known to be more complicated; they have a graded response rather than the simple thresholding of the McCulloch–Pitts model, work in continuous time, and can perform more general non-linear functions of their inputs, for example, logical functions. Nevertheless, the McCulloch–Pitts model has been extremely influential in the development of artificial neural networks.

Feed-forward neural networks can equally be seen as a way to parametrize a fairly general non-linear function. Such networks *are* rather general: Cybenko (1989), Funahashi (1989), Hornik, Stinchcombe and White (1989) and later authors have shown that neural networks with linear output units can approximate any continuous function f uniformly on compact sets, by increasing the size of the hidden layer.

The approximation results are non-constructive, and in practice the weights have to be chosen to minimize some fitting criterion, for example, least squares

$$ E = \sum_p \| t^p - y^p \|^2 $$

where t^p is the target and y^p the output for the pth example pattern. Other measures have been proposed, including for $y \in [0, 1]$ 'maximum likelihood' (in fact minus the logarithm of a conditional likelihood) or equivalently the Kullback–Leibler distance, which amount to minimizing

$$ E = \sum_p \sum_k \left[t_k^p \log \frac{t_k^p}{y_k^p} + (1 - t_k^p) \log \frac{1 - t_k^p}{1 - y_k^p} \right] \qquad (8.14) $$

This is half the deviance for a binary logistic model with linear predictor given by (8.12) or (8.13). For a multinomial log-linear model with K classes we can use a neural network with K outputs and the negative conditional log-likelihood

$$ E = \sum_p \sum_k -t_k^p \log p_k^p, \qquad p_k^p = \frac{e^{y_k^p}}{\sum_{c=1}^{K} e^{y_c^p}} \qquad (8.15) $$

since exactly one of the targets t_k^p will be one (for the class which occurred) and the others all zero. This is often known by the pretentious name of 'softmax' fitting. Since there is some redundancy in (8.15) we may set one output to zero (but usually do not).

One way to ensure that f is smooth is to restrict the class of estimates, for example, by using a limited number of spline knots. Another way is *regularization* in which the fit criterion is altered to

$$ E + \lambda C(f) $$

with a penalty C on the 'roughness' of f. *Weight decay*, specific to neural networks, uses as penalty the sum of squares of the weights w_{ij}. (This only makes sense if the inputs are rescaled to range about $[0, 1]$ to be comparable with the outputs of internal units.) The use of weight decay seems both to help the optimization process and to avoid over-fitting. Arguments in Ripley (1993, 1994) based on a Bayesian interpretation suggest $\lambda \approx 10^{-4} - 10^{-2}$ depending on the degree of fit expected, for least-squares fitting to variables of range one and $\lambda \approx 0.01 - 0.1$ for the entropy fit.

Software to fit feed-forward neural networks with a single hidden layer but allowing skip-layer connections (as in (8.13)) is provided in our library section nnet. The format of the call to the fitting function nnet is

```
nnet(formula, data, weights, size, Wts, linout = F, entropy = F,
     softmax = F, skip = F, rang = 0.7, decay = 0, maxit = 100,
     trace = T)
```

The non-standard arguments are as follows.

size	number of units in the hidden layer.
Wts	optional initial vector for w_{ij}.
linout	logical for linear output units.
entropy	logical for entropy rather than least-squares fit.
softmax	logical for log-probability models.
skip	logical for links from inputs to outputs.
rang	if Wts is missing, use random weights from runif(n,-rang, rang).
decay	parameter λ.
maxit	maximum of iterations for the optimizer.
Hess	should the Hessian matrix at the solution be returned?
trace	logical for output from the optimizer. Very reassuring!

There are predict, print and summary methods for neural networks, and a function nnet.Hess to compute the Hessian with respect to the weight parameters and so check if a secure local minimum has been found. For our rock example we have

```
> library(nnet)
> attach(rock)
> area1 <- area/10000; peri1 <- peri/10000
> rock1 <- data.frame(perm, area = area1, peri = peri1, shape)
> rock.nn <- nnet(log(perm) ~ area + peri + shape, rock1,
      size = 3, decay = 1e-3, linout = T, skip = T,
      maxit = 1000, Hess = T)
# weights:  19
initial  value 1092.816748
iter  10 value 32.272454
    . . . .
final  value 14.069537
converged

> summary(rock.nn)
a 3-3-1 network with 19 weights
options were - skip-layer connections  linear output units
  decay=0.001
  b->h1 i1->h1 i2->h1 i3->h1
   1.21   8.74 -15.00  -3.45
```

```
      b->h2 i1->h2 i2->h2 i3->h2
      9.50   -4.34 -12.66   2.48
      b->h3 i1->h3 i2->h3 i3->h3
      6.20   -7.63 -10.97   3.12
       b->o   h1->o   h2->o   h3->o   i1->o   i2->o   i3->o
      7.74  20.17  -7.68  -7.02 -12.95 -15.19   6.92
> sum((log(perm) - predict(rock.nn))^2)
[1] 12.231
> detach()
> eigen(rock.nn@Hessian, T)$values      # rock.nn$Hessian in R
  [1] 9.1533e+02 1.6346e+02 1.3521e+02 3.0368e+01 7.3914e+00
  [6] 3.4012e+00 2.2879e+00 1.0917e+00 3.9823e-01 2.7867e-01
 [11] 1.9953e-01 7.5159e-02 3.2513e-02 2.5950e-02 1.9077e-02
 [16] 1.0834e-02 6.8937e-03 3.7671e-03 2.6974e-03
```

(There are several solutions and a random starting point, so your results may well differ.) The quoted values include the weight decay term. The eigenvalues of the Hessian suggest that a secure local minimum has been achieved. In the summary the b refers to the bias unit (input unit 0), and i, h and o to input, hidden and bias units.

To view the fitted surface for the rock dataset we can use essentially the same code as we used for the fits by ppr.

```
Xp <- expand.grid(area = seq(0.1, 1.2, 0.05),
                  peri = seq(0, 0.5, 0.02), shape = 0.2)
trellis.device()
rock.grid <- cbind(Xp, fit = predict(rock.nn, Xp))
## S: Trellis 3D plot
wireframe(fit ~ area + peri, rock.grid,
          screen = list(z = 160, x = -60),
          aspect = c(1, 0.5), drape = T)
## R: use persp, see page 242
```

An example: The cpus data

To use the nnet software effectively it is essential to scale the problem. A preliminary run with a linear model demonstrates that we get essentially the same results as the conventional approach to linear models.

```
attach(cpus2)
cpus3 <- data.frame(syct = syct-2, mmin = mmin-3, mmax = mmax-4,
                    cach = cach/256, chmin = chmin/100,
                    chmax = chmax/100, perf = perf)
detach()

test.cpus <- function(fit)
  sqrt(sum((log10(cpus3[-cpus.samp, "perf"]) -
            predict(fit, cpus3[-cpus.samp,]))^2)/109)
cpus.nn1 <- nnet(log10(perf) ~ ., cpus3[cpus.samp,],
                 linout = T, skip = T, size = 0)
test.cpus(cpus.nn1)
[1] 0.21295
```

We now consider adding non-linear terms to the model.

```
cpus.nn2 <- nnet(log10(perf) ~ ., cpus3[cpus.samp,], linout = T,
                 skip = T, size = 4, decay = 0.01, maxit = 1000)
final  value 2.332258
test.cpus(cpus.nn2)
[1] 0.20968
cpus.nn3 <- nnet(log10(perf) ~ ., cpus3[cpus.samp,], linout = T,
                 skip = T, size = 10, decay = 0.01, maxit = 1000)
final  value 2.338387
test.cpus(cpus.nn3)
[1] 0.20645
cpus.nn4 <- nnet(log10(perf) ~ ., cpus3[cpus.samp,], linout = T,
                 skip = T, size = 25, decay = 0.01, maxit = 1000)
final  value 2.332207
test.cpus(cpus.nn4)
[1] 0.20933
```

This demonstrates that the degree of fit is almost completely controlled by the amount of weight decay rather than the number of hidden units (provided there are sufficient). We have to be able to choose the amount of weight decay *without* looking at the test set. To do so we borrow the ideas of Chapter 12, by using cross-validation and by averaging across multiple fits.

```
CVnn.cpus <- function(formula, data = cpus3[cpus.samp, ],
     size = c(0, 4, 4, 10, 10),
     lambda = c(0, rep(c(0.003, 0.01), 2)),
     nreps = 5, nifold = 10, ...)
{
  CVnn1 <- function(formula, data, nreps = 1, ri,  ...)
  {
    truth <- log10(data$perf)
    res <- numeric(length(truth))
    cat("  fold")
    for (i in sort(unique(ri))) {
      cat(" ", i,  sep = "")
      for(rep in 1:nreps) {
        learn <- nnet(formula, data[ri !=i,], trace = F, ...)
        res[ri == i] <- res[ri == i] +
                          predict(learn, data[ri == i,])
      }
    }
    cat("\n")
    sum((truth - res/nreps)^2)
  }
  choice <- numeric(length(lambda))
  ri <- sample(nifold, nrow(data), replace = T)
  for(j in seq(along = lambda)) {
    cat("  size =", size[j], "decay =", lambda[j], "\n")
    choice[j] <- CVnn1(formula, data, nreps = nreps, ri = ri,
                       size = size[j], decay = lambda[j], ...)
```

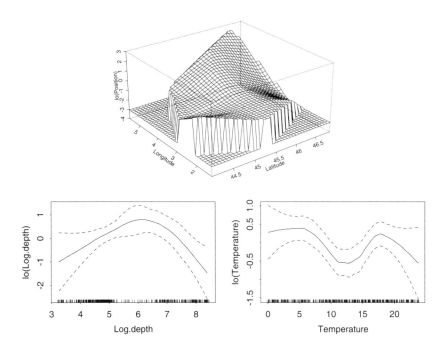

Figure 8.12: Terms of a gam fit to presence of mackerel eggs with lo terms for spatial position, depth and surface temperature. Based on Bowman and Azzalini (1997, Fig. 8.7).

```
        }
     cbind(size = size, decay = lambda, fit = sqrt(choice/100))
   }
   CVnn.cpus(log10(perf) ~ ., data = cpus3[cpus.samp,],
             linout = T, skip = T, maxit = 1000)
         size decay     fit
   [1,]     0 0.000 0.20256
   [2,]     4 0.003 0.18311
   [3,]     4 0.010 0.17837
   [4,]    10 0.003 0.19622
   [5,]    10 0.010 0.18322
```

This took about 75 seconds. The cross-validated results seem rather insensitive to the choice of model. The non-linearity does not seem justified.

More extensive examples of the use of neural networks are given in Chapter 12.

8.11   Conclusions

We have considered a large, perhaps bewildering, variety of extensions to linear regression. These can be thought of as belonging to two classes, the 'black-box' fully automatic and maximally flexible routines represented by projection pursuit regression and neural networks, and the small steps under full user control of parametric and (to a less extent) additive models. Although the latter may gain in interpretation, as we saw in the `rock` example they may not be general enough, and this is a common experience. They may also be sufficiently flexible to be difficult to interpret; Figure 8.12 shows a biologically implausible `gam` fit that is probably exhibiting the equivalent for additive models of collinearity.

What is best for any particular problem depends on its aim, in particular whether prediction or interpretation is paramount. The methods of this chapter are powerful tools with very little distribution theory to support them, so it is very easy to over-fit and over-explain features of the data. Be warned!

Chapter 9

Tree-Based Methods

The use of tree-based models may be unfamiliar to statisticians, although researchers in other fields have found trees to be an attractive way to express knowledge and aid decision-making. Keys such as Figure 9.1 are common in botany and in medical decision-making, and provide a way to encapsulate and structure the knowledge of experts to be used by less-experienced users. Notice how this tree uses both categorical variables and splits on continuous variables. (It is a tree, and readers are encouraged to draw it.)

The automatic construction of decision trees dates from work in the social sciences by Morgan and Sonquist (1963) and Morgan and Messenger (1973). In statistics Breiman *et al.* (1984) had a seminal influence both in bringing the work to the attention of statisticians and in proposing new algorithms for constructing trees. At around the same time decision-tree induction was beginning to be used in the field of *machine learning*, notably by Quinlan (1979, 1983, 1986, 1993), and in engineering (Henrichon and Fu, 1969; Sethi and Sarvarayudu, 1982). Whereas there is now an extensive literature in machine learning (see Murthy, 1998), further statistical contributions are sparse. The introduction within S of tree-based models described by Clark and Pregibon (1992) made the methods much more freely available through function `tree` and its support functions. The library section `rpart` (Therneau and Atkinson, 1997) provides a faster and more tightly-packaged set of S functions for fitting trees to data, which we describe here.

Ripley (1996, Chapter 7) gives a comprehensive survey of the subject, with proofs of the theoretical results.

Constructing trees may be seen as a type of variable selection. Questions of interaction between variables are handled automatically, and so is monotonic transformation of the x variables. These issues are reduced to which variables to divide on, and how to achieve the split.

Figure 9.1 is a *classification* tree since its endpoint is a factor giving the species. Although this is the most common use, it is also possible to have *regression* trees in which each terminal node gives a predicted value, as shown in Figure 9.2 for our dataset `cpus`.

Much of the machine learning literature is concerned with logical variables and correct decisions. The endpoint of a tree is a (labelled) partition of the space $\mathcal{X}$ of possible observations. In logical problems it is assumed that there *is* a partition of the space $\mathcal{X}$ that will correctly classify all observations, and the task is to

1.	Leaves subterete to slightly flattened, plant with bulb	2.
	Leaves flat, plant with rhizome	4.
2.	Perianth-tube > 10 mm	**I. × hollandica**
	Perianth-tube < 10 mm	3.
3.	Leaves evergreen	**I. xiphium**
	Leaves dying in winter	**I. latifolia**
4.	Outer tepals bearded	**I. germanica**
	Outer tepals not bearded	5.
5.	Tepals predominately yellow	6.
	Tepals blue, purple, mauve or violet	8.
6.	Leaves evergreen	**I. foetidissima**
	Leaves dying in winter	7.
7.	Inner tepals white	**I. orientalis**
	Tepals yellow all over	**I. pseudocorus**
8.	Leaves evergreen	**I. foetidissima**
	Leaves dying in winter	9.
9.	Stems hollow, perianth-tube 4–7mm	**I. sibirica**
	Stems solid, perianth-tube 7–20mm	10.
10.	Upper part of ovary sterile	11.
	Ovary without sterile apical part	12.
11.	Capsule beak 5–8mm, 1 rib	**I. enstata**
	Capsule beak 8–16mm, 2 ridges	**I. spuria**
12.	Outer tepals glabrous, many seeds	**I. versicolor**
	Outer tepals pubescent, 0–few seeds	**I. × robusta**

Figure 9.1: Key to British species of the genus *Iris*. Simplified from Stace (1991, p. 1140), by omitting parts of his descriptions.

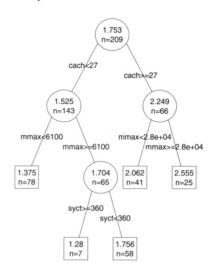

Figure 9.2: A regression tree for the cpu performance data on $\log_{10}$ scale. The values in each node are the prediction for the node and the number of cases reaching the node.

Table 9.1: Example decisions for the space shuttle autolander problem.

stability	error	sign	wind	magnitude	visibility	decision
any	any	any	any	any	no	auto
xstab	any	any	any	any	yes	noauto
stab	LX	any	any	any	yes	noauto
stab	XL	any	any	any	yes	noauto
stab	MM	nn	tail	any	yes	noauto
any	any	any	any	Out of range	yes	noauto
stab	SS	any	any	Light	yes	auto
stab	SS	any	any	Medium	yes	auto
stab	SS	any	any	Strong	yes	auto
stab	MM	pp	head	Light	yes	auto
stab	MM	pp	head	Medium	yes	auto
stab	MM	pp	tail	Light	yes	auto
stab	MM	pp	tail	Medium	yes	auto
stab	MM	pp	head	Strong	yes	noauto
stab	MM	pp	tail	Strong	yes	auto

find a tree to describe it succinctly. A famous example of Donald Michie (for example, Michie, 1989) is whether the space shuttle pilot should use the autolander or land manually (Table 9.1). Some enumeration will show that the decision has been specified for 253 out of the 256 possible observations. Some cases have been specified twice. This body of expert opinion needed to be reduced to a simple decision aid, as shown in Figure 9.3. (There are several comparably simple trees that summarize this table.)

Note that the botanical problem is treated as if it were a logical problem, although there will be occasional specimens that do not meet the specification for their species.

9.1 Partitioning Methods

The ideas for classification and regression trees are quite similar, but the terminology differs. Classification trees are more familiar and it is a little easier to justify the tree-construction procedure, so we consider them first.

Classification trees

We have already noted that the endpoint for a tree is a partition of the space $\mathcal{X}$, and we compare trees by how well that partition corresponds to the correct decision rule for the problem. In logical problems the easiest way to compare partitions is to count the number of errors, or, if we have a prior over the space $\mathcal{X}$, to compute the probability of error.

In statistical problems the distributions of the classes over $\mathcal{X}$ usually overlap, so there is no partition that completely describes the classes. Then for each cell of

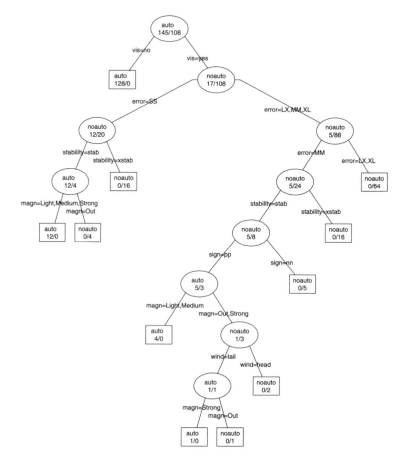

Figure 9.3: A decision tree for shuttle autolander problem. The numbers m/n denote the proportion of training cases reaching that node classified into each class (`auto`/`noauto`). This figure was drawn by `post.rpart`.

the partition there will be a probability distribution over the classes, and the Bayes decision rule will choose the class with highest probability. This corresponds to assessing partitions by the overall probability of misclassification. Of course, in practice we do not have the whole probability structure, but a training set of n classified examples that we assume are an independent random sample. Then we can estimate the misclassification rate by the proportion of the training set that is misclassified.

Almost all current tree-construction methods use a one-step lookahead. That is, they choose the next split in an optimal way, without attempting to optimize the performance of the whole tree. (This avoids a combinatorial explosion over future choices, and is akin to a very simple strategy for playing a game such as chess.) However, by choosing the right measure to optimize at each split, we can ease future splits. It does not seem appropriate to use the misclassification rate to choose the splits.

What class of splits should we allow? Both Breiman *et al.*'s CART methodology and the S functions only allow binary splits, which avoids one difficulty in comparing splits, that of normalization by size. For a continuous variable x_j the allowed splits are of the form $x_j < t$ versus $x_j \geqslant t$. For ordered factors the splits are of the same type. For general factors the levels are divided into two classes. (Note that for L levels there are 2^L possible splits, and if we disallow the empty split and ignore the order, there are still $2^{L-1} - 1$. For ordered factors there are only $L - 1$ possible splits.) Some algorithms[1] allow a linear combination of continuous variables to be split, and Boolean combinations to be formed of binary variables.

The justification for the original S methodology is to view the tree as providing a probability model (hence the title 'tree-based models' of Clark and Pregibon, 1992). At each node i of a classification tree we have a probability distribution p_{ik} over the classes. The partition is given by the *leaves* of the tree (also known as terminal nodes). Each case in the training set is assigned to a leaf, and so at each leaf we have a random sample n_{ik} from the multinomial distribution specified by p_{ik}.

We now condition on the observed variables x_i in the training set, and hence we know the numbers n_i of cases assigned to every node of the tree, in particular to the leaves. The conditional likelihood is then proportional to

$$\prod_{\text{cases } j} p_{[j]y_j} = \prod_{\text{leaves } i} \prod_{\text{classes } k} p_{ik}^{n_{ik}}$$

where $[j]$ denotes the leaf assigned to case j. This allows us to define a deviance for the tree as

$$D = \sum_i D_i, \qquad D_i = -2 \sum_k n_{ik} \log p_{ik}$$

as a sum over leaves.

Now consider splitting node s into nodes t and u. This changes the probability model within node s, so the reduction in deviance for the tree is

$$D_s - D_t - D_u = 2 \sum_k \left[n_{tk} \log \frac{p_{tk}}{p_{sk}} + n_{uk} \log \frac{p_{uk}}{p_{sk}} \right]$$

Since we do not know the probabilities, we estimate them from the proportions in the split node, obtaining

$$\hat{p}_{tk} = \frac{n_{tk}}{n_t}, \qquad \hat{p}_{uk} = \frac{n_{uk}}{n_u}, \qquad \hat{p}_{sk} = \frac{n_t \hat{p}_{tk} + n_u \hat{p}_{uk}}{n_s} = \frac{n_{sk}}{n_s}$$

so the reduction in deviance is

$$D_s - D_t - D_u = 2 \sum_k \left[n_{tk} \log \frac{n_{tk} n_s}{n_{sk} n_t} + n_{uk} \log \frac{n_{uk} n_s}{n_{sk} n_u} \right]$$

---

[1] Including CART but excluding those considered here.

$$= 2\left[\sum_k n_{tk} \log n_{tk} + n_{uk} \log n_{uk} - n_{sk} \log n_{sk}\right.$$

$$\left. + n_s \log n_s - n_t \log n_t - n_u \log n_u\right]$$

This gives a measure of the value of a split. Note that it is size-biased; there is more value in splitting leaves with large numbers of cases.

The tree construction process takes the maximum reduction in deviance over all allowed splits of all leaves, to choose the next split. (Note that for continuous variates the value depends only on the split of the ranks of the observed values, so we may take a finite set of splits.) The tree construction continues until the number of cases reaching each leaf is small (by default $n_i < 20$ in rpart, $n_i < 10$ in tree) or the leaf is homogeneous enough. Note that as all leaves not meeting the stopping criterion will eventually be split, an alternative view is to consider splitting any leaf and choose the best allowed split (if any) for that leaf, proceeding until no further splits are allowable.

This justification for the value of a split follows Ciampi *et al.* (1987) and Clark & Pregibon, but differs from most of the literature on tree construction. The more common approach is to define a measure of the impurity of the distribution at a node, and choose the split that most reduces the average impurity. Two common measures are the *entropy* or *information* $\sum p_{ik} \log p_{ik}$ and the *Gini index*

$$\sum_{j \neq k} p_{ij} p_{ik} = 1 - \sum_k p_{ik}^2$$

As the probabilities are unknown, they are estimated from the node proportions. With the entropy measure, the average impurity differs from D by a constant factor, so the tree construction process is the same, except perhaps for the stopping rule. Breiman *et al.* preferred the Gini index, which is the default in rpart.

Regression trees

The prediction for a regression tree is constant over each cell of the partition of $\mathcal{X}$ induced by the leaves of the tree. The deviance is defined as

$$D = \sum_{\text{cases } j} (y_j - \mu_{[j]})^2$$

and so clearly we should estimate the constant μ_i for leaf i by the mean of the values of the training-set cases assigned to that node. Then the deviance is the sum over leaves of D_i, the corrected sum of squares for cases within that node, and the value (impurity reduction) of a split is the reduction in the residual sum of squares.

The obvious probability model (and that proposed by Clark & Pregibon) is to take a normal $N(\mu_i, \sigma^2)$ distribution within each leaf. Then D is the usual scaled deviance for a Gaussian GLM. However, the distribution at internal nodes of the tree is then a mixture of normal distributions, and so D_i is only appropriate at the leaves. The tree-construction process has to be seen as a hierarchical refinement of probability models, very similar to forward variable selection in regression.

Missing values

One attraction of tree-based methods is the ease with which missing values can be handled. Consider the botanical key of Figure 9.1. We need to know about only a small subset of the 10 observations to classify any case, and part of the art of constructing such trees is to avoid observations that will be difficult or missing in some of the species (or as in 'capsules', for some of the cases). A general strategy is to 'drop' a case down the tree as far as it will go. If it reaches a leaf we can predict y for it. Otherwise we use the distribution at the node reached to predict y, as shown in Figures 9.2 and 9.3, which have predictions at all nodes.

An alternative strategy is used by many botanical keys and can be seen at nodes 9 and 12 of Figure 9.1. A list of characteristics is given, the most important first, and a decision made from those observations that are available. This is codified in the method of *surrogate splits* in which surrogate rules are available at non-terminal nodes to be used if the splitting variable is unobserved. Another attractive strategy is to split cases with missing values, and pass part of the case down each branch of the tree (Ripley, 1996, p. 232).

The default behaviour of `rpart` is to find surrogate splits during tree construction, and use them if missing values are found during prediction. This can be changed by the option `usesurrogate = 0` to stop cases as soon as a missing attribute is encountered. A further choice is what do to if *all* surrogates are missing: option `usesurrogate = 1` stops whereas `usesurrogate = 2` (the default) sends the case in the majority direction.

Function `predict.tree` allows a choice of case-splitting or stopping (the default) governed by the logical argument `split`.

Cutting trees down to size

With 'noisy' data, that is when the distributions for the classes overlap, it is quite possible to grow a tree that fits the training set well, but that has adapted too well to features of that subset of $\mathcal{X}$. Similarly, regression trees can be too elaborate and over-fit the training data. We need an analogue of variable selection in regression.

The established methodology is cost-complexity *pruning*, first introduced by Breiman *et al.* (1984). They considered rooted subtrees of the tree $\mathcal{T}$ grown by the construction algorithm, that is the possible results of snipping off terminal subtrees on $\mathcal{T}$. The pruning process chooses one of the rooted subtrees. Let R_i be a measure evaluated at the leaves, such as the deviance or the number of errors, and let R be the value for the tree, the sum over the leaves of R_i. Let the size of the tree be the number of leaves. Then Breiman *et al.* showed that the set of rooted subtrees of $\mathcal{T}$ that minimize[2] the cost-complexity measure

$$R_\alpha = R + \alpha \, \text{size}$$

is itself nested. That is, as we increase α we can find the optimal trees by a sequence of snip operations on the current tree (just like pruning a real tree). This

---

[2]For given α there may be more than one minimizing tree; ties are resolved by size, there being a minimizing tree contained in all the others.

produces a sequence of trees from the size of $\mathcal{T}$ down to the tree $\mathcal{T}_\emptyset$ with just the root node, but it may prune more than one node at a time. (Short proofs of these assertions are given by Ripley, 1996, Chapter 7. The tree $\mathcal{T}$ is not necessarily optimal for $\alpha = 0$.)

We need a good way to choose the degree of pruning. If a separate validation set is available, we can predict on that set, and compute the deviance versus α for the pruned trees. This will often have a minimum within the range of trees considered, and we can choose the smallest tree whose deviance is close to the minimum.

If no validation set is available we can make one by splitting the training set. Suppose we split the training set into 10 (roughly) equally sized parts. We can then use 9 parts to grow the tree and test it on the tenth. This can be done in 10 ways, and we can average the results. This is known as (10-fold) *cross-validation*.

9.2   Implementation in `rpart`

The simplest way to use tree-based methods is via the library section `rpart`[3] by Terry Therneau and Beth Atkinson (Therneau and Atkinson, 1997). The underlying philosophy is of one function, `rpart`, that both grows a tree and computes the optimal pruning for all α; although there is a function `prune.rpart`, it merely further prunes the tree at points already determined by the call to `rpart`, which has itself done some pruning. It is also possible to print a pruned tree by giving a pruning parameter to `print.rpart`. By default `rpart` runs a 10-fold cross-validation and the results are stored in the `rpart` object to allow the user to choose the degree of pruning at a later stage. Since all the work is done in a C function the calculations are fast.

The `rpart` system was designed to be easily extended to new types of responses. We only consider the following types, selected by the argument `method`.

`"anova"` A regression tree, with the impurity criterion the reduction in sum of squares on creating a binary split of the data at that node. The criterion $R(T)$ used for pruning is the mean square error of the predictions of the tree on the current dataset (that is, the residual mean square).

`"class"` A classification tree, with a categorical (factor) response and default impurity criterion the Gini index. The deviance-based approach corresponds to the entropy index, selected by the argument `parms = list(split="information")`. The pruning criterion $R(T)$ is the predicted loss, normally the error rate.

If the `method` argument is missing an appropriate type is inferred from the response variable in the formula.

We first consider a regression tree for our `cpus` data discussed on page 177. The model is specified by a model formula with terms separated by `+`; interactions make no sense for trees. The argument `cp` is α divided by the number

---

[3]Available from `statlib`; see page 464. Versions (sometimes later) are available from `http://www.mayo.edu/hsr/Sfunc.html`. Package `rpart` should be part of all R installations.

$R(T_\emptyset)$ for the root tree: in setting it we are specifying the smallest value of α we wish to consider.

```
> library(rpart)
> set.seed(123)
> cpus.rp <- rpart(log10(perf) ~ ., cpus[ , 2:8], cp = 1e-3)
> cpus.rp  # gives a large tree not shown here.
> print(cpus.rp, cp = 0.01) # default pruning
node), split, n, deviance, yval
      * denotes terminal node

 1) root 209 43.116000 1.7533
   2) cach<27 143 11.791000 1.5246
     4) mmax<6100 78  3.893700 1.3748
       8) mmax<1750 12  0.784250 1.0887 *
       9) mmax>=1750 66  1.948700 1.4268 *
     5) mmax>=6100 65  4.045200 1.7044
      10) syct>=360 7  0.129080 1.2797 *
      11) syct<360 58  2.501200 1.7557
        22) chmin<5.5 46  1.226200 1.6986 *
        23) chmin>=5.5 12  0.550710 1.9745 *
   3) cach>=27 66  7.642600 2.2488
     6) mmax<28000 41  2.341400 2.0620
      12) cach<96.5 34  1.592000 2.0081
        24) mmax<11240 14  0.424620 1.8266 *
        25) mmax>=11240 20  0.383400 2.1352 *
      13) cach>=96.5 7  0.171730 2.3236 *
     7) mmax>=28000 25  1.522900 2.5552
      14) cach<56 7  0.069294 2.2684 *
      15) cach>=56 18  0.653510 2.6668 *
```

This shows the predicted value (yval) and deviance within each node. We can plot the full tree by

```
plot(cpus.rp, uniform = T); text(cpus.rp, digits = 3)
```

There are lots of options to produce prettier trees; see help(plot.rpart) for details.

Note that the tree has yet not been pruned to final size. We use printcp and plotcp to extract the information stored in the rpart object. (All the errors are proportions of $R(T_\emptyset)$, the error for the root tree.)

```
> printcp(cpus.rp)
Regression tree:
    ....

       CP nsplit rel error xerror   xstd
1 0.54927      0     1.000 1.005 0.0972
2 0.08934      1     0.451 0.480 0.0487
3 0.08763      2     0.361 0.427 0.0433
4 0.03282      3     0.274 0.322 0.0322
5 0.02692      4     0.241 0.306 0.0306
6 0.01856      5     0.214 0.278 0.0294
```

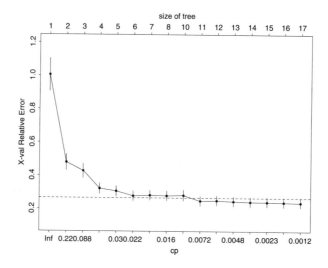

Figure 9.4: Plot by `plotcp` of the `rpart` object `cpus.rp`.

7	0.01680	6	0.195	0.281	0.0292
8	0.01579	7	0.179	0.279	0.0289
9	0.00946	9	0.147	0.281	0.0322
10	0.00548	10	0.138	0.247	0.0289
11	0.00523	11	0.132	0.250	0.0289
12	0.00440	12	0.127	0.245	0.0287
13	0.00229	13	0.123	0.242	0.0284
14	0.00227	14	0.120	0.241	0.0282
15	0.00141	15	0.118	0.240	0.0282
16	0.00100	16	0.117	0.238	0.0279

```
> plotcp(cpus.rp)
```

The columns `xerror` and `xstd` are random, depending on the random partition used in the 10-fold cross-validation that has been computed within `rpart`.

The complexity parameter may then be chosen to minimize `xerror`. An alternative procedure is to use the 1-SE rule, the largest value with `xerror` within one standard deviation of the minimum. In this case the 1-SE rule gives $0.238 + 0.0279$, so we choose line 10, a tree with 10 splits and hence 11 leaves.[4] This is easier to see graphically in Figure 9.4, where we take the leftmost pruning point with value below the line.

We can examine the pruned tree (Figure 9.5) by

```
> cpus.rp1 <- prune(cpus.rp, cp = 0.006)
> print(cpus.rp1, digits = 3)  # not shown
> plot(cpus.rp1, branch = 0.4, uniform = T)
> text(cpus.rp1, digits = 3)
```

---

[4]The number of leaves is always one more than the number of splits.

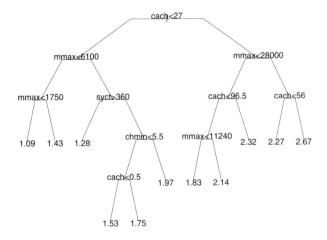

Figure 9.5: Plot of the `rpart` object `cpus.rp1`.

Forensic glass

The forensic glass dataset `fgl` has six classes, and allows us to illustrate classification trees. We can form a large tree and examine possible pruning points by

```
> set.seed(123)
> fgl.rp <- rpart(type ~ ., fgl, cp = 0.001)
> plotcp(fgl.rp)
> printcp(fgl.rp)

Classification tree:
rpart(formula = type ~ ., data = fgl, cp = 0.001)

Variables actually used in tree construction:
[1] Al Ba Ca Fe Mg Na RI

Root node error: 138/214 = 0.645

        CP nsplit rel error xerror   xstd
1 0.2065      0    1.000  1.000 0.0507
2 0.0725      2    0.587  0.594 0.0515
3 0.0580      3    0.514  0.587 0.0514
4 0.0362      4    0.457  0.551 0.0507
5 0.0326      5    0.420  0.536 0.0504
6 0.0109      7    0.355  0.478 0.0490
7 0.0010      9    0.333  0.500 0.0495
```

This suggests (Figure 9.6) a tree of size 8, plotted in Figure 9.7 by

```
fgl.rp2 <- prune(fgl.rp, cp = 0.02)
plot(fgl.rp2, uniform = T); text(fgl.rp2, use.n = T)
```

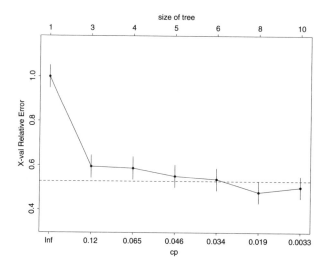

Figure 9.6: Plot by `plotcp` of the `rpart` object `fgl.rp`.

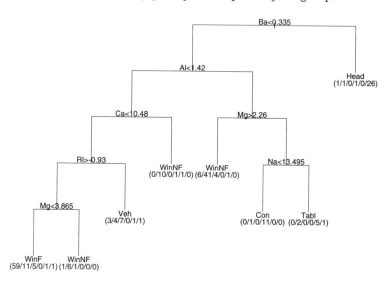

Figure 9.7: Plot of the `rpart` object `fgl.rp2`. The numbers below each leaf are the frequencies of each class, in the order of the factor levels.

and which can be printed out by

```
> fgl.rp2
node), split, n, loss, yval, (yprob)
      * denotes terminal node

 1) root 214 138 WinNF (0.33 0.36 0.079 0.061 0.042 0.14)
   2) Ba< 0.335 185 110 WinNF (0.37 0.41 0.092 0.065 0.049 0.016)
```

```
  4) Al< 1.42 113   50 WinF (0.56 0.27 0.12 0.0088 0.027 0.018)
    8) Ca< 10.48 101   38 WinF (0.62 0.21 0.13 0 0.02 0.02)
     16) RI>=-0.93 85   25 WinF (0.71 0.2 0.071 0 0.012 0.012)
        32) Mg< 3.865 77   18 WinF (0.77 0.14 0.065 0 0.013 0.013) *
        33) Mg>=3.865 8    2 WinNF (0.12 0.75 0.12 0 0 0) *
      17) RI< -0.93 16    9 Veh (0.19 0.25 0.44 0 0.062 0.062) *
    9) Ca>=10.48 12    2 WinNF (0 0.83 0 0.083 0.083 0) *
  5) Al>=1.42 72   28 WinNF (0.083 0.61 0.056 0.15 0.083 0.014)
   10) Mg>=2.26 52   11 WinNF (0.12 0.79 0.077 0 0.019 0) *
   11) Mg< 2.26 20    9 Con (0 0.15 0 0.55 0.25 0.05)
     22) Na< 13.495 12    1 Con (0 0.083 0 0.92 0 0) *
     23) Na>=13.495 8    3 Tabl (0 0.25 0 0 0.62 0.12) *
 3) Ba>=0.335 29    3 Head (0.034 0.034 0 0.034 0 0.9) *
```

The (yprob) give the distribution by class within each node.

The summary method, summary.rpart, produces voluminous output (which can be diverted to a file *via* its file argument).

```
> summary(fgl.rp2)
Call:
rpart(formula = type ~ ., data = fgl, cp = 0.001)
  n= 214

        CP nsplit rel error  xerror     xstd
1 0.206522      0  1.00000 1.00000 0.050729
2 0.072464      2  0.58696 0.59420 0.051536
3 0.057971      3  0.51449 0.58696 0.051414
4 0.036232      4  0.45652 0.55072 0.050729
5 0.032609      5  0.42029 0.53623 0.050419
6 0.020000      7  0.35507 0.47826 0.048957


Node number 1: 214 observations,    complexity param=0.20652
  predicted class=WinNF  expected loss=0.64486
    class counts:    70    76    17    13     9    29
   probabilities: 0.327 0.355 0.079 0.061 0.042 0.136
  left son=2 (185 obs) right son=3 (29 obs)
  Primary splits:
      Ba < 0.335  to the left,  improve=26.045, (0 missing)
      Mg < 2.695  to the right, improve=21.529, (0 missing)
      Al < 1.775  to the left,  improve=20.043, (0 missing)
      Na < 14.065 to the left,  improve=17.505, (0 missing)
      K  < 0.055  to the right, improve=14.617, (0 missing)
  Surrogate splits:
      Al < 1.92   to the left,  agree=0.935, adj=0.517, (0 split)
      Na < 14.22  to the left,  agree=0.902, adj=0.276, (0 split)
      Mg < 0.165  to the right, agree=0.883, adj=0.138, (0 split)
      Ca < 6.56   to the right, agree=0.883, adj=0.138, (0 split)
      K  < 0.055  to the right, agree=0.879, adj=0.103, (0 split)

Node number 2: 185 observations,    complexity param=0.20652
  predicted class=WinNF  expected loss=0.59459
```

```
     class counts:    69    75    17    12     9     3
   probabilities: 0.373 0.405 0.092 0.065 0.049 0.016
 left son=4 (113 obs) right son=5 (72 obs)
 Primary splits:
     Al < 1.42    to the left,  improve=16.0860, (0 missing)
     RI < -0.845 to the right, improve=12.2360, (0 missing)
     Mg < 2.56    to the right, improve=11.8230, (0 missing)
     Ca < 8.325  to the left,  improve=10.5980, (0 missing)
     K  < 0.01    to the right, improve= 7.3524, (0 missing)
 Surrogate splits:
     RI < -0.845 to the right, agree=0.757, adj=0.375, (0 split)
     Ca < 8.235  to the right, agree=0.746, adj=0.347, (0 split)
     K  < 0.625  to the left,  agree=0.730, adj=0.306, (0 split)
     Mg < 2.4    to the right, agree=0.665, adj=0.139, (0 split)
     Si < 73.095 to the left,  agree=0.627, adj=0.042, (0 split)

Node number 3: 29 observations
   predicted class=Head   expected loss=0.10345
     class counts:     1     1     0     1     0    26
   probabilities: 0.034 0.034 0.000 0.034 0.000 0.897
   ....
```

The initial table is that given by `printcp`. The summary method gives the top few (up to 1 + maxcompete, default five) splits and their reduction in impurity, plus up to maxsurrogate (default five) surrogates.

This is an example in which the choice of impurity index does matter, so let us also try the entropy index (Gini being the default).

```
> set.seed(123)
> fgl.rp3 <- rpart(type ~ ., fgl, cp = 0.001,
                   parms = list(split="information"))
> plotcp(fgl.rp3)
> printcp(fgl.rp3)
   ....
       CP nsplit rel error xerror   xstd
1 0.19565      0     1.000  1.000 0.0507
2 0.07971      2     0.609  0.688 0.0527
3 0.05797      3     0.529  0.667 0.0525
4 0.05072      4     0.471  0.645 0.0522
5 0.03382      5     0.420  0.558 0.0509
6 0.00725      8     0.319  0.529 0.0503
7 0.00362      9     0.312  0.493 0.0494
8 0.00100     11     0.304  0.493 0.0494
> fgl.rp4 <- prune(fgl.rp3, cp = 0.03)
> plot(fgl.rp4, uniform = T); text(fgl.rp4, use.n = T)
```

Note from Figure 9.8 that the root split chosen for the entropy measure was a leading contender given in `summary(fgl.rp2)` for the Gini measure.

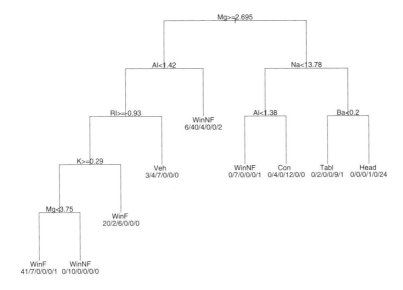

Figure 9.8: Plot of the rpart object fgl.rp4.

Plots

There are plot methods for use on a standard graphics devices (plot.rpart and text.rpart), plus a method for post for plots in PostScript. Note that post.rpart is just a wrapper for calls to plot.rpart and text.rpart on a postscript device.

The function plot.rpart has a wide range of options to choose the layout of the plotted tree. Let us consider some examples

```
plot(cpus.rp, branch = 0.6, compress = T, uniform = T)
text(cpus.rp, digits = 3, all = T, use.n = T)
```

The argument branch controls the slope of the branches. Arguments uniform and compress control whether the spacing reflects the importance of the splits (by default it does) and whether a compact style is used. The call to text.rpart may have additional arguments all which gives the value at all nodes (not just the leaves) and use.n which if true gives the numbers of cases reaching the node (and for classification trees the number of each class).

Fine control

The function rpart.control is usually used to collect arguments for the control parameter of rpart, but they can also be passed directly to rpart. Help is provided by help(rpart.control).

The parameter minsplit gives the smallest node that will be considered for a split; this defaults to 20. Parameter minbucket is the minimum number of observations in a daughter node, which defaults to 7 (minsplit/3, rounded up).

If a split does not result in a branch T_t with $R(T_t)$ at least $\mathtt{cp} \times |T_t| \times R(T_\emptyset)$ it is not considered further. This is a form of 'pre-pruning'; the tree presented has been pruned to this value and the knowledge that this will happen can be used to stop tree growth.[5] In many of our examples the minimum of $\mathtt{xerror}$ occurs for values of $\mathtt{cp}$ less than 0.01 (the default), so we choose a smaller value.

The number of cross-validations is controlled by parameter $\mathtt{xval}$, default 10. This can be set to zero at early stages of exploration, since this will produce a very significant speedup.

Missing data

If both control parameters $\mathtt{maxsurrogate}$ and $\mathtt{usesurrogate}$ are positive, $\mathtt{rpart}$ uses surrogate splits for prediction as described on page 257.

The default $\mathtt{na.action}$ during training is $\mathtt{na.rpart}$, which excludes cases only if the response or *all* the explanatory variables are missing. (This looks like a sub-optimal choice, as cases with missing response are useful for finding surrogate variables.)

When missing values are encountered in considering a split they are ignored and the probabilities and impurity measures are calculated from the non-missing values of that variable. Surrogate splits are then used to allocate the missing cases to the daughter nodes.

Surrogate splits are chosen to match as well as possible the primary split (viewed as a binary classification), and retained provided they send at least two cases down each branch, and agree as well as the rule of following the majority. The measure of agreement is the number of cases that are sent the same way, possibly after swapping 'left' and 'right' for the surrogate. (By default, missing values on the surrogate are ignored, so this measure is biased towards surrogate variables with few missing values. Changing the control parameter $\mathtt{surrogatestyle}$ to one uses instead the percentage of non-missing cases sent the same way.)

9.3   Implementation in $\mathtt{tree}$

Although we recommend that $\mathtt{rpart}$ be used, the $\mathtt{tree}$ function has some advantages, mainly in showing the process in more detail and in allowing case-splitting of missing values (page 257). The current pruning and prediction functions for $\mathtt{tree}$ were written by BDR, and an alternative tree-growing function $\mathtt{Tree}$ is available in BDR's library section $\mathtt{Tree}$[6] that also allows the Gini index to be used. (The $\mathtt{tree}$ function in the R package $\mathtt{tree}$ is a variant of $\mathtt{Tree}$.)

Again consider a regression tree for our $\mathtt{cpus}$ data. For $\mathtt{tree}$, growing and pruning are separate processes, so first we grow it:

---

[5]If $R(T_t) \geqslant 0$, splits of nodes with $R(t) < \mathtt{cp}R(T_\emptyset)$ will always be pruned.

[6]At http://www.stats.ox.ac.uk/pub/S.

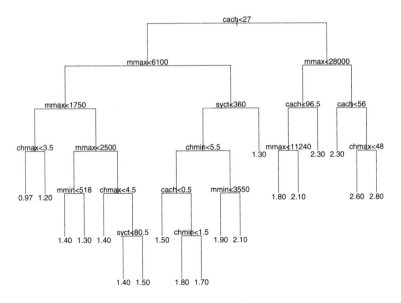

Figure 9.9: Plot of the `tree` object `cpus.ltr`.

```
> cpus.ltr <- tree(log10(perf) ~ ., data = cpus[, 2:8])
> summary(cpus.ltr)

Regression tree:
tree(formula = log10(perf) ~ ., data = cpus[, 2:8])
Number of terminal nodes:  19
Residual mean deviance:  0.0239 = 4.55 / 190
    ....
> cpus.ltr
 # a large tree, omitted here
> plot(cpus.ltr, type="u");  text(cpus.ltr)
```

The tree is shown in Figure 9.9.

We can now consider pruning the tree, using the cross-validation function `cv.tree` to find a suitable pruning point; we rather arbitrarily choose a tree of size 10 (Figure 9.10).

```
> set.seed(321)
> plot(cv.tree(cpus.ltr, , prune.tree))
> cpus.ltr1 <- prune.tree(cpus.ltr, best = 10)
> plot(cpus.ltr1, type = "u");   text(cpus.ltr1, digits = 3)
```

Now consider the forensic glass dataset `fgl`. First we grow a large tree.

```
> fgl.tr <- tree(type ~ ., fgl)
> summary(fgl.tr)
Classification tree:
tree(formula = type ~ ., data = fgl)
Number of terminal nodes:  24
```

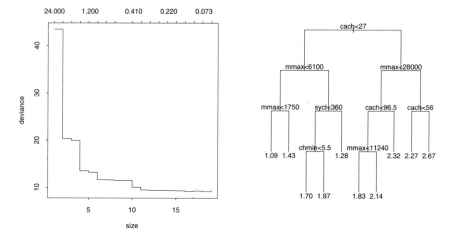

Figure 9.10: Plots of the cross-validation sequence and the pruned tree of size 10 for the cpus dataset.

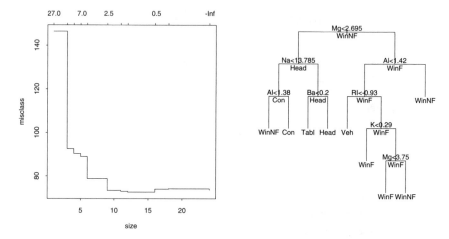

Figure 9.11: The cross-validation sequence and the pruned tree of size 9 for the `fgl` dataset.

```
Residual mean deviance:  0.649 = 123 / 190
Misclassification error rate: 0.145 = 31 / 214
> plot(fgl.tr);  text(fgl.tr, all = T, cex = 0.5)
```

We will then use cross-validation on error rates (not the default). Because this is pretty unstable over choice of split, we average over five 10-fold cross-validations.

```
> set.seed(123)
> fgl.cv <- cv.tree(fgl.tr,, prune.misclass)
> for(i in 2:5)  fgl.cv$dev <- fgl.cv$dev +
        cv.tree(fgl.tr,, prune.misclass)$dev
```

```
> fgl.cv$dev <- fgl.cv$dev/5
> fgl.cv
$size:
 [1] 24 18 16 12 11  9  6  5  4  3  1
$dev:
 [1]   73.6  74.2  74.0  72.6  73.0  73.4  78.8  89.0  90.4  92.6
[11] 146.4
> plot(fgl.cv)
```

which suggests that a pruned tree of size 9 suffices, as might one of size 6 (see Figure 9.11).

```
fgl.tr1 <- prune.misclass(fgl.tr, best = 9)
plot(fgl.tr1, type = "u");  text(fgl.tr1, all = T)
```

Chapter 10

Random and Mixed Effects

Models with *mixed effects* contain both *fixed* and *random* effects. Fixed effects are what we have been considering up to now; the only source of randomness in our models arises from regarding the cases as independent random samples. Thus in regression we have an additive measurement error that we assume is independent between cases, and in a GLM we observe independent binomial, Poisson, gamma … random variates whose mean is a deterministic function of the explanatory variables.

Random effects arise when we have more than one observation on one experimental unit (or clusters of similar experimental units). Because we expect the units to vary independently, we will have correlated observations within a unit/cluster. Rather than model the correlations directly, we consider the variates that vary with the units to be *random effects*. Perhaps the most common example is having multiple measurements on a single subject. Since these tend to occur through time, this area is also known as *repeated measures* or studies of *longitudinal* or *panel* data. Because we have observations within units, and the units may themselves be in clusters, social scientists tend to refer to this as *multilevel* analysis (Goldstein, 1995; Snijders and Bosker, 1999).

Another application of mixed effects models is to the classic nested experimental designs such as split-plot designs.

Sometimes the main interest is in the sizes of the random effects as measured by the variances of the various random variables, so yet another name is *variance components* (Searle *et al.*, 1992).

The approach of mixed models is sometimes known as *conditional* or *subject-specific* modelling: the regression coefficients apply to each individual but not necessarily to the population. Contrast this with the *marginal* approach represented by the *generalized estimating equations* that we consider in Section 10.5 where the regression coefficients apply to the population but not necessarily to individuals.

The main tool for fitting mixed models in S is the `nlme3` software of Pinheiro, Bates and collaborators, described in Pinheiro and Bates (2000). This is part[1] of S-PLUS, and a recommended package for R. You might need to use

---

[1] It is a separate chapter that is normally installed and attached. The latest version can be found at `http://nlme.stat.wisc.edu`.

`library(nlme3)` or `library(nlme)` to use it.

Davidian and Giltinan (1995) and Vonesh and Chinchilli (1997) are other references for these models.

10.1 Linear Models

The simplest case where a mixed effects model might be considered is when there are two stages of sampling. At the first stage units are selected at random from a population and at the second stage several measurements are made on each unit sampled in the first stage. For example, we were given data on 141 patients measured in 1997 and 1999. A naïve regression would have 282 observations, assumed independent. This is likely to give reasonable estimates of the regression coefficients, but to overstate their significance. An alternative would be to average the two measurements, but clearly the two-stage approach is the most satisfying one.

Prater's *gasoline* data (Prater, 1956) were originally used to build an estimation equation for yield of the refining process. The possible predictor variables were the specific gravity, vapour pressure and ASTM 10% point measured on the crude oil itself and the volatility of the desired product measured as the ASTM endpoint. The dataset is included in `MASS` as `petrol`. Several authors[2] have noted that the three measurements made on the crude oil occur in only 10 discrete sets and have inferred that only 10 crude oil samples were involved and several yields with varying ASTM endpoints were measured on each crude oil sample. We adopt this view here.

A natural way to inspect the data is to plot the yield against the ASTM endpoint for each sample, for example, using Trellis graphics by

```
xyplot(Y ~ EP | No, data = petrol,
    xlab = "ASTM end point (deg. F)",
    ylab = "Yield as a percent of crude",
    panel = function(x, y) {
      panel.grid()
      m <- sort.list(x)
      panel.xyplot(x[m], y[m], type = "b", cex = 0.5)
    })
```

Figure 10.1 shows a fairly consistent and linear rise in yield with the ASTM endpoint but with some variation in intercept that may correspond to differences in the covariates measured on the crude oil itself. In fact the intercepts appear to be steadily decreasing and since the samples are arranged so that the value of V10 is increasing this suggests a significant regression on V10 with a negative coefficient.

If we regard the 10 crude oil samples as fixed, the most general model we might consider would have separate simple linear regressions of Y on EP for each crude oil sample. First we centre the determining variables so that the origin is in the centre of the design space, thus making the intercepts simpler to interpret.

---

[2]Including Daniel and Wood (1980) and Hand *et al.* (1994).

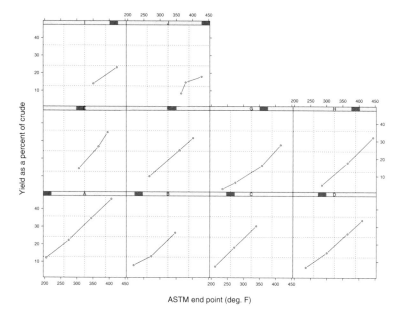

Figure 10.1: Prater's gasoline data: yield versus ASTM endpoint within samples.

```
> Petrol <- petrol
> names(Petrol)
[1] "No"  "SG"  "VP"  "V10" "EP"  "Y"
> Petrol[, 2:5] <- scale(Petrol[, 2:5], scale = F)
> pet1.lm <- lm(Y ~ No/EP - 1, Petrol)
> matrix(round(coef(pet1.lm), 2), 2, 10, byrow = T,
        dimnames = list(c("b0", "b1"), levels(Petrol$No)))

       A     B     C     D     E     F     G     H     I    J
b0 32.75 23.62 28.99 21.13 19.82 20.42 14.78 12.68 11.62 6.18
b1  0.17  0.15  0.18  0.15  0.23  0.16  0.14  0.17  0.13 0.13
```

There is a large variation in intercepts, and some variation in slopes. We test if the slopes can be considered constant; as they can we adopt the simpler parallel-line regression model.

```
> pet2.lm <- lm(Y ~ No - 1 + EP, Petrol)
> anova(pet2.lm, pet1.lm)
      Terms RDf    RSS  Df Sum of Sq F Value    Pr(F)
 No - 1 + EP   21 74.132
   No/EP - 1   12 30.329   9    43.803  1.9257  0.1439
```

We still need to explain the variation in intercepts; we try a regression on SG, VP and V10.

```
> pet3.lm <- lm(Y ~ SG + VP + V10 + EP, Petrol)
> anova(pet3.lm, pet2.lm)
```

```
              Terms RDf    RSS  Df Sum of Sq F Value    Pr(F)
 SG + VP + V10 + EP  27 134.80
         No - 1 + EP  21  74.13   6    60.672  2.8645 0.033681
```

(Notice that SG, VP and V10 are constant within the levels of No so these two models are genuinely nested.) The result suggests that differences between intercepts are *not* adequately explained by such a regression.

A promising way of generalizing the model is to assume that the 10 crude oil samples form a random sample from a population where the intercepts after regression on the determining variables depend on the sample. The model we investigate has the form

$$y_{ij} = \mu + \zeta_i + \beta_1 \, \mathrm{SG}_i + \beta_2 \, \mathrm{VP}_i + \beta_3 \, \mathrm{V10}_i + \beta_4 \, \mathrm{EP}_{ij} + \epsilon_{ij}$$

where i denotes the sample and j the observation on that sample, and $\zeta_i \sim N(0, \sigma_1^2)$ and $\epsilon_{ij} \sim N(0, \sigma^2)$, independently. This allows for the possibility that predictions from the fitted equation will need to cope with two sources of error, one associated with the sampling process for the crude oil sample and the other with the measurement process within the sample itself.

The fitting function for linear mixed effects models is lme. It has many arguments; the main ones are the first three that specify the fixed effects model, the random effects including the factor(s) defining the groups over which the random effects vary, both as formulae, and the usual data frame. For our example we can specify random intercepts for group No by

```
> pet3.lme <- lme(Y ~ SG + VP + V10 + EP,
                 random = ~ 1 | No, data = Petrol)
> summary(pet3.lme)
Linear mixed-effects model fit by REML
      AIC     BIC  logLik
  166.38 175.45 -76.191

Random effects:
         (Intercept) Residual
StdDev:      1.4447   1.8722

Fixed effects: Y ~ SG + VP + V10 + EP
              Value Std.Error DF t-value p-value
(Intercept) 19.707   0.56827 21  34.679  <.0001
         SG  0.219   0.14694  6   1.493  0.1860
         VP  0.546   0.52052  6   1.049  0.3347
        V10 -0.154   0.03996  6  -3.860  0.0084
         EP  0.157   0.00559 21  28.128  <.0001
 . . . .
```

The estimate of the residual variance within groups is $\hat{\sigma}^2 = 1.8722^2 = 3.51$ and the variance of the random effects between groups is $\hat{\sigma}_1^2 = 1.444^2 = 2.09$.

The default estimation method is REML,[3] in which the parameters in the variance structure (such as the variance components) are estimated by maximizing the

---

[3]REsidual (or REduced or REstricted) Maximum Likelihood

marginal likelihood of the residuals[4] from a least-squares fit of the linear model, and then the fixed effects are estimated by maximum likelihood assuming that the variance structure is known, which amounts to fitting by generalized least squares. With such a small sample and so few groups the difference between REML and maximum likelihood estimation is appreciable, especially in the variance estimates and hence the standard error estimates.

```
> pet3.lme <- update(pet3.lme, method = "ML")
> summary(pet3.lme)
Linear mixed-effects model fit by maximum likelihood
      AIC    BIC  logLik
  149.38 159.64 -67.692

Random effects:
         (Intercept) Residual
StdDev:     0.92889   1.8273

Fixed effects: Y ~ SG + VP + V10 + EP
             Value Std.Error DF t-value p-value
(Intercept) 19.694   0.47815 21  41.188  <.0001
         SG  0.221   0.12282  6   1.802  0.1216
         VP  0.549   0.44076  6   1.246  0.2590
        V10 -0.153   0.03417  6  -4.469  0.0042
         EP  0.156   0.00587 21  26.620  <.0001
```

At this point we can test if the random effects lead to a better fit. Function lme will not fit without random effects, but we can compare with a model fitted by lm,

```
> anova(pet3.lme, pet3.lm)
           Model df    AIC    BIC  logLik   Test L.Ratio p-value
pet3.lme       1  7 149.38 159.64 -67.692
 pet3.lm       2  6 148.83 157.63 -68.415 1 vs 2  1.4475  0.2289
```

which suggests that the random effects are negligibly small. Nevertheless, we carry on.

Both tables of fixed effects estimates suggest that only V10 and EP may be useful predictors (although the t-value columns assume that the asymptotic theory is appropriate, in particular that the estimates of variance components are accurate). We can check this by fitting it as a submodel and performing a likelihood ratio test. For testing hypotheses on the fixed effects the estimation method *must* be set to ML to ensure the test compares likelihoods based on the same data.

```
> pet4.lme <- update(pet3.lme, fixed = Y ~ V10 + EP)
> anova(pet4.lme, pet3.lme)
           Model df    AIC    BIC  logLik   Test L.Ratio p-value
pet4.lme       1  5 149.61 156.94 -69.806
pet3.lme       2  7 149.38 159.64 -67.692 1 vs 2  4.2285  0.1207
```

---

[4]This is a reasonable procedure since taking the residuals from *any* generalized least-squares fit will lead to the same marginal likelihood; McCullagh and Nelder (1989, pp. 247, 282–3).

```
> fixed.effects(pet4.lme)
(Intercept)       V10        EP
     19.652  -0.21081  0.15759
> coef(pet4.lme)
   (Intercept)       V10        EP
A       21.054  -0.21081  0.15759
B       18.338  -0.21081  0.15759
C       21.486  -0.21081  0.15759
   . . . .
```

Note how the `coef` method combines the slopes (the fixed effects) and the (random effect) intercept for each sample, whereas the coefficients of the fixed effects may be extracted by the function `fixed.effects` (short form `fixef`). The 'estimates' being given for the random effects are in fact BLUPs, *best linear unbiased predictors* (see the review paper by Robinson, 1991).

The `AIC` column gives (as in Section 6.8) minus twice the log-likelihood plus twice the number of parameters, and `BIC` refers to the criterion of Schwarz (1978) with penalty $\log n$ times the number of parameters.

Finally we check if we need both random regression intercepts and slopes on EP, by fitting the model

$$y_{ij} = \mu + \zeta_i + \beta_3 \, \mathrm{V10}_i + (\beta_4 + \eta_i) \, \mathrm{EP}_{ij} + \epsilon_{ij}$$

where (ζ_i, η_i) and ϵ_{ij} are independent, but ζ_i and η_i can be correlated.

```
> pet5.lme <- update(pet4.lme, random = ~ 1 + EP | No)
> anova(pet4.lme, pet5.lme)
          Model df    AIC     BIC  logLik    Test   L.Ratio
pet4.lme      1  5 149.61  156.94 -69.806
pet5.lme      2  7 153.61  163.87 -69.805 1 vs 2 0.0025194
          p-value
pet4.lme
pet5.lme  0.9987
```

The test is non-significant, leading us to retain the simpler model. Notice that the degrees of freedom for testing this hypothesis are $7 - 5 = 2$. The additional two parameters are the variance of the random slopes and the covariance between the random slopes and random intercepts.

Prediction and fitted values

A little thought must be given to fitted and predicted values for a mixed effects model. Do we want the prediction for a new data point in an existing group, or from a new group; in other words, which of the random effects are the same as when we fitted? This is handled by the argument `level`. For our final model `pet4.lme` the fitted values might be

$$\hat{\mu} + \hat{\beta}_3 \, \mathrm{V10}_i + \hat{\beta}_4 \, \mathrm{EP}_{ij} \qquad \text{and} \qquad \hat{\mu} + \hat{\zeta}_i + \hat{\beta}_3 \, \mathrm{V10}_i + \hat{\beta}_4 \, \mathrm{EP}_{ij}$$

given by `level = 0` and `level = 1` (the default in this example). Random effects are set either to zero or to their BLUP values.

A multi-level study

Snijders and Bosker (1999, Example 4.1) use as a running example a study of 2287 eighth-grade pupils (aged about 11) in 133 classes in 131 schools in the Netherlands. The response used is the score on a language test. We can prepare a data frame by centring the variables, for IQ centring within classes and including the class averages.

```
nl1 <- nlschools; attach(nl1)
classMeans <- tapply(IQ, class, mean)
nl1$IQave <- classMeans[as.character(class)]
detach()
cen <- c("IQ", "IQave", "SES")
nl1[cen] <- scale(nl1[cen], center = T, scale = F)
```

Snijders and Bosker work up to a model (their Table 5.4) with fixed effects for social-economic status, the average IQ of the class[5] and the interaction between pupils' IQ and whether they are taught in a multi-grade class. There is a random intercept, and a random slope for IQ within class. We can obtain similar results by

```
> options(contrasts = c("contr.treatment", "contr.poly"))
> nl.lme <- lme(lang ~ IQ*COMB + IQave + SES,
                random = ~ IQ | class, data = nl1)
> summary(nl.lme)
    ....
Random effects:
             StdDev   Corr
(Intercept) 2.79627 (Inter
         IQ 0.42276 -0.576
   Residual 6.25764

Fixed effects: lang ~ IQ * COMB + IQave + SES
              Value Std.Error   DF t-value p-value
(Intercept) 41.415   0.35620 2151  116.27  <.0001
         IQ  2.111   0.09627 2151   21.93  <.0001
       COMB -1.667   0.58903  130   -2.83  0.0054
      IQave  0.854   0.32396  130    2.64  0.0094
        SES  0.157   0.01465 2151   10.71  <.0001
    IQ:COMB  0.467   0.17541 2151    2.66  0.0079
    ....
```

Note that we can learn similar things from two fixed-effects models, within and between classes, the latter weighted by class size.

```
> summary(lm(lang ~ IQ*COMB + SES + class, data = nl1,
             singular.ok = T), cor = F)
Coefficients:
             Value Std. Error  t value Pr(>|t|)
         IQ  2.094   0.083      25.284   0.000
```

[5]Although this is what the text says they used, their code used the *school* average.

```
           SES   0.168   0.015      10.891   0.000
        IQ:COMB   0.402   0.162       2.478   0.013
    ....
> nl2 <- cbind(aggregate(nl1[c(1,7)], list(class = nl1$class), mean),
            unique(nl1[c("class", "COMB", "GS")]))
> summary(lm(lang ~ IQave + COMB, data = nl2, weights = GS))
Coefficients:
              Value Std. Error  t value Pr(>|t|)
(Intercept)  41.278    0.370    111.508    0.000
      IQave   3.906    0.294     13.265    0.000
       COMB  -1.624    0.586     -2.770    0.006
```

When worrying about the random effects, perhaps we should not overlook that
IQ is a measurement with perhaps as much measurement error as the response
lang, so perhaps a measurement-error model (Fuller, 1987) should be used.

A longitudinal study

Diggle, Liang and Zeger (1994) give an example of repeated measurements on
the log-size[6] of 79 Sitka spruce trees, 54 of which were grown in ozone-enriched
chambers and 25 of which were controls. The size was measured five times in
1988, at roughly monthly intervals.[7]

We consider a general curve on the five times for each group, then an
overall general curve plus a linear difference between the two groups. Taking
ordered(Time) parametrizes the curve by polynomial components.

```
> sitka.lme <- lme(size ~ treat*ordered(Time),
    random = ~1 | tree, data = Sitka, method = "ML")
> Sitka <- Sitka  # make a local copy for S-PLUS
> attach(Sitka)
> Sitka$treatslope <- Time * (treat == "ozone")
> detach()
> sitka.lme2 <- update(sitka.lme,
    fixed = size ~ ordered(Time) + treat + treatslope)
> anova(sitka.lme, sitka.lme2)
          Model df    AIC     BIC   logLik   Test L.Ratio p-value
sitka.lme   1 12 30.946 78.693 -3.4732
sitka.lme2  2  9 25.434 61.244 -3.7172 1 vs 2 0.48813  0.9215

# fitted curves
> matrix(fitted(sitka.lme2, level = 0)[c(301:305, 1:5)],
        2, 5, byrow = T,
    dimnames = list(c("control", "ozone"), unique(Sitka$Time)))
          152    174    201    227    258
control 4.1640 4.6213 5.0509 5.4427 5.6467
  ozone 4.0606 4.4709 4.8427 5.1789 5.3167
```

The first model just fits the means within each group at each time, but the simpler
model needs to take the error structure into account.

---

[6]By convention this is log height plus twice log diameter.

[7]The time is given in days since 1 January 1988.

Covariance structures for random effects and residuals

It is time to specify precisely what models `lme` considers. For a single level of random effects, let i index the groups and j the measurements on each group. Then the linear mixed-effects model is

$$y_{ij} = \mu_{ij} + z_{ij}\zeta_i + \epsilon_{ij}, \qquad \mu_{ij} = x_{ij}\beta \qquad (10.1)$$

Here x and z are row vectors of explanatory variables associated with each measurement. The ϵ_{ij} are independent for each group but possibly correlated within groups, so

$$\mathrm{var}(\epsilon_{ij}) = \sigma^2 g(\mu_{ij}, z_{ij}, \theta), \qquad \mathrm{corr}(\epsilon_{ij}) = \Gamma(\alpha) \qquad (10.2)$$

Thus the variances are allowed to depend on the means and other covariates. The random effects ζ_i are independent of the ϵ_{ij} with a variance matrix

$$\mathrm{var}(\zeta_i) = D(\alpha_\zeta) \qquad (10.3)$$

Almost all examples will be much simpler than this, and by default $g \equiv 1$, $\Gamma = I$ and D is unrestricted.

Full details of how to specify the components of a linear mixed-effects model can be obtained by `help(lme)`. It is possible for users to supply their own functions to compute many of these components. The free parameters $(\sigma^2, \theta, \alpha, \alpha_\zeta)$ in the specification of the error distribution are estimated, by REML by default.

We could try to allow for the possible effect of serial correlation in the `Sitka` data by specifying a continuous-time $AR(1)$ model for the within-tree measurements, but that fails to optimize over the correlation. This is illustrated later for a non-linear model.

10.2   Classic Nested Designs

Classic applications of random effects in experimental designs have a high degree of balance. The function `raov` in S-PLUS may be used for balanced designs with only random effects, and gives a conventional analysis including the estimation of variance components. The S-PLUS function `varcomp` is more general, and may be used to estimate variance components for balanced or unbalanced mixed models.

To illustrate these functions we take a portion of the data from our data frame `coop`. This is a cooperative trial in analytical chemistry taken from Analytical Methods Committee (1987). Seven specimens were sent to six laboratories, each three times a month apart for duplicate analysis. The response is the concentration of (unspecified) analyte in g/kg. We use the data from Specimen 1 shown in Table 10.1.

The purpose of the study was to assess components of variation in cooperative trials. For this purpose, the laboratories and batches are regarded as random. A model for the response for laboratory i, batch j and duplicate k is

$$y_{ijk} = \mu + \xi_i + \beta_{ij} + \epsilon_{ijk}$$

Table 10.1: Analyte concentrations (g/kg) from six laboratories and three batches.

	Laboratory					
Batch	1	2	3	4	5	6
1	0.29	0.40	0.40	0.9	0.44	0.38
	0.33	0.40	0.35	1.3	0.44	0.39
2	0.33	0.43	0.38	0.9	0.45	0.40
	0.32	0.36	0.32	1.1	0.45	0.46
3	0.34	0.42	0.38	0.9	0.42	0.72
	0.31	0.40	0.33	0.9	0.46	0.79

where ξ, β and ϵ are independent random variables with zero means and variances σ_L^2, σ_B^2 and σ_e^2 respectively. For l laboratories, b batches and $r = 2$ duplicates a nested analysis of variance gives:

Source of variation	Degrees of freedom	Sum of squares	Mean square	E(MS)
Between laboratories	$l - 1$	$br \sum_i (\bar{y}_i - \bar{y})^2$	MS_L	$br\sigma_L^2 + r\sigma_B^2 + \sigma_e^2$
Batches within laboratories	$l(b - 1)$	$r \sum_{ij} (\bar{y}_{ij} - \bar{y}_i)^2$	MS_B	$r\sigma_B^2 + \sigma_e^2$
Replicates within batches	$lb(r - 1)$	$\sum_{ijk} (y_{ij} - \bar{y}_{ij})^2$	MS_e	σ_e^2

So the unbiased estimators of the variance components are

$$\hat{\sigma}_L^2 = (br)^{-1}(MS_L - MS_B), \quad \hat{\sigma}_B^2 = r^{-1}(MS_B - MS_e), \quad \hat{\sigma}_e^2 = MS_e$$

The model is fitted in the same way as an analysis of variance model with `raov` replacing `aov`:

```
> summary(raov(Conc ~ Lab/Bat, data = coop, subset = Spc=="S1"))
              Df Sum of Sq Mean Sq Est. Var.
        Lab    5    1.8902 0.37804  0.060168
Bat %in% Lab  12    0.2044 0.01703  0.005368
   Residuals  18    0.1134 0.00630  0.006297
```

The same variance component estimates can be found using `varcomp`, but as this allows mixed models we need first to declare which factors are random effects using the `is.random` function. All factors in a data frame are declared to be random effects by

```
> coop <- coop  # make a local copy
> is.random(coop) <- T
```

If we use `Spc` as a factor it may need to be treated as fixed. Individual factors can also have their status declared in the same way. When used as an unassigned expression `is.random` reports the status of (all the factors in) its argument:

```
> is.random(coop$Spc) <- F
> is.random(coop)
 Lab Spc Bat
   T   F   T
```

We can now estimate the variance components:

```
> varcomp(Conc ~ Lab/Bat, data = coop, subset = Spc=="S1")
Variances:
        Lab Bat %in% Lab Residuals
  0.060168     0.0053681 0.0062972
```

The fitted values for such a model are technically the grand mean, but the `fitted` method function calculates them as if it were a fixed effects model, so giving the BLUPs. Residuals are calculated as differences of observations from the BLUPs. An examination of the residuals and fitted values points up laboratory 4 batches 1 and 2 as possibly suspect, and these will inflate the estimate of σ_e^2. Variance component estimates are known to be very sensitive to aberrant observations; we can get some check on this by repeating the analysis with a 'robust' estimating method. The result is very different:

```
> varcomp(Conc ~ Lab/Bat, data = coop, subset = Spc=="S1",
          method = c("winsor", "minque0"))
Variances:
        Lab Bat %in% Lab Residuals
  0.0040043   -0.00065681 0.0062979
```

(A related robust analysis is given in Analytical Methods Committee, 1989b.)

The possible estimation methods[8] for `varcomp` are `"minque0"` for minimum norm quadratic estimators (the default), `"ml"` for maximum likelihood and `"reml"` for residual (or reduced or restricted) maximum likelihood. Method `"winsor"` specifies that the data are 'cleaned' before further analysis. As in our example, a second method may be specified as the one to use for estimation of the variance components on the cleaned data. The method used, Winsorizing, is discussed in Section 5.5; unfortunately, the tuning constant is not adjustable, and is set for rather mild data cleaning.

Multistratum models

Multistratum models occur where there is more than one source of random variation in an experiment, as in a split-plot experiment, and have been most used in agricultural field trials. The sample information and the model for the means of the observations may be partitioned into so-called *strata*. In orthogonal experiments such as split-plot designs each parameter may be estimated in one and only one stratum, but in non-orthogonal experiments such as a balanced incomplete

---

[8]See, for example, Rao (1971a,b) and Rao and Kleffe (1988).

block design with 'random blocks', treatment contrasts are estimable both from the between-block and the within-block strata. Combining the estimates from two strata is known as 'the recovery of interblock information', for which the interested reader should consult a comprehensive reference such as Scheffé (1959, pp. 170ff). Non-orthogonal experiments are best analysed by `lme`.

The details of the computational scheme employed for multistratum experiments are given by Heiberger (1989).

A split-plot experiment

Our example was first used[9] by Yates (1935) and is discussed further in Yates (1937). The experiment involved varieties of oats and manure (nitrogen), conducted in six blocks of three whole plots. Each whole plot was divided into four subplots. Three varieties of oats were used in the experiment with one variety being sown in each whole plot (this being a limitation of the seed drill used), while four levels of manure (0, 0.2, 0.4 and 0.6 cwt per acre) were used, one level in each of the four subplots of each whole plot.

The data are shown in Table 10.2, where the blocks are labelled I–VI, V_i denotes the ith variety and N_j denotes the jth level of nitrogen. The dataset is available in `MASS` as the data frame `oats`, with variables `B`, `V`, `N` and `Y`. Since the levels of `N` are ordered, it is appropriate to create an ordered factor:

```
oats <- oats  # make a local copy: needed in S-PLUS
oats$Nf <- ordered(oats$N, levels = sort(levels(oats$N)))
```

Table 10.2: A split-plot field trial of oat varieties.

		N_1	N_2	N_3	N_4			N_1	N_2	N_3	N_4
	V_1	111	130	157	174		V_1	74	89	81	122
I	V_2	117	114	161	141	IV	V_2	64	103	132	133
	V_3	105	140	118	156		V_3	70	89	104	117
	V_1	61	91	97	100		V_1	62	90	100	116
II	V_2	70	108	126	149	V	V_2	80	82	94	126
	V_3	96	124	121	144		V_3	63	70	109	99
	V_1	68	64	112	86		V_1	53	74	118	113
III	V_2	60	102	89	96	VI	V_2	89	82	86	104
	V_3	89	129	132	124		V_3	97	99	119	121

The strata for the model we use here are:

1. a 1-dimensional stratum corresponding to the total of all observations,

2. a 5-dimensional stratum corresponding to contrasts between block totals,

---

[9] It has also been used by many authors since; see, for example, John (1971, §5.7).

3. a 12-dimensional stratum corresponding to contrasts between variety (or equivalently whole plot) totals within the same block, and

4. a 54-dimensional stratum corresponding to contrasts within whole plots.

(We use the term 'contrast' here to mean a linear function with coefficients adding to zero, and thus representing a comparison.)

Only the overall mean is estimable within stratum 1 and since no degrees of freedom are left for error, this stratum is suppressed in the printed summaries (although a component corresponding to it is present in the fitted-model object). Stratum 2 has no information on any treatment effect. Information on the V main effect is only available from stratum 3, and the N main effect and N × V interaction are only estimable within stratum 4.

Multistratum models may be fitted using aov, and are specified by a model formula of the form

$$response \sim mean.formula + \texttt{Error}(strata.formula)$$

In our example the *strata.formula* is B/V, specifying strata 2 and 3; the fourth stratum is included automatically as the "within" stratum, the residual stratum from the strata formula.

The fitted-model object is of class "aovlist", which is a list of fitted-model objects corresponding to the individual strata. Each component is an object of class aov (although in some respects incomplete). The appropriate display function is summary, which displays separately the ANOVA tables for each stratum (except the first).

```
> oats.aov <- aov(Y ~ Nf*V + Error(B/V), data = oats, qr = T)
> summary(oats.aov)

Error: B
            Df Sum of Sq Mean Sq F Value Pr(F)
Residuals   5     15875  3175.1

Error: V %in% B
            Df Sum of Sq Mean Sq F Value   Pr(F)
V            2    1786.4  893.18  1.4853 0.27239
Residuals   10    6013.3  601.33

Error: Within
            Df Sum of Sq Mean Sq F Value  Pr(F)
Nf           3     20020  6673.5 37.686  0.0000
Nf:V         6       322    53.6  0.303  0.9322
Residuals   45      7969   177.1
```

There is a clear nitrogen effect, but no evidence of variety differences nor interactions. Since the levels of Nf are quantitative, it is natural to consider partitioning the sums of squares for nitrogen and its interactions into polynomial components. Here the details of factor coding become important. Since the levels of nitrogen are equally spaced, and we chose an ordered factor, the default contrast matrix is appropriate. We can obtain an analysis of variance table with the degrees of freedom for nitrogen partitioned into components by giving an additional argument to the `summary` function:

```
> summary(oats.aov, split = list(Nf = list(L = 1, Dev = 2:3)))
    ....
Error: Within
                 Df Sum of Sq Mean Sq  F Value    Pr(F)
          Nf   3     20020      6673    37.69 0.00000
       Nf: L   1     19536     19536   110.32 0.00000
     Nf: Dev   2       484       242     1.37 0.26528
        Nf:V   6       322        54     0.30 0.93220
     Nf:V: L   2       168        84     0.48 0.62476
   Nf:V: Dev   4       153        38     0.22 0.92786
   Residuals  45      7969       177
```

The `split` argument is a named list with the name of each component that of some factor appearing in the model. Each component is itself a list with components of the form `name = int.seq` where `name` is the name for the table entry and `int.seq` is an integer sequence giving the contrasts to be grouped under that name in the ANOVA table.

Residuals in multistratum analyses: Projections

Residuals and fitted values from the individual strata are available in the usual way by accessing each component as a fitted-model object. Thus `fitted(oats.aov[[4]])` and `resid(oats.aov[[4]])` are vectors of length 54 representing fitted values and residuals from the last stratum, based on 54 orthonormal linear functions of the original data vector. It is not possible to associate them uniquely with the plots of the original experiment.

The function `proj` takes a fitted model object and finds the projections of the original data vector onto the subspaces defined by each line in the analysis of variance tables (including, for multistratum objects, the suppressed table with the grand mean only). The result is a list of matrices, one for each stratum, where the column names for each are the component names from the analysis of variance tables. As the argument `qr = T` has been set when the model was initially fitted, this calculation is considerably faster. Diagnostic plots computed by

```
plot(fitted(oats.aov[[4]]), studres(oats.aov[[4]]))
abline(h = 0, lty = 2)
oats.pr <- proj(oats.aov)
qqnorm(oats.pr[[4]][,"Residuals"], ylab = "Stratum 4 residuals")
qqline(oats.pr[[4]][,"Residuals"])
```

show nothing amiss.

Tables of means and components of variance

A better appreciation of the results of the experiment comes from looking at the estimated marginal means. The function `model.tables` calculates tables of the effects, means or residuals and optionally their standard errors; tables of means

are really only well defined and useful for balanced designs, and the standard errors calculated are for differences of means. We now refit the model omitting the
V:N interaction, but retaining the V main effect (since that might be considered a
blocking factor). Since the `model.tables` calculation also requires the projections, we should fit the model with either `qr = T` or `proj = T` set to avoid later
refitting.

```
> oats.aov <- aov(Y ~ N + V + Error(B/V), data = oats, qr = T)
> model.tables(oats.aov, type = "means", se = T)
Tables of means
 N
 0.0cwt 0.2cwt 0.4cwt 0.6cwt
  79.39  98.89  114.2  123.4
 V
 Golden.rain Marvellous Victory
       104.5      109.8   97.63

Standard errors for differences of means
            N       V
         4.25  7.0789
replic. 18.00 24.0000
```

When interactions are present the table is a little more complicated.

Finally we use `varcomp` to estimate the variance components associated with
the lowest three strata. To do this we need to declare B to be a random factor:

```
> is.random(oats$B) <- T
> varcomp(Y ~ N + V + B/V, data = oats)
Variances:
       B V %in% B Residuals
 214.48    109.69    162.56
    ....
```

In this simple balanced design the estimates are the same as those obtained by
equating the residual mean squares to their expectations. The result suggests that
the main blocking scheme has been reasonably effective.

Relationship to `lme` *models*

The `raov` and multistratum models are also linear mixed effects models, and
showing how they might be fitted with `lme` allows us to demonstrate some further
features of that function. In both cases we need more than one level of random
effects, which we specify by a nested model in the `random` formula.

For the cooperative trial `coop` we can use

```
> lme(Conc ~ 1, random = ~1 | Lab/Bat, data = coop,
      subset = Spc=="S1")
Random effects:
 Formula:  ~ 1 | Lab
         (Intercept)
StdDev:     0.24529
```

```
Formula:  ~ 1 | Bat %in% Lab
          (Intercept) Residual
StdDev:     0.073267 0.079355
```

As the design is balanced, these REML estimates of standard deviations agree
with the default minque variance estimates from `varcomp` on page 281.

The final model we used for the `oats` dataset corresponds to the linear model

$$y_{bnv} = \mu + \alpha_n + \beta_v + \eta_b + \zeta_{bv} + \epsilon_{bnv}$$

with three random terms corresponding to plots, subplots and observations. We
can code this in `lme` by

```
> options(contrasts = c("contr.treatment", "contr.poly"))
> summary(lme(Y ~ N + V, random = ~1 | B/V, data = oats))
Random effects:
 Formula:  ~ 1 | B
          (Intercept)
StdDev:      14.645

 Formula:  ~ 1 | V %in% B
          (Intercept) Residual
StdDev:      10.473     12.75

Fixed effects: Y ~ N + V
                Value Std.Error DF t-value p-value
(Intercept)    79.917    8.2203 51   9.722  <.0001
     NO.2cwt   19.500    4.2500 51   4.588  <.0001
     NO.4cwt   34.833    4.2500 51   8.196  <.0001
     NO.6cwt   44.000    4.2500 51  10.353  <.0001
 VMarvellous    5.292    7.0788 10   0.748  0.4720
    VVictory   -6.875    7.0788 10  -0.971  0.3544
     . . . .
```

The variance components are estimated as $14.645^2 = 214.48$, $10.4735^2 = 109.69$ and $12.75^2 = 162.56$. The standard errors for treatment differences also
agree with those found by `aov`.

10.3 Non-Linear Mixed Effects Models

Non-linear mixed effects models are fitted by the function `nlme`. Most of its many
arguments are specified in the same way as for `lme`, and the class of models that
can be fitted is based on those described by equations (10.1) to (10.3) on page 279.
The difference is that

$$y_{ij} = \mu_{ij} + \epsilon_{ij}, \qquad \mu_{ij} = f(\boldsymbol{x}_{ij}, \boldsymbol{\beta}, \boldsymbol{\zeta}_i) \tag{10.4}$$

so the conditional mean is a non-linear function specified by giving both fixed
and random parameters. Which parameters are fixed and which are random is

specified by the arguments `fixed` and `random`, each of which is a set of formulae with left-hand side the parameter name. Parameters can be in both sets, and
random effects will have mean zero unless they are. For example, we can specify
a simple exponential growth curve for each tree in the `Sitka` data with random
intercept and asymptote by

```
options(contrasts = c("contr.treatment", "contr.poly"))
sitka.nlme <- nlme(size ~ A + B * (1 - exp(-(Time-100)/C)),
    fixed = list(A ~ treat, B ~ treat, C ~ 1),
    random = A + B ~ 1 | tree, data = Sitka,
    start = list(fixed = c(2, 0, 4, 0, 100)), verbose = T)
```

Here the shape of the curve is taken as common to all trees. It is necessary to
specify starting values for all the fixed parameters (in the order they occur) and
starting values can be specified for other parameters. Note the way the formula
for `random` is expressed; this is one of a variety of forms described on the help
page for `nlme`.

It is not usually possible to find the exact likelihood of such a model, as the
non-linearity prevents integration over the random effects. Various linearization
schemes have been proposed; `nlme` uses the strategy of Lindstrom and Bates
(1990). This replaces the integration by joint maximization over the parameters
and random effects ζ_i, then uses linearization about the conditional modes of
the random effects to estimate the variance parameters. Since the integration is
replaced by evaluation at the mode, the log-likelihood (and hence AIC and BIC
values) is often a poor approximation. REML fitting is not well defined, and
`nlme` defaults to (approximate) maximum likelihood.

As `nlme` calls `lme` iteratively and `lme` fits can be slow, `nlme` fits can be very
slow (and also memory-intensive). Adding the argument `verbose = T` helps to
monitor progress. Finding good starting values can help a great deal. Our starting
values were chosen by inspection of the growth curves.

```
> summary(sitka.nlme)
      AIC      BIC  logLik
  -96.283 -60.473  57.142

Random effects:
                 StdDev   Corr
A.(Intercept) 0.83558  A.(Int
B.(Intercept) 0.81953  -0.69
     Residual 0.10298

Fixed effects: list(A ~ treat, B ~ treat, C ~ 1)
               Value Std.Error  DF t-value p-value
A.(Intercept)  2.303    0.1995 312  11.542  <.0001
      A.treat  0.175    0.2117 312   0.826  0.4093
B.(Intercept)  3.921    0.1808 312  21.687  <.0001
      B.treat -0.565    0.2156 312  -2.618  0.0093
            C 81.734    4.7231 312  17.305  <.0001
  ....
```

The 't value' for a difference in slope by treatment is convincing. We could use
anova against a model with fixed = list(A ~ 1, B ~ 1, C ~ 1), but that
would be undermined by the approximate nature of the likelihood.

For this model we can get a convincing fit with a continuous-time $AR(1)$
model for the within-tree measurements

```
> summary(update(sitka.nlme,
                 corr = corCAR1(0.95, ~Time | tree)))
      AIC    BIC logLik
  -104.51 -64.72 62.254

Random effects:
                StdDev   Corr
A.(Intercept) 0.81609 A.(Int
B.(Intercept) 0.76066 -0.674
     Residual 0.13066

Correlation Structure: Continuous AR(1)
    Phi
 0.9675

Fixed effects: list(A ~ treat, B ~ treat, C ~ 1)
               Value Std.Error  DF t-value p-value
A.(Intercept)  2.312    0.2052 312  11.267  <.0001
      A.treat  0.171    0.2144 312   0.796  0.4265
B.(Intercept)  3.892    0.1813 312  21.466  <.0001
      B.treat -0.564    0.2162 312  -2.607  0.0096
            C 80.875    5.2888 312  15.292  <.0001
```

The correlation is in units of days; at the average spacing between observations of
26.5 days the estimated correlation is $0.9675^{26.5} \approx 0.42$. A good starting value
for the correlation is needed to persuade nlme to fit this parameter.

Blood pressure in rabbits

We also consider the data in Table 10.3 described in Ludbrook (1994).[10] To quote
from the paper:

> Five rabbits were studied on two occasions, after treatment with saline
> (control) and after treatment with the 5-HT$_3$ antagonist MDL 72222. After
> each treatment ascending doses of phenylbiguanide (PBG) were injected in-
> travenously at 10 minute intervals and the responses of mean blood pressure
> measured. The goal was to test whether the cardiogenic chemoreflex elicited
> by PBG depends on the activation of 5-HT$_3$ receptors.

The response is the *change* in blood pressure relative to the start of the experiment.
The dataset is a data frame Rabbit in MASS.

There are three strata of variation:

1. between animals,

---

[10]We are grateful to Professor Ludbrook for supplying us with the numerical data.

Table 10.3: Data from a blood pressure experiment with five rabbits.

Treatment	Rabbit	Dose of Phenylbiguanide (μg)					
		6.25	12.5	25	50	100	200
Placebo	1	0.50	4.50	10.00	26.00	37.00	32.00
	2	1.00	1.25	4.00	12.00	27.00	29.00
	3	0.75	3.00	3.00	14.00	22.00	24.00
	4	1.25	1.50	6.00	19.00	33.00	33.00
	5	1.50	1.50	5.00	16.00	20.00	18.00
MDL 72222	1	1.25	0.75	4.00	9.00	25.00	37.00
	2	1.40	1.70	1.00	2.00	15.00	28.00
	3	0.75	2.30	3.00	5.00	26.00	25.00
	4	2.60	1.20	2.00	3.00	11.00	22.00
	5	2.40	2.50	1.50	2.00	9.00	19.00

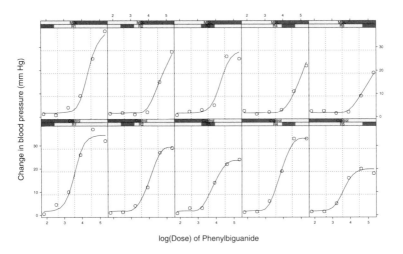

Figure 10.2: Data from a cardiovascular experiment using five rabbits on two occasions: on control and with treatment. On each occasion the animals are given increasing doses of phenylbiguanide at about 10 minutes apart. The response is the change in blood pressure. The fitted curve is derived from a model on page 292.

2. within animals between occasions and

3. within animals within occasions.

We can specify this by having the random effects depend on `Animal/Run`. We take `Treatment` to be a factor on which the fixed effects in the model might depend.

A plot of the data in Figure 10.2 shows the blood pressure rising with the dose of PBG on each occasion generally according to a sigmoid curve. Evidently the main effect of the treatment is to suppress the change in blood pressure at lower doses of PBG, hence translating the response curve a distance to the right. There is a great deal of variation, though, so there may well be other effects as well.

Ludbrook suggests a four-parameter logistic response function in $\log(\text{dose})$ (as is commonly used in this context) and we adopt his suggestion. This function has the form

$$f(\alpha, \beta, \lambda, \theta, x) = \alpha + \frac{\beta - \alpha}{1 + \exp[(x - \lambda)/\theta]}$$

Notice that $f(\alpha, \beta, \lambda, \theta, x)$ and $f(\beta, \alpha, \lambda, -\theta, x)$ are identically equal in x; to resolve this lack of identification we require that θ be positive. For our example this makes α the right asymptote, β the left asymptote (or 'baseline'), λ the $\log(\text{dose})$ (LD50) leading to a response exactly halfway between asymptotes and θ an abscissa scale parameter determining the rapidity of ascent.

We start by considering the two treatment groups separately (when there are only two strata of variation). We guess some initial values then use an `nls` fit to refine them.

```
Fpl <- deriv(~ A + (B-A)/(1 + exp((log(d) - ld50)/th)),
    c("A","B","ld50","th"), function(d, A, B, ld50, th) {})

st <- coef(nls(BPchange ~ Fpl(Dose, A, B, ld50, th),
            start = c(A = 25, B = 0, ld50 = 4, th = 0.25),
            data = Rabbit))
Rc.nlme <- nlme(BPchange ~ Fpl(Dose, A, B, ld50, th),
        fixed = list(A ~ 1, B ~ 1, ld50 ~ 1, th ~ 1),
        random = A + ld50 ~ 1 | Animal, data = Rabbit,
        subset = Treatment == "Control",
        start = list(fixed = st))
Rm.nlme <- update(Rc.nlme, subset = Treatment=="MDL")
```

We may now look at the results of the separate analyses.

```
> Rc.nlme
  Log-likelihood: -66.502
  Fixed: list(A ~ 1, B ~ 1, ld50 ~ 1, th ~ 1)
       A      B    ld50      th
  28.333 1.513 3.7743 0.28962

Random effects:
          StdDev  Corr
       A 5.76874 A
    ld50 0.17952 0.112
Residual 1.36737

> Rm.nlme
  Log-likelihood: -65.422
  Fixed: list(A ~ 1, B ~ 1, ld50 ~ 1, th ~ 1)
       A      B    ld50      th
  27.521 1.7839 4.5257 0.24236

Random effects:
```

```
            StdDev    Corr
      A 5.36554 A
   ld50 0.19004 -0.594
Residual 1.44162
```

This suggests that the random variation of ld50 between animals in each group
is small. The separate results suggest a combined model in which the distribution
of the random effects does not depend on the treatment. Initially we allow all the
parameter means to differ by group.

```
> c1 <- c(28, 1.6, 4.1, 0.27, 0)
> R.nlme1 <- nlme(BPchange ~ Fpl(Dose, A, B, ld50, th),
      fixed = list(A ~ Treatment, B ~ Treatment,
                   ld50 ~ Treatment, th ~ Treatment),
      random = A + ld50 ~ 1 | Animal/Run, data = Rabbit,
      start = list(fixed = c1[c(1, 5, 2, 5, 3, 5, 4, 5)]))
> summary(R.nlme1)
    AIC    BIC  logLik
  292.62 324.04 -131.31

Random effects:
 Level: Animal
                    StdDev    Corr
   A.(Intercept) 4.606326 A.(Int
ld50.(Intercept) 0.062593 -0.165

 Level: Run %in% Animal
                    StdDev   Corr
   A.(Intercept) 3.24882 A.(Int
ld50.(Intercept) 0.17072 -0.348
        Residual 1.41128

Fixed effects:
                   Value Std.Error DF t-value p-value
   A.(Intercept)  28.326    2.7802 43  10.188  <.0001
     A.Treatment  -0.727    2.5184 43  -0.288  0.7744
   B.(Intercept)   1.525    0.5155 43   2.958  0.0050
     B.Treatment   0.261    0.6460 43   0.405  0.6877
ld50.(Intercept)   3.778    0.0955 43  39.577  <.0001
  ld50.Treatment   0.747    0.1286 43   5.809  <.0001
  th.(Intercept)   0.290    0.0323 43   8.957  <.0001
    th.Treatment  -0.047    0.0459 43  -1.019  0.3137
```

Note that most of the random variation in ld50 is between occasions rather than
between animals whereas, for A both strata of variation are important but the
larger effect is that between animals. The table suggests that only ld50 depends
on the treatment, confirmed by

```
> R.nlme2 <- update(R.nlme1,
      fixed = list(A ~ 1, B ~ 1, ld50 ~ Treatment, th ~ 1),
      start = list(fixed = c1[c(1:3, 5, 4)]))
```

```
> anova(R.nlme2, R.nlme1)
          Model df    AIC    BIC  logLik    Test L.Ratio p-value
R.nlme2       1 12 287.29 312.43 -131.65
R.nlme1       2 15 292.62 324.04 -131.31 1 vs 2 0.66971  0.8803
> summary(R.nlme2)
   ....
```

	Value	Std.Error	DF	t-value	p-value
A	28.170	2.4908	46	11.309	<.0001
B	1.667	0.3069	46	5.433	<.0001
ld50.(Intercept)	3.779	0.0921	46	41.040	<.0001
ld50.Treatment	0.759	0.1217	46	6.234	<.0001
th	0.271	0.0227	46	11.964	<.0001

```
   ....
```

Finally we display the results and a spline approximation to the (BLUP) fitted curve.

```
xyplot(BPchange ~ log(Dose) | Animal * Treatment, Rabbit,
    xlab = "log(Dose) of Phenylbiguanide",
    ylab = "Change in blood pressure (mm Hg)",
    subscripts = T, aspect = "xy", panel =
        function(x, y, subscripts) {
            panel.grid()
            panel.xyplot(x, y)
            sp <- spline(x, fitted(R.nlme2)[subscripts])
            panel.xyplot(sp$x, sp$y, type = "l")
        })
```

The result, shown on Figure 10.2 on page 289, seems to fit the data very well.

10.4 Generalized Linear Mixed Models

Generalized linear mixed models (GLMMs) are simply stated: they are generalized linear models in which the linear predictor contains random effects. In symbols, if we have observations indexed by j on m units i,

$$E\left[Y_{ij} \mid \zeta_i\right] = g^{-1}(\eta_{ij}), \qquad \eta_{ij} = x_{ij}\beta + z_{ij}\zeta_i \qquad (10.5)$$

(for link function g) so conditionally on the random effects ζ the standard GLM applies. We complete the specification by giving $Y_{ij} \mid \zeta_i$ a conditional distribution[11] from an exponential family. Thus the likelihood is

$$L(\beta; Y) = \prod_i \int e^{\sum_j \ell(\theta; Y_{ij})} \, p(\zeta_i) \, d\zeta_i$$

where $\ell(\theta_i; Y_{ij})$ is the summand in (7.2), and $\theta_i = (\gamma')^{-1}(\mu_i) = (\gamma')^{-1}(g^{-1}(\eta_i))$ depends on β and ζ_i.

---

[11]Or, for a quasi-likelihood model specifying the variance as a function of the mean.

It is almost universal to assume that the random effects are normally distributed, with zero mean and variance matrix D, so the likelihood becomes

$$L(\beta; Y) \propto |D|^{-m/2} \prod_i \int \exp \kappa_i(\zeta_i)\, d\zeta_i, \quad \kappa_i(\zeta_i) = \sum_j \ell(\theta; Y_{ij}) + \zeta_i^T D^{-1} \zeta_i$$

The integrals complicate likelihood-based model fitting, and efficient ways to fit GLMMs are a research topic. All of the methods are approximate, either based on theoretical approximations (*ad hoc* methods) or numerical approximations to integrals.

Both Diggle *et al.* (1994) and Laird (1996) provide good overviews of the inference issues underlying this section and the next.

Running examples

We consider two examples, one of binary observations and one of counts.

The `bacteria` dataset records observations on weeks 0, 2, 4, 6 and 11 on 50 subjects (but 30 observations are missing). The response is presence/absence of a particular bacteria. There are basically three treatments, a placebo and an active drug with and without extra effort to ensure that it was taken. It is expected that the response declines with time, perhaps in a way that depends on the treatment. A basic GLM analysis is

```
> bacteria <- bacteria # needed in S-PLUS
> contrasts(bacteria$trt) <- structure(contr.sdif(3),
      dimnames = list(NULL, c("drug", "encourage")))
> summary(glm(y ~ trt * week, binomial, data = bacteria),
         cor = F)
Coefficients:
                      Value Std. Error  t value
      (Intercept)  1.975474   0.300296  6.57841
         trtdrug -0.998471   0.694168 -1.43837
    trtencourage  0.838648   0.734645  1.14157
            week -0.118137   0.044581 -2.64994
     trtdrugweek -0.017218   0.105647 -0.16298
trtencourageweek -0.070434   0.109625 -0.64250

> summary(glm(y ~ trt + week, binomial, data = bacteria),
         cor = F)
Coefficients:
                  Value Std. Error t value
  (Intercept)  1.96017   0.296867  6.6029
     trtdrug -1.10667   0.424942 -2.6043
trtencourage  0.45502   0.427608  1.0641
        week -0.11577   0.044127 -2.6237
```

which suggests that the week effect does not depend on the treatment.

Epileptic seizures

Thall and Vail (1990) give a dataset on two-week seizure counts for 59 epileptics. The number of seizures was recorded for a baseline period of 8 weeks, and then patients were randomly assigned to a treatment group or a control group. Counts were then recorded for four successive two-week periods. The subject's age is the only covariate. Thall and Vail (1990, p. 665) state that the patients were subsequently crossed-over to the other treatment; had the full data been made available a more penetrating analysis would be possible.

These data have been analysed again by Breslow and Clayton (1993), Diggle *et al.* (1994) and others. There are two ways to look at this problem. Breslow and Clayton treat the baseline counts as a fixed predictor, and use a Poisson log-linear model with predictors lbase, the log of the baseline divided by four,[12] log of age and the interaction of lbase and treatment. (The logged regressors were centred here.) A GLM analysis would give

```
> summary(glm(y ~ lbase*trt + lage + V4, family = poisson,
              data = epil), cor = F)
Coefficients:
                Value Std. Error  t value
(Intercept)   1.89792   0.042583  44.5695
      lbase   0.94862   0.043585  21.7649
        trt  -0.34587   0.060962  -5.6736
       lage   0.88759   0.116453   7.6219
         V4  -0.15977   0.054574  -2.9276
  lbase:trt   0.56153   0.063497   8.8435


Residual Deviance: 869.07 on 230 degrees of freedom
```

so there is evidence of considerable over-dispersion.

A second approach is that there are five periods of observation in two 8-week intervals, before and during the study. We can summarize the data for a GLM analysis by

```
epil <- epil # needed in S-PLUS
epil2 <- epil[epil$period == 1, ]
epil2["period"] <- rep(0, 59); epil2["y"] <- epil2["base"]
epil["time"] <- 1; epil2["time"] <- 4
epil2 <- rbind(epil, epil2)
epil2$pred <- unclass(epil2$trt) * (epil2$period > 0)
epil2$subject <- factor(epil2$subject)
epil3 <- aggregate(epil2, list(epil2$subject, epil2$period > 0),
    function(x) if(is.numeric(x)) sum(x) else x[1])
epil3$pred <- factor(epil3$pred,
    labels = c("base", "placebo", "drug"))
contrasts(epil3$pred) <- structure(contr.sdif(3),
    dimnames = list(NULL, c("placebo-base", "drug-placebo")))
summary(glm(y ~ pred + factor(subject) + offset(log(time)),
            family = poisson, data = epil3), cor = F)
```

---

[12]For comparibility with the two-week periods

```
Coefficients:
                 Value Std. Error   t value
  (Intercept)  1.743838   0.022683  76.87852
  placebo-base  0.108719   0.046911   2.31755
  drug-placebo -0.101602   0.065070  -1.56143

Residual Deviance: 303.16 on 57 degrees of freedom
```

This suggests only a small drug effect. Ramsey and Schafer (1997, Chapter 22) discuss this example in more detail, but the wide range in variation *within* patients for the four treatment periods shows that a Poisson model is inappropriate.

```
> glm(y ~ factor(subject), family = poisson, data = epil)
     ....
Degrees of Freedom: 236 Total; 177 Residual
Residual Deviance: 399.29
```

Conditional inference

For special case of a single random effect, a per-unit intercept, we can perform conditional inference treating the intercepts ζ_i as ancillary parameters rather than as random variables. (This corresponds to the traditional analysis of block effects in experimental designs as fixed effects.) For a Poisson GLMM the conditional inference is a multinomial log-linear model taking as fixed the total count for each subject, equivalent to our Poisson GLM analysis for epil3 viewed as a surrogate Poisson model.

For binary GLMMs conditional inference results (Diggle *et al.*, 1994, §.9.2.1) in a conditional likelihood which may be fitted by coxph. For example, for the bacteria data (where the treatment effects are confounded with subjects)

```
> # R: library(survival)
> bacteria$Time <- rep(1, nrow(bacteria))
> coxph(Surv(Time, unclass(y)) ~ week + strata(ID),
        data = bacteria, method = "exact")

        coef exp(coef) se(coef)     z     p
week -0.163      0.85   0.0547 -2.97 0.003
```

We use a model 'observed' on time interval $[0, 1]$ with 'success' corresponding to a death at time 1. We can use this model to explore the non-linearity of the change through time

```
> coxph(Surv(Time, unclass(y)) ~ factor(week) + strata(ID),
        data = bacteria, method = "exact")

                  coef exp(coef) se(coef)      z     p
factor(week)2    0.198     1.219    0.724  0.274 0.780
factor(week)4   -1.421     0.242    0.667 -2.131 0.033
factor(week)6   -1.661     0.190    0.682 -2.434 0.015
factor(week)11  -1.675     0.187    0.678 -2.471 0.013
```

```
Likelihood ratio test=15.4  on 4 df, p=0.00385  n= 220

> coxph(Surv(Time, unclass(y)) ~ I(week > 2) + strata(ID),
        data = bacteria, method = "exact")

                coef exp(coef) se(coef)      z       p
I(week > 2) -1.67      0.188    0.482 -3.47 0.00053
```

and conclude that the main effect is after 2 weeks. Re-fitting this as a GLM gives

```
> fit <- glm(y ~ trt + I(week > 2), binomial, data = bacteria)
> summary(fit, cor = F)

                    Value Std. Error t value
   (Intercept)   2.24791    0.35471  6.3374
       trtdrug  -1.11867    0.42824 -2.6123
  trtencourage   0.48146    0.43265  1.1128
   I(week > 2)  -1.29482    0.40924 -3.1640

Residual Deviance: 199.18 on 216 degrees of freedom
```

Note that this appears to be under-dispersed even though we have not yet allowed for between-subject variation. However, that is illusory, as the residual deviance has mean below the number of degrees of freedom in most binary GLMs (page 209). It is better to look at the Pearson chi-squared statistic

```
> sum(residuals(fit, type = "pearson")^2)
[1] 223.46
```

Numerical integration

To perform maximum-likelihood estimation, we need to be able to compute the likelihood. We have to integrate over the random effects. This can be a formidable task; Evans and Swartz (2000) is a book-length review of the available techniques.

The approaches fall into two camps. For the simple interpretation of a GLMM as having a subject-specific intercept, we only have to do a one-dimensional integration for each subject, and numerical quadrature can be used. For normally-distributed random effects it is usual to use a Gauss–Hermite scheme (Evans and Swartz, 2000, §5.3.4; Monahan, 2001, §10.3). This is implemented in our function glmmNQ.

Conversely, for an elaborate specification of random effects the only effective integration schemes seem to be those based on Markov Chain Monte Carlo ideas (Gamerman, 1997a; Robert and Casella, 1999). Clayton (1996) and Gamerman (1997b) present MCMC schemes specifically for GLMMs; GLMMGibbs is an R package by Jonathan Myles and David Clayton implementing MCMC for canonical links and a single subject-specific random effect. We can apply glmm from GLMMGibbs to the epilepsy problem.[13] Each run takes a few minutes; we run

---

[13] It does not currently handle many Bernoulli GLMMs. We had mixed success with the bacteria example.

Table 10.4: Summary results for GLMM and marginal models. The parameters in the marginal (the outer two) and the conditional models are not strictly comparable. The differences in the standard errors are mainly attributable to the extent to which under- and over-dispersion is considered.

Bacteria data	GLM	Numerical integration	PQL	GEE
(Intercept)	2.24 (0.35)	2.85 (0.53)	2.75 (0.38)	2.26 (0.35)
drug	−1.12 (0.42)	−1.37 (0.69)	−1.24 (0.64)	−1.11 (0.59)
encourage	0.48 (0.43)	0.58 (0.71)	0.49 (0.67)	0.48 (0.51)
I(week > 2)	−1.29 (0.41)	−1.62 (0.48)	−1.61 (0.36)	−1.32 (0.36)
σ		1.30 (0.41)	1.41	

Epilepsy data	GLM	Numerical integration	PQL	GEE
(Intercept)	1.90 (0.04)	1.88 (0.28)	1.87 (0.11)	1.89 (0.11)
lbase	0.95 (0.04)	0.89 (0.14)	0.88 (0.13)	0.95 (0.10)
trt	−0.34 (0.06)	−0.34 (0.17)	−0.31 (0.15)	−0.34 (0.18)
lage	0.89 (0.12)	0.47 (0.37)	0.53 (0.35)	0.90 (0.28)
V4	−0.16 (0.15)	−0.16 (0.05)	−0.16 (0.08)	−0.16 (0.07)
lbase:trt	0.56 (0.06)	0.34 (0.22)	0.34 (0.20)	0.56 (0.17)
σ		0.55 (0.07)	0.44	
(Intercept)	3.48 (0.02)	3.20 (0.19)	3.21 (0.10)	3.13 (0.03)
placebo - base	0.09 (0.04)	0.11 (0.05)	0.11 (0.10)	0.11 (0.11)
drug - placebo	−0.07 (0.05)	−0.11 (0.06)	−0.11 (0.13)	−0.10 (0.15)
σ		0.80 (0.08)	0.73	

100 000 iterations and record 1 in 100. There is a plot method to examine the results.

```
> library(GLMMGibbs)
# declare a random intercept for each subject
> epil$subject <- Ra(data = factor(epil$subject))
> glmm(y ~ lbase*trt + lage + V4 + subject, family = poisson,
        data = epil, keep = 100000, thin = 100)

> epil3$subject <- Ra(data = factor(epil3$subject))
> glmm(y ~ pred + subject, family = poisson,
        data = epil3, keep = 100000, thin = 100)
```

PQL methods

There are several closely related approximations that are most often known as PQL (Breslow and Clayton, 1993) for *penalized quasi-likelihood*.

Schall (1991) proposed a conceptually simple algorithm that reduces to maximum-likelihood estimation for Gaussian GLMMs. Recall that the IWLS algorithm for GLMs works by linearizing the solution at the current estimate of the mean μ, and regressing the working values z with working weights W. Schall proposed using mixed-effects linear fitting instead of least-square-fitting to update the fixed effects and hence the estimate of the mean. Now the mixed-effects fit depends on the current estimates of the variance components. Then for a linear problem the MLEs and REML estimates of the variances can be found by scaling the sums of squares of the BLUPs and the residuals, and Schall proposed to use the estimates from the linearization inside a loop.

An alternative derivation (Breslow and Clayton, 1993) is to apply Laplace's method (Evans and Swartz, 2000, p. 62) to the integrals over ζ_i expanding κ as a quadratic Taylor expansion about its maximum. Wolfinger and O"Connell (1993) give related methods.

This approach is closely related to that used by `nlme`. A GLM can also be considered as a non-linear model with non-linearity determined by the link function, and with the variance specified as a function of the mean. (Indeed, for a quasi-likelihood fit, that is all that is specified.) So we can consider fitting a GLMM by specifying a `nlme` model with single non-linear function of a linear predictor, and a variance function as a function of the mean. That is equivalent to PQL up to details in the approximations. We supply a wrapper `glmmPQL` function in `MASS` that implements linearization about the BLUPs. We used

```
> summary(glmmPQL(y ~ trt + I(week> 2), random = ~ 1 | ID,
                  family = binomial, data = bacteria))
> summary(glmmPQL(y ~ lbase*trt + lage + V4,
                  random = ~ 1 | subject,
                  family = poisson, data = epil))
> summary(glmmPQL(y ~ pred, random = ~1 | subject,
                  family = poisson, data = epil3))
```

The function `glme` of Pinheiro's GLME library section[14] for S-PLUS is a more closely integrated version of the same approach.

Over-dispersion

We have to consider more carefully residual over- or under-dispersion. In many GLMM analyses one would expect the random effects to account for most of the marginal over-dispersion, but as the epilepsy example shows, this can be unrealistic. By default `glmmPQL` allows the dispersion to be estimated whereas `glme` fixes the dispersion. The results shown in Table 10.4 are for estimated dispersion, and seem much closer to the results by numerical integration (even though the latter assumes the binomial or Poisson model).

---

[14]From `http://nlme.stat.wisc.edu`.

10.5   GEE Models

Consider once again the simplest scenario of multiple observations y_{ij} on a set of units. If we take a mixed-effects model, say

$$y_{ij} = \mu_{ij} + z_{ij}\zeta_i + \epsilon_{ij}, \qquad \mu_{ij} = x_{ij}\,\beta$$

then y_{ij} are a correlated collection of random variates, marginally with a multivariate normal distribution with mean μ_{ij} and a computable covariance matrix. Maximum-likelihood fitting of a `lme` model is just maximizing the marginal multivariate-normal likelihood.

We can consider a marginal model which is a GLM for Y_{ij}, that is

$$Y_{ij} \sim \text{family}(g^{-1}(\eta_{ij})), \qquad \eta_{ij} = x_{ij}\,\beta \tag{10.6}$$

but with the assumption of a correlation structure for Y_{ij}. Laird (1996, §4.4) and McCullagh and Nelder (1989, Chapter 9) consider more formal models which can give rise to correlated observations from GLMs, but these models are often known as GEE models since they are most often fitted by *generalized estimating equations* (Liang and Zeger, 1986; Diggle *et al.*, 1994). This is another theoretical approximation that is asymptotically unbiased under mild conditions.

Now consider the analogous GLMM: Y_{ij} are once again correlated, but unless the link g is linear their means are no longer $\mu_{ij} = g^{-1}(\eta_{ij})$. Note that the mean for a GLMM with random intercepts and a log link differs from a marginal model only by a shift in intercept (Diggle *et al.*, 1994, §7.4) for

$$E\,Y_{ij} = E\exp(x_{ij}\,\beta + \zeta_i) = g^{-1}(x_{ij}\,\beta) \times E\,e^{\zeta_1}$$

but of course the marginal distribution of the GLMM fit is not Poisson.

We can fit GEE models by the `gee` and `yags` library sections of Vincent Carey.[15]

```
> library(yags)
> attach(bacteria)
> summary(yags(unclass(y) - 1 ~ trt + I(week > 2),
              family = binomial,
              id = ID, corstr = "exchangeable"))
    . . . .
Estimated Scale Parameter:  1.04

    ....Working Correlation Parameter(s)
[1] 0.136

> attach(epil)
> summary(yags(y ~ lbase*trt + lage + V4, family = poisson,
              id = subject, corstr = "exchangeable"))
    . . . .
```

---

[15]`http://www.biostat.harvard.edu/~carey/.`

```
Estimated Scale Parameter:   4.42
Working Correlation Parameter(s)
[1,] 0.354

> library(gee)
> options(contrasts = c("contr.sum", "contr.poly"))
> summary(gee(y ~ pred + factor(subject), family = poisson,
      id = subject,data = epil3,  corstr = "exchangeable"))
```

It is important to realize that the parameters β in (10.5) and (10.6) may have different interpretations. In a GLMM β measures the effect of the covariates for an individual; in a GEE model it measures the effect in the population, usually smaller. If a GLMM is a good model, then often so is a GEE model but for substantially different parameter estimates (Zeger *et al.*, 1988). For a log-linear model only the intercept will differ, but for logistic models the slopes will be attenuated. Since at least in these two cases the effect is known, Zeger *et al.* point out that GLMM models can also be fitted by GEE methods, provided the latter are extended to estimate the variance components of the random effects. Since we have an estimate of the variance components we can estimate that the population slopes will be attenuated by a factor of about $\sqrt{1 + 0.346\sigma^2} \approx 1.3$ in the bacteria example. (The number 0.346 comes from Zeger *et al.*, 1988.) This agrees fairly well with the numerical integration results in Table 10.4.

Chapter 11

Exploratory Multivariate Analysis

Multivariate analysis is concerned with datasets that have more than one response variable for each observational or experimental unit. The datasets can be summarized by data matrices X with n rows and p columns, the rows representing the observations or cases, and the columns the variables. The matrix can be viewed either way, depending on whether the main interest is in the relationships between the cases or between the variables. Note that for consistency we represent the variables of a case by the *row* vector x.

The main division in multivariate methods is between those methods that assume a given structure, for example, dividing the cases into groups, and those that seek to discover structure from the evidence of the data matrix alone (nowadays often called *data mining*, see for example Hand *et al.*, 2001). Methods for known structure are considered in Chapter 12.

In pattern-recognition terminology the distinction is between *supervised* and *unsupervised* methods. One of our examples is the (in)famous iris data collected by Anderson (1935) and given and analysed by Fisher (1936). This has 150 cases, which are stated to be 50 of each of the three species *Iris setosa, I. virginica* and *I. versicolor*. Each case has four measurements on the length and width of its petals and sepals. *A priori* this seems a supervised problem, and the obvious questions are to use measurements on a future case to classify it, and perhaps to ask how the variables vary among the species. (In fact, Fisher used these data to test a genetic hypothesis which placed *I. versicolor* as a hybrid two-thirds of the way from *I. setosa* to *I. virginica*.) However, the classification of species is uncertain, and similar data have been used to identify species by grouping the cases. (Indeed, Wilson (1982) and McLachlan (1992, §6.9) consider whether the iris data can be split into subspecies.)

Krzanowski (1988) and Mardia, Kent and Bibby (1979) are two general references on multivariate analysis. For pattern recognition we follow Ripley (1996), which also has a computationally-informed account of multivariate analysis.

Most of the emphasis in the literature and in this chapter is on continuous measurements, but we do look briefly at multi-way discrete data in Section 11.4.

Colour can be used very effectively to differentiate groups in the plots of this chapter, on screen if not on paper. The code given here uses both colours and symbols, but you may prefer to use only one of these to differentiate groups. (The colours used are chosen for use on a trellis device.)

In R many of the functions use in this chapter are in the standard package `mva` which is attached by default.

Running example: *Leptograpsus variegatus* crabs

Mahon (see Campbell and Mahon, 1974) recorded data on 200 specimens of *Leptograpsus variegatus* crabs on the shore in Western Australia. This occurs in two colour forms, blue and orange, and he collected 50 of each form of each sex and made five physical measurements. These were the carapace (shell) length `CL` and width `CW`, the size of the frontal lobe `FL` and rear width `RW`, and the body depth `BD`. Part of the authors' thesis was to establish that the two colour forms were clearly differentiated morphologically, to support classification as two separate species.

The data are physical measurements, so a sound initial strategy is to work on log scale. This has been done throughout.

11.1   Visualization Methods

The simplest way to examine multivariate data is via a *pairs* or *scatterplot matrix* plot, enhanced to show the groups as discussed in Section 4.5. Pairs plots are a set of two-dimensional projections of a high-dimension point cloud.

However, pairs plots can easily miss interesting structure in the data that depends on three or more of the variables, and genuinely multivariate methods explore the data in a less coordinate-dependent way. Many of the most effective routes to explore multivariate data use dynamic graphics such as exploratory projection pursuit (for example, Huber, 1985; Friedman, 1987; Jones and Sibson, 1987 and Ripley, 1996) which chooses 'interesting' rotations of the point cloud. These are available through interfaces to the package XGobi[1] for machines running X11.[2] A successor to XGobi, GGobi,[3] is under development.

Many of the other visualization methods can be viewed as projection methods for particular definitions of 'interestingness'.

Principal component analysis

Principal component analysis (PCA) has a number of different interpretations. The simplest is a projection method finding projections of maximal variability. That is, it seeks linear combinations of the columns of X with maximal (or minimal) variance. Because the variance can be scaled by rescaling the combination, we constrain the combinations to have unit length (which is true of projections).

Let S denote the covariance matrix of the data X, which is defined[4] by

$$nS = (X - n^{-1}11^T X)^T (X - n^{-1}11^T X) = (X^T X - n\overline{x}\,\overline{x}^T)$$

---

[1] http://www.research.att.com/areas/stat/xgobi/

[2] On UNIX *and* on Windows: a Windows port of XGobi is available at http://www.stats.ox.ac.uk/pub/SWin.

[3] http://www.ggobi.org.

[4] A divisor of $n - 1$ is more conventional, but `princomp` calls `cov.wt`, which uses n.

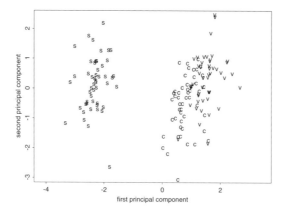

Figure 11.1: First two principal components for the log-transformed `iris` data.

where $\bar{x} = \mathbf{1}^T X/n$ is the row vector of means of the variables. Then the sample variance of a linear combination $x a$ of a row vector x is $a^T \Sigma a$ and this is to be maximized (or minimized) subject to $\|a\|^2 = a^T a = 1$. Since Σ is a non-negative definite matrix, it has an eigendecomposition

$$\Sigma = C^T \Lambda C$$

where Λ is a diagonal matrix of (non-negative) eigenvalues in decreasing order. Let $b = Ca$, which has the same length as a (since C is orthogonal). The problem is then equivalent to maximizing $b^T \Lambda b = \sum \lambda_i b_i^2$ subject to $\sum b_i^2 = 1$. Clearly the variance is maximized by taking b to be the first unit vector, or equivalently taking a to be the column eigenvector corresponding to the largest eigenvalue of Σ. Taking subsequent eigenvectors gives combinations with as large as possible variance that are uncorrelated with those that have been taken earlier. The ith principal component is then the ith linear combination picked by this procedure. (It is only determined up to a change of sign; you may get different signs in different implementations of S.)

The first k principal components span a subspace containing the 'best' k-dimensional view of the data. It has a maximal covariance matrix (both in trace and determinant). It also best approximates the original points in the sense of minimizing the sum of squared distances from the points to their projections. The first few principal components are often useful to reveal structure in the data. The principal components corresponding to the smallest eigenvalues are the most nearly constant combinations of the variables, and can also be of interest.

Note that the principal components depend on the scaling of the original variables, and this will be undesirable except perhaps if (as in the `iris` data) they are in comparable units. (Even in this case, correlations would often be used.) Otherwise it is conventional to take the principal components of the *correlation* matrix, implicitly rescaling all the variables to have unit sample variance.

The function `princomp` computes principal components. The argument `cor` controls whether the covariance or correlation matrix is used (via rescaling the variables).

```
> # S: ir <- rbind(iris[,,1], iris[,,2], iris[,,3])
> # R: data(iris3); ir <- rbind(iris3[,,1], iris3[,,2], iris3[,,3])
> ir.species <- factor(c(rep("s", 50), rep("c", 50), rep("v", 50)))
> (ir.pca <- princomp(log(ir), cor = T))
Standard deviations:
 Comp.1  Comp.2 Comp.3  Comp.4
 1.7125 0.95238 0.3647 0.16568

    . . . .

> summary(ir.pca)
Importance of components:
                          Comp.1  Comp.2   Comp.3    Comp.4
     Standard deviation 1.71246 0.95238 0.364703 0.1656840
 Proportion of Variance 0.73313 0.22676 0.033252 0.0068628
  Cumulative Proportion 0.73313 0.95989 0.993137 1.0000000
> plot(ir.pca)
> loadings(ir.pca)
          Comp.1 Comp.2 Comp.3 Comp.4
Sepal L.   0.504  0.455  0.709  0.191
Sepal W.  -0.302  0.889 -0.331
Petal L.   0.577        -0.219 -0.786
Petal W.   0.567        -0.583  0.580
> ir.pc <- predict(ir.pca)
> eqscplot(ir.pc[, 1:2], type = "n",
      xlab = "first principal component",
      ylab = "second principal component")
> text(ir.pc[, 1:2], labels = as.character(ir.species),
      col = 3 + codes(ir.species))
```

In the terminology of this function, the *loadings* are columns giving the linear combinations a for each principal component, and the *scores* are the data on the principal components. The plot (not shown) is the screeplot, a barplot of the variances of the principal components labelled by $\sum_{i=1}^{j} \lambda_i / \text{trace}(\Sigma)$. The result of loadings is rather deceptive, as small entries are suppressed in printing but will be insignificant only if the correlation matrix is used, and that is *not* the default. The predict method rotates to the principal components.

As well as a data matrix x, the function princomp can accept data via a model formula with an empty left-hand side or as a variance or correlation matrix specified by argument covlist, of the form output by cov.wt and cov.rob (see page 336). Using the latter is one way to robustify principal component anal-
S+ ysis. (S-PLUS has princompRob in library section robust, using covRob.)

Figure 11.1 shows the first two principal components for the iris data based on the covariance matrix, revealing the group structure if it had not already been known. *A warning:* principal component analysis will reveal the gross features of the data, which may already be known, and is often best applied to residuals after the known structure has been removed. As we discovered in Figure 4.13 on page 96, animals come in varying sizes and two sexes!

```
> lcrabs <- log(crabs[, 4:8])
> crabs.grp <- factor(c("B", "b", "O", "o")[rep(1:4, each = 50)])
> (lcrabs.pca <- princomp(lcrabs))
Standard deviations:
  Comp.1   Comp.2   Comp.3   Comp.4    Comp.5
 0.51664 0.074654 0.047914 0.024804 0.0090522
> loadings(lcrabs.pca)
    Comp.1 Comp.2 Comp.3 Comp.4 Comp.5
FL   0.452  0.157  0.438 -0.752  0.114
RW   0.387 -0.911
CL   0.453  0.204 -0.371        -0.784
CW   0.440        -0.672         0.591
BD   0.497  0.315  0.458  0.652  0.136
> lcrabs.pc <- predict(lcrabs.pca)
> dimnames(lcrabs.pc) <- list(NULL, paste("PC", 1:5, sep = ""))
```

(As the data on log scale *are* very comparable, we did not rescale the variables to unit variance.) The first principal component had by far the largest standard deviation, with coefficients that show it to be a 'size' effect. A plot of the second and third principal components shows an almost total separation into forms (Figure 4.13 and 4.14 on pages 96 and 97) on the third PC, the second PC distinguishing sex. The coefficients of the third PC show that it is contrasting overall size with FL and BD.

One ancillary use of principal component analysis is to *sphere* the data. After transformation to principal components, the coordinates are uncorrelated, but with different variances. Sphering the data amounts to rescaling each principal component to have unit variance, so the variance matrix becomes the identity. If the data were a sample from a multivariate normal distribution the point cloud would look spherical, and many measures of 'interestingness' in exploratory projection pursuit look for features in sphered data. Borrowing a term from time series, sphering is sometimes known as *pre-whitening*.

There are two books devoted solely to principal components, Jackson (1991) and Jolliffe (1986), which we think overstates its value as a technique.

Exploratory projection pursuit

Using projection pursuit in XGobi or GGobi allows us to examine the data much more thoroughly. Try one of

```
library(xgobi)
xgobi(lcrabs, colors = c("SkyBlue", "SlateBlue", "Orange",
      "Red")[rep(1:4, each = 50)])
xgobi(lcrabs, glyphs = 12 + 5*rep(0:3, each = 50))
```

A result of optimizing by the 'holes' index is shown in Figure 11.2.

Distance methods

This is a class of methods based on representing the cases in a low-dimensional Euclidean space so that their proximity reflects the similarity of their variables.

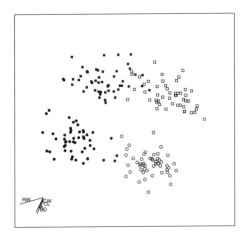

Figure 11.2: Projection pursuit view of the `crabs` data. Males are coded as filled symbols, females as open symbols, the blue colour form as squares and the orange form as circles.

We can think of 'squeezing' a high-dimensional point cloud into a small number of dimensions (2, perhaps 3) whilst preserving as well as possible the inter-point distances.

To do so we have to produce a measure of (dis)similarity. The function `dist` uses one of four distance measures between the points in the p-dimensional space of variables; the default is Euclidean distance. Distances are often called *dissimilarities*. Jardine and Sibson (1971) discuss several families of similarity and dissimilarity measures. For categorical variables most dissimilarities are measures of agreement. The *simple matching coefficient* is the proportion of categorical variables on which the cases differ. The *Jaccard coefficient* applies to categorical variables with a preferred level. It is the proportion of such variables with one of the cases at the preferred level in which the cases differ. The `binary` method of `dist` is of this family, being the Jaccard coefficient if all non-zero levels are preferred. Applied to logical variables on two cases it gives the proportion of variables in which only one is true among those that are true on at least one case.

R   The function `daisy` (in package `cluster` in R) provides a more general way to compute dissimilarity matrices. The main extension is to variables that are not on interval scale, for example, ordinal, log-ratio and asymmetric binary variables. There are many variants of these coefficients; Kaufman and Rousseeuw (1990, §2.5) provide a readable summary and recommendations, and Cox and Cox (2001, Chapter 2) provide a more comprehensive catalogue.

The most obvious of the distance methods is *multidimensional scaling* (MDS), which seeks a configuration in $\mathbb{R}^d$ such that distances between the points best match (in a sense to be defined) those of the distance matrix. We start with the classical form of multidimensional scaling, which is also known as *principal coordinate analysis*. For the `iris` data we can use:

```
ir.scal <- cmdscale(dist(ir), k = 2, eig = T)
ir.scal$points[, 2] <- -ir.scal$points[, 2]
```

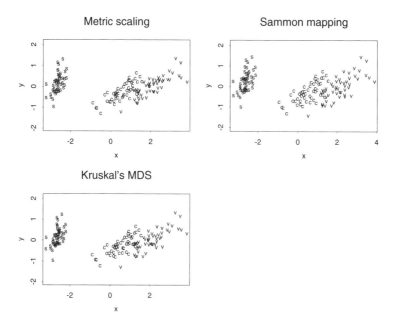

Figure 11.3: Distance-based representations of the `iris` data. The top left plot is by multidimensional scaling, the top right by Sammon's non-linear mapping, the bottom left by Kruskal's isotonic multidimensional scaling. Note that each is defined up to shifts, rotations and reflections.

```
eqscplot(ir.scal$points, type = "n")
text(ir.scal$points, labels = as.character(ir.species),
     col = 3 + codes(ir.species), cex = 0.8)
```

where care is taken to ensure correct scaling of the axes (see the top left plot of Figure 11.3). Note that a configuration can be determined only up to translation, rotation and reflection, since Euclidean distance is invariant under the group of rigid motions and reflections. (We chose to reflect this plot to match later ones.) An idea of how good the fit is can be obtained by calculating a measure[5] of 'stress':

```
> distp <- dist(ir)
> dist2 <- dist(ir.scal$points)
> sum((distp - dist2)^2)/sum(distp^2)
[1] 0.001747
```

which shows the fit is good. Using classical multidimensional scaling with a Euclidean distance as here is equivalent to plotting the first k principal components (without rescaling to correlations).

Another form of multidimensional scaling is Sammon's (1969) non-linear mapping, which given a dissimilarity d on n points constructs a k-dimensional

---

[5]There are many such measures.

configuration with distances $\tilde{d}$ to minimize a weighted 'stress'

$$E_{\text{Sammon}}(d, \tilde{d}) = \frac{1}{\sum_{i \neq j} d_{ij}} \sum_{i \neq j} \frac{(d_{ij} - \tilde{d}_{ij})^2}{d_{ij}}$$

by an iterative algorithm implemented in our function sammon. We have to drop duplicate observations to make sense of $E(d, \tilde{d})$; running sammon will report which observations are duplicates.[6] Figure 11.4 was produced by

```
ir.sam <- sammon(dist(ir[-143,]))
eqscplot(ir.sam$points, type = "n")
text(ir.sam$points, labels = as.character(ir.species[-143]),
     col = 3 + codes(ir.species), cex = 0.8)
```

Contrast this with the objective for classical MDS applied to a Euclidean configuration of points (but not in general), which minimizes

$$E_{\text{classical}}(d, \tilde{d}) = \sum_{i \neq j} [d_{ij}^2 - \tilde{d}_{ij}^2] \Big/ \sum_{i \neq j} d_{ij}^2$$

The Sammon function puts much more stress on reproducing small distances accurately, which is normally what is needed.

A more thoroughly non-metric version of multidimensional scaling goes back to Kruskal and Shepard in the 1960s (see Cox and Cox, 2001 and Ripley, 1996). The idea is to choose a configuration to minimize

$$STRESS^2 = \sum_{i \neq j} [\theta(d_{ij}) - \tilde{d}_{ij}]^2 \Big/ \sum_{i \neq j} \tilde{d}_{ij}^2$$

over both the configuration of points and an increasing function θ. Now the location, rotation, reflection and scale of the configuration are all indeterminate. This is implemented in function isoMDS which we can use by

```
ir.iso <- isoMDS(dist(ir[-143,]))
eqscplot(ir.iso$points, type = "n")
text(ir.iso$points, labels = as.character(ir.species[-143]),
     col = 3 + codes(ir.species), cex = 0.8)
```

The optimization task is difficult and this can be quite slow.

MDS plots of the crabs data tend to show just large and small crabs, so we have to remove the dominant effect of size. We used the carapace area as a good measure of size, and divided all measurements by the square root of the area. It is also necessary to account for the sex differences, which we can do by analysing each sex separately, or by subtracting the mean for each sex, which we did:

```
cr.scale <- 0.5 * log(crabs$CL * crabs$CW)
slcrabs <- lcrabs - cr.scale
cr.means <- matrix(0, 2, 5)
```

---

[6]In S we would use (1:150)[duplicated(ir)] .

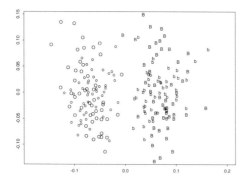

Figure 11.4: Sammon mapping of `crabs` data adjusted for size and sex. Males are coded as capitals, females as lower case, colours as the initial letter of blue or orange.

Figure 11.5: Isotonic multidimensional scaling representation of the `fgl` data. The groups are plotted by the initial letter, except F for window float glass, and N for window non-float glass. Small dissimilarities correspond to small distances on the plot and conversely.

```
cr.means[1,] <- colMeans(slcrabs[crabs$sex == "F", ])
cr.means[2,] <- colMeans(slcrabs[crabs$sex == "M", ])
dslcrabs <- slcrabs - cr.means[as.numeric(crabs$sex), ]
lcrabs.sam <- sammon(dist(dslcrabs))
eqscplot(lcrabs.sam$points, type = "n", xlab = "", ylab = "")
text(lcrabs.sam$points, labels = as.character(crabs.grp))
```

The MDS representations can be quite different in examples such as our dataset `fgl` that do not project well into a small number of dimensions; Figure 11.5 shows a non-metric MDS plot. (We omit one of an identical pair of fragments.)

```
fgl.iso <- isoMDS(dist(as.matrix(fgl[-40, -10])))
eqscplot(fgl.iso$points, type = "n", xlab = "", ylab = "", axes = F)
# either
for(i in seq(along = levels(fgl$type))) {
  set <- fgl$type[-40] == levels(fgl$type)[i]
  points(fgl.iso$points[set,], pch = 18, cex = 0.6, col = 2 + i)}
# S: key(text = list(levels(fgl$type), col = 3:8))
# or
text(fgl.iso$points,
```

```
        labels = c("F", "N", "V", "C", "T", "H")[fgl$type[-40]],
        cex = 0.6)
fgl.iso3 <- isoMDS(dist(as.matrix(fgl[-40, -10]))), k = 3)
# S: brush(fgl.iso3$points)
fgl.col <- c("SkyBlue", "SlateBlue", "Orange", "Orchid",
            "Green", "HotPink")[fgl$type]
xgobi(fgl.iso3$points, colors = fgl.col)
```

This dataset fits much better into three dimensions, but that poses a challenge of viewing the results in some S environments. The optimization can be displayed dynamically in XGvis, part of XGobi.

Self-organizing maps

All multidimensional scaling algorithms are slow, not least because they work with all the distances between pairs of points and so scale at least as $O(n^2)$ and often worse. Engineers have looked for methods to find maps from many more than hundreds of points, of which the best known is 'Self-Organizing Maps' (Kohonen, 1995). Kohonen describes his own motivation as:

> 'I just wanted an algorithm that would effectively map similar patterns (pattern vectors close to each other in the input signal space) onto contiguous locations in the output space.' (p. VI)

which is the same aim as most variants of MDS. However, he interpreted 'contiguous' *via* a rectangular or hexagonal 2-D lattice of representatives[7] m_j, with representatives at nearby points on the grid that are more similar than those that are widely separated. Data points are then assigned to the nearest representative (in Euclidean distance). Since Euclidean distance is used, pre-scaling of the data is important.

Kohonen's SOM is a family of algorithms with no well-defined objective to be optimized, and the results can be critically dependent on the initialization and the values of the tuning constants used. Despite this high degree of arbitrariness, the method scales well (it is at worst linear in n) and often produces useful insights in datasets whose size is way beyond MDS methods (for example, Roberts and Tarassenko, 1995).

If all the data are available at once (as will be the case in S applications), the preferred method is *batch SOM* (Kohonen, 1995, §3.14). For a single iteration, assign all the data points to representatives, and then update all the representatives by replacing each by the mean of all data points assigned to that representative or one of its neighbours (possibly using a distance-weighted mean). The algorithm proceeds iteratively, shrinking the neighbourhood radius to zero over a small number of iterations. Figure 11.6 shows the result of one run of the following code.

```
library(class)
gr <- somgrid(topo = "hexagonal")
crabs.som <- batchSOM(lcrabs, gr, c(4, 4, 2, 2, 1, 1, 1, 0, 0))
plot(crabs.som)
```

---

[7]Called 'codes' or a 'codebook' in some of the literature.

Figure 11.6: Batch SOM applied to the `crabs` dataset. The left plot is a `stars` plot of the representatives, and the right plot shows the assignments of the original points, coded as in 11.4 and placed randomly within the circle. (Plots from R.)

```
bins <- as.numeric(knn1(crabs.som$code, lcrabs, 0:47))
plot(crabs.som$grid, type = "n")
symbols(crabs.som$grid$pts[, 1], crabs.som$grid$pts[, 2],
        circles = rep(0.4, 48), inches = F, add = T)
text(crabs.som$grid$pts[bins, ] + rnorm(400, 0, 0.1),
     as.character(crabs.grp))
```

The initialization used is to select a random subset of the data points. Different runs give different patterns but do generally show the gradation for small to large animals shown in the left panel[8] of Figure 11.6.

Traditional SOM uses an on-line algorithm, in which examples are presented in turn until convergence, usually by sampling from the dataset. Whenever an example x is presented, the closest representative m_j is found. Then

$$m_i \leftarrow m_i + \alpha[x - m_i] \qquad \text{for all neighbours } i \ .$$

Both the constant α and the definition of 'neighbour' change with time. This can be explored *via* function `SOM`, for example,

```
crabs.som2 <- SOM(lcrabs, gr); plot(crabs.som2)
```

See Murtagh and Hernández-Pajares (1995) for another statistical assessment.

Biplots

The biplot (Gabriel, 1971) is a method to represent both the cases and variables. We suppose that X has been centred to remove column means. The biplot represents X by two sets of vectors of dimensions n and p producing a rank-2 approximation to X. The best (in the sense of least squares) such approximation is given by replacing Λ in the singular value decomposition of X by D, a diagonal matrix setting $\lambda_3, \ldots$ to zero, so

$$X \approx \tilde{X} = [\boldsymbol{u}_1 \, \boldsymbol{u}_2] \begin{bmatrix} \lambda_1 & 0 \\ 0 & \lambda_2 \end{bmatrix} \begin{bmatrix} \boldsymbol{v}_1^T \\ \boldsymbol{v}_2^T \end{bmatrix} = GH^T$$

---

[8]In S-PLUS the `stars` plot will be drawn on a rectangular grid.

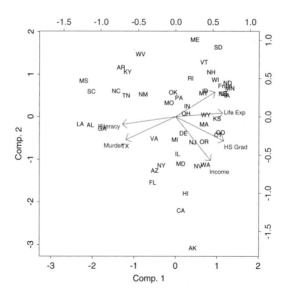

Figure 11.7: Principal component biplot of the part of the state.x77 data. Distances between states represent Mahalanobis distance, and inner products between variables represent correlations. (The arrows extend 80% of the way along the variable's vector.)

where the diagonal scaling factors can be absorbed into G and H in a number of ways. For example, we could take

$$G = n^{a/2} \left[\boldsymbol{u}_1 \, \boldsymbol{u}_2 \right] \begin{bmatrix} \lambda_1 & 0 \\ 0 & \lambda_2 \end{bmatrix}^{1-\lambda}, \qquad H = n^{-a/2} \left[\boldsymbol{v}_1 \, \boldsymbol{v}_2 \right] \begin{bmatrix} \lambda_1 & 0 \\ 0 & \lambda_2 \end{bmatrix}^{\lambda}$$

The biplot then consists of plotting the $n + p$ two-dimensional vectors that form the rows of G and H. The interpretation is based on inner products between vectors from the two sets, which give the elements of $\widetilde{X}$. For $\lambda = a = 0$ this is just a plot of the first two principal components and the projections of the variable axes.

The most popular choice is $\lambda = a = 1$ (which Gabriel, 1971, calls the *principal component biplot*). Then G contains the first two principal components scaled to unit variance, so the Euclidean distances between the rows of G represent the Mahalanobis distances (page 334) between the observations and the inner products between the rows of H represent the covariances between the (possibly scaled) variables (Jolliffe, 1986, pp. 77–8); thus the lengths of the vectors represent the standard deviations.

Figure 11.7 shows a biplot with $\lambda = 1$, obtained by[9]

```
library(MASS, first = T)      # enhanced biplot.princomp
# R: data(state)
state <- state.x77[, 2:7]; row.names(state) <- state.abb
```

---

[9]An enhanced version of biplot.princomp from MASS is used.

```
biplot(princomp(state, cor = T), pc.biplot = T, cex = 0.7,
       expand = 0.8)
```

We specified a rescaling of the original variables to unit variance. (There are additional arguments `scale`, which specifies λ, and `expand`, which specifies a scaling of the rows of H relative to the rows of G, both of which default to 1.)

Gower and Hand (1996) in a book-length discussion of biplots criticize conventional plots such as Figure 11.7. In particular they point out that the axis scales are not at all helpful. Notice the two sets of scales. That on the lower and left axes refers to the values of the rows of G. The upper/right scale is for the values of the rows of H which are shown as arrows.

Independent component analysis

Independent component analysis (ICA) was named by Comon (1994), and has since become a 'hot' topic in data visualization; see the books Lee (1998); Hyvärinen *et al.* (2001) and the expositions by Hyvärinen and Oja (2000) and Hastie *et al.* (2001, §14.6).

ICA looks for rotations of sphered data that have approximately independent coordinates. This will be true (in theory) for all rotations of samples from multivariate normal distributions, so ICA is of most interest for distributions that are far from normal.

The original context for ICA was 'unmixing' of signals. Suppose there are $k \leqslant p$ independent sources in a data matrix S, and we observe the p linear combinations $X = SA$ with mixing matrix A. The 'unmixing' problem is to recover S. Clearly there are identifiability problems: we cannot recover the amplitudes or the labels of the signals, so we may as well suppose that the signals have unit variances. Unmixing is often illustrated by the problem of listening to just one speaker at a party. Note that this is a 'no noise' model: all the randomness is assumed to come from the signals.

Suppose the data X have been sphered; by assumption S is sphered and so X has variance $A^T A$ and we look for an orthogonal matrix A. Thus ICA algorithms can be seen as exploratory projection pursuit in which the measure of interestingness emphasises independence (not just uncorrelatedness), say as the sum of the entropies of the projected coordinates. Like most projection pursuit indices, approximations are used for speed, and that proposed by Hyvärinen and Oja (2000) is implemented is the R package `fastICA`.[10] We can illustrate this for the `crabs` data, where the first and fourth signals shown in Figure 11.8 seem to pick out the two colour forms and two sexes respectively.

```
library(fastICA)
nICA <- 4
crabs.ica <- fastICA(crabs[, 4:8], nICA)
Z <- crabs.ica$S
par(mfrow = c(2, nICA))
for(i in 1:nICA) boxplot(split(Z[, i], crabs.grp))
```

---

[10]By Jonathan Marchini. Also ported to S-PLUS.

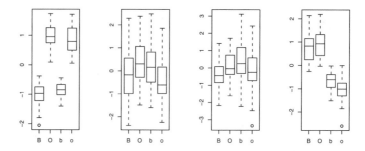

Figure 11.8: Boxplots of four 'signals' recovered by ICA from the `crabs` data.

There is a lot of arbitrariness in the use of ICA, in particular in choosing the number of signals. We might have expected to need two here, when the results are much less impressive.

Glyph representations

There is a wide range of ways to trigger multiple perceptions of a figure, and we can use these to represent each of a moderately large number of rows of a data matrix by an individual figure. Perhaps the best known of these are Chernoff's faces (Chernoff, 1973, implemented in the S-PLUS function `faces`; there are other versions by Bruckner, 1978 and Flury and Riedwyl, 1981) and the star plots as implemented in the function `stars` (see Figure 11.6), but Wilkinson (1999, Chapter 3) gives many more.

These glyph plots do depend on the ordering of variables and perhaps also their scaling, and they do rely on properties of human visual perception. So they have rightly been criticised as subject to manipulation, and one should be aware of the possibility that the effect may differ by viewer.[11] Nevertheless they can be very effective as tools for private exploration.

As an example, a stars plot for the `state.x77` dataset with variables in the order showing up in the biplot of Figure 11.7 can be drawn by

```
# S: stars(state.x77[, c(7, 4, 6, 2, 5, 3)], byrow = T)
# R: stars(state.x77[, c(7, 4, 6, 2, 5, 3)], full = FALSE,
          key.loc = c(10, 2))
```

Parallel coordinate plots

Parallel coordinates plots (Inselberg, 1984; Wegman, 1990) join the same points across a set of parallel axes. We can show the `state.x77` dataset in the order showing up in the biplot of Figure 11.7 by

```
parcoord(state.x77[, c(7, 4, 6, 2, 5, 3)])
```

---

[11]Especially if colour is involved; it is amazingly common to overlook the prevalence of red–green colour blindness.

Figure 11.9: R version of `stars` plot of the `state.x77` dataset.

Such plots are often too 'busy' without a means of interaction to identify observations, sign-change and reorder variables, brush groups and so on (as is possible in XGobi and GGobi). As an example of a revealing parallel coordinate plot try

```
parcoord(log(ir)[, c(3, 4, 2, 1)], col = 1 + (0:149)%/%50)
```

on a device which can plot colour.

11.2   Cluster Analysis

Cluster analysis is concerned with discovering groupings among the cases of our n by p matrix. A comprehensive general reference is Gordon (1999); Kaufman and Rousseeuw (1990) give a good introduction and their methods are available in S-PLUS and in package `cluster` for R Clustering methods can be clustered in many different ways; here is one.

- Agglomerative hierarchical methods (`hclust`, `agnes`, `mclust`).

  - Produces a set of clusterings, usually one with k clusters for each $k = n, \ldots, 2$, successively amalgamating groups.

  - Main differences are in calculating group–group dissimilarities from point–point dissimilarities.

  – Computationally easy.

- Optimal partitioning methods (`kmeans`, `pam`, `clara`, `fanny`).

  – Produces a clustering for fixed K.

  – Need an initial clustering.

  – Lots of different criteria to optimize, some based on probability models.

  – Can have distinct 'outlier' group(s).

- Divisive hierarchical methods (`diana`, `mona`).

  – Produces a set of clusterings, usually one for each $k = 2, \ldots, K \ll n$.

  – Computationally nigh-impossible to find optimal divisions (Gordon, 1999, p. 90).

  – Most available methods are *monothetic* (split on one variable at each stage).

Do not assume that 'clustering' methods are the best way to discover interesting groupings in the data; in our experience the visualization methods are often far more effective. There are many different clustering methods, often giving different answers, and so the danger of over-interpretation is high.

Many methods are based on a measure of the similarity or dissimilarity between cases, but some need the data matrix itself. A *dissimilarity coefficient* d is symmetric ($d(A, B) = d(B, A)$), non-negative and $d(A, A)$ is zero. A similarity coefficient has the scale reversed. Dissimilarities may be *metric*

$$d(A, C) \leqslant d(A, B) + d(B, C)$$

or *ultrametric*

$$d(A, B) \leqslant \max\big(d(A, C), d(B, C)\big)$$

but need not be either. We have already seen several dissimilarities calculated by `dist` and `daisy`.

Ultrametric dissimilarities have the appealing property that they can be represented by a *dendrogram* such as those shown in Figure 11.10, in which the dissimilarity between two cases can be read from the height at which they join a single group. Hierarchical clustering methods can be thought of as approximating a dissimilarity by an ultrametric dissimilarity. Jardine and Sibson (1971) argue that one method, single-link clustering, uniquely has all the desirable properties of a clustering method. This measures distances between clusters by the dissimilarity of the closest pair, and agglomerates by adding the shortest possible link (that is, joining the two closest clusters). Other authors disagree, and Kaufman and Rousseeuw (1990, §5.2) give a different set of desirable properties leading uniquely to their preferred method, which views the dissimilarity between clusters as the average of the dissimilarities between members of those clusters. Another popular method is complete-linkage, which views the dissimilarity between clusters as the maximum of the dissimilarities between members.

The function `hclust` implements these three choices, selected by its `method` argument which takes values `"compact"` (the default, for complete-linkage,

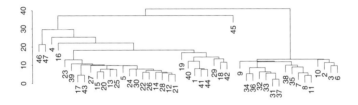

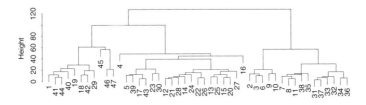

Figure 11.10: Dendrograms for the socio-economic data on Swiss provinces computed by single-link clustering (top) and divisive clustering (bottom).

R   called "complete" in R), "average" and "connected" (for single-linkage, called "single" in R). Function agnes also has these (with the R names) and others.

The S dataset[12] swiss.x gives five measures of socio-economic data on Swiss provinces about 1888, given by Mosteller and Tukey (1977, pp. 549–551). The data are percentages, so Euclidean distance is a reasonable choice. We use single-link clustering:

```
# S: h <- hclust(dist(swiss.x), method = "connected")
# R: data(swiss); swiss.x <- as.matrix(swiss[, -1])
# R: h <- hclust(dist(swiss.x), method = "single")
plclust(h)
cutree(h, 3)
# S: plclust( clorder(h, cutree(h, 3) ))
```

The hierarchy of clusters in a dendrogram is obtained by cutting it at different heights. The first plot suggests three main clusters, and the remaining code re-orders the dendrogram to display (see Figure 11.10) those clusters more clearly. Note that there appear to be two main groups, with the point 45 well separated from them.

Function diana performs *divisive* clustering, in which the clusters are repeat-edly subdivided rather than joined, using the algorithm of Macnaughton-Smith *et al.* (1964). Divisive clustering is an attractive option when a grouping into a few large clusters is of interest. The lower panel of Figure 11.10 was produced by pltree(diana(swiss.x)).

---

[12] In R the numbers are slightly different, and the provinces has been given names.

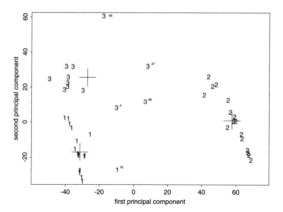

Figure 11.11: The Swiss provinces data plotted on its first two principal components. The labels are the groups assigned by K-means; the crosses denote the group means. Five points are labelled with smaller symbols.

Partitioning methods

The K-means clustering algorithm (MacQueen, 1967; Hartigan, 1975; Hartigan and Wong, 1979) chooses a pre-specified number of cluster centres to minimize the within-class sum of squares from those centres. As such it is most appropriate to continuous variables, suitably scaled. The algorithm needs a starting point, so we choose the means of the clusters identified by group-average clustering. The clusters *are* altered (cluster 3 contained just point 45), and are shown in principal-component space in Figure 11.11. (Its standard deviations show that a two-dimensional representation is reasonable.)

```
h <- hclust(dist(swiss.x), method = "average")
initial <- tapply(swiss.x, list(rep(cutree(h, 3),
   ncol(swiss.x)), col(swiss.x)), mean)
dimnames(initial) <- list(NULL, dimnames(swiss.x)[[2]])
km <- kmeans(swiss.x, initial)
(swiss.pca <- princomp(swiss.x))
Standard deviations:
  Comp.1 Comp.2 Comp.3 Comp.4 Comp.5
  42.903 21.202  7.588 3.6879 2.7211
   ....
swiss.px <- predict(swiss.pca)
dimnames(km$centers)[[2]] <- dimnames(swiss.x)[[2]]
swiss.centers <- predict(swiss.pca, km$centers)
eqscplot(swiss.px[, 1:2], type = "n",
        xlab = "first principal component",
        ylab = "second principal component")
text(swiss.px[, 1:2], labels = km$cluster)
points(swiss.centers[,1:2], pch = 3, cex = 3)
identify(swiss.px[, 1:2], cex = 0.5)
```

By definition, K-means clustering needs access to the data matrix and uses Euclidean distance. We can apply a similar method using only dissimilarities if we confine the cluster centres to the set of given examples. This is known as the k-medoids criterion (of Vinod, 1969) implemented in pam and clara. Using pam picks provinces 29, 8 and 28 as cluster centres.

```
> library(cluster)              # needed in R only
> swiss.pam <- pam(swiss.px, 3)
> summary(swiss.pam)
Medoids:
      Comp. 1   Comp. 2 Comp. 3 Comp. 4  Comp. 5
[1,] -29.716  18.22162  1.4265 -1.3206  0.95201
[2,]  58.609   0.56211  2.2320 -4.1778  4.22828
[3,] -28.844 -19.54901  3.1506  2.3870 -2.46842
Clustering vector:
 [1] 1 2 2 1 3 2 2 2 2 2 2 3 3 3 3 3 1 1 1 3 3 3 3 3 3 3 3 3
[29] 1 3 2 2 2 2 2 2 2 2 1 1 1 1 1 1 1 1 1
 . . . .
> eqscplot(swiss.px[, 1:2], type = "n",
           xlab = "first principal component",
           ylab = "second principal component")
> text(swiss.px[,1:2], labels = swiss.pam$clustering)
> points(swiss.pam$medoid[,1:2], pch = 3, cex = 5)
```

The function fanny implements a 'fuzzy' version of the k-medoids criterion. Rather than point i having a membership of just one cluster v, its membership is partitioned among clusters as positive weights u_{iv} summing to one. The criterion then is

$$\min_{(u_{iv})} \sum_v \frac{\sum_{i,j} u_{iv}^2 u_{jv}^2 d_{ij}}{2 \sum_i u_{iv}^2}.$$

For our running example we find

```
> fanny(swiss.px, 3)
 iterations objective
        16    354.01
Membership coefficients:
       [,1]     [,2]     [,3]
[1,] 0.725016 0.075485 0.199499
[2,] 0.189978 0.643928 0.166094
[3,] 0.191282 0.643596 0.165123
 . . . .
Closest hard clustering:
 [1] 1 2 2 1 3 2 2 2 2 2 2 3 3 3 3 3 1 1 1 3 3 3 3 3 3 3 3 3
[29] 1 3 2 2 2 2 2 2 2 2 1 1 1 1 1 1 1 1 1
```

The 'hard' clustering is formed by assigning each point to the cluster for which its membership coefficient is highest.

Other partitioning methods are based on the idea that the data are independent samples from a series of group populations, but the group labels have been lost, so the data can be regarded as from a mixture distribution. The idea is then to find the

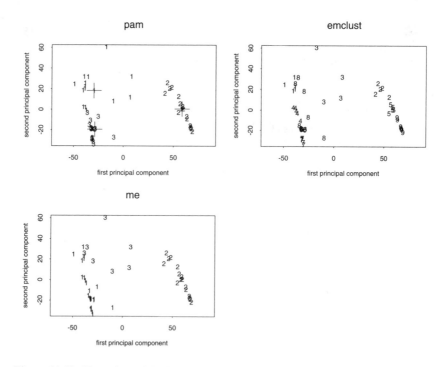

Figure 11.12: Clusterings of the Swiss provinces data by pam with three clusters (with the medoids marked by crosses), me with three clusters and emclust with up to nine clusters (it chose nine).

mixture distribution, usually as a mixture of multivariate normals, and to assign points to the component for which their posterior probability of membership is highest.

S+   S-PLUS has functions mclust, mclass and mreloc based on 'maximum-likelihood' clustering in which the mixture parameters and the classification are optimized simultaneously. Later work in the mclust library section[13] uses sounder methods in which the mixtures are fitted first. Nevertheless, fitting normal mixtures is a difficult problem, and the results obtained are often heavily dependent on the initial configuration supplied.

K-means clustering can be seen as 'maximum-likelihood' clustering where the clusters are assumed all to be spherically symmetric multivariate normals with the same spread. The modelName argument to the mclust functions allows a wider choice of normal mixture components, including "EI" (equal spherical) "VI" (spherical, differing by component), "EEE" (same elliptical), "VEV" (same shape elliptical, variable size and orientation) and "VVV" (arbitrary components).

Library section mclust provides hierarchical clustering via functions hc and mhclass. Then for a given number k of clusters the fitted mixture can be optimized by calling me (which here does not change the classification).

---

[13] Available at http://www.stat.washington.edu/fraley/mclust/ and for R from CRAN.

```
library(mclust)
h <- hc(modelName = "VVV", swiss.x)
(mh <- as.vector(mhclass(h, 3)))
 [1] 1 2 2 3 1 2 2 2 2 2 2 1 1 1 1 1 1 1 1 1 1 1 1 1 1 1 1 1
[29] 1 1 2 2 2 2 2 2 2 2 1 1 1 1 1 1 3 3 3
z <- me(modelName = "VVV", swiss.x,   z = 0.5*(unmap(mh)+1/3))
eqscplot(swiss.px[, 1:2], type = "n",
            xlab = "first principal component",
            ylab = "second principal component")
text(swiss.px[, 1:2], labels = max.col(z$z))
```

Function `EMclust` automates the whole cluster process, including choosing the number of clusters and between different `modelName`'s. It chooses lots of clusters (see Figure 11.12).

```
> vals <- EMclust(swiss.x) # all possible models, 0:9 clusters.
> (sm <- summary(vals, swiss.x))
> eqscplot(swiss.px[, 1:2], type = "n",
            xlab = "first principal component",
            ylab = "second principal component")
> text(swiss.px[, 1:2], labels = sm$classification)
```

Another possibility is to use function `EMclustN` to fit a cluster model including a background 'noise' term, that is a component that is a uniform Poisson process, controlled by argument `noise`.

11.3   Factor Analysis

Principal component analysis looks for linear combinations of the data matrix X that are uncorrelated and of high variance. Independent component analysis seeks linear combinations that are independent. *Factor analysis* seeks linear combinations of the variables, called *factors*, that represent underlying fundamental quantities of which the observed variables are expressions. The examples tend to be controversial ones such as 'intelligence' and 'social deprivation', the idea being that a small number of factors might explain a large number of measurements in an observational study. Such factors are to be inferred from the data.

We can think of both the factors of factor analysis and the signals of independent component analysis as *latent variables*, unobserved variables on each experimental unit that determine the patterns in the observations. The difference is that it is not the factors that are assumed to be independent, but rather the observations conditional on the factors.

The factor analysis model for a single common factor f is

$$x = \mu + \lambda f + u \tag{11.1}$$

where λ is a vector known as the *loadings* and u is a vector of *unique* (or *specific*) factors for that observational unit. To help make the model identifiable, we assume that the factor f has mean zero and variance one, and that u has mean zero and unknown *diagonal* covariance matrix Ψ. For $k < p$ common factors we have a vector f of common factors and a loadings matrix Λ, and

$$x = \mu + \Lambda f + u \qquad (11.2)$$

where the components of f have unit variance and are uncorrelated and f and u are taken to be uncorrelated. Note that *all* the correlations amongst the variables in x must be explained by the common factors; if we assume joint normality the observed variables x will be conditionally independent given f.

Principal component analysis also seeks a linear subspace like Λf to explain the data, but measures the lack of fit by the sum of squares of the u_i. Since factor analysis allows an arbitrary diagonal covariance matrix Ψ, its measure of fit of the u_i depends on the problem and should be independent of the units of measurement of the observed variables. (Changing the units of measurement of the observations does not change the common factors if the loadings and unique factors are re-expressed in the new units.)

Equation (11.2) and the conditions on f express the covariance matrix Σ of the data as

$$\Sigma = \Lambda\Lambda^T + \Psi \qquad (11.3)$$

Conversely, if (11.3) holds, there is a k-factor model of the form (11.2). Note that the common factors $G^T f$ and loadings matrix ΛG give rise to the same model for Σ, for any $k \times k$ orthogonal matrix G. Choosing an appropriate G is known as choosing a *rotation*. All we can achieve statistically is to fit the space spanned by the factors, so choosing a rotation is a way to choose an interpretable basis for that space. Note that if

$$s = \tfrac{1}{2}p(p+1) - [p(k+1) - \tfrac{1}{2}k(k-1)] = \tfrac{1}{2}(p-k)^2 - \tfrac{1}{2}(p+k) < 0$$

we would expect an infinity of solutions to (11.3). This value is known as the *degrees of freedom*, and comes from the number of elements in Σ minus the number of parameters in Ψ and Λ (taking account of the rotational freedom in Λ since only $\Lambda\Lambda^T$ is determined). Thus it is usual to assume $s \geqslant 0$; for $s = 0$ there may be a unique solution, no solution or an infinity of solutions (Lawley and Maxwell, 1971, pp. 10–11).

The variances of the original variables are decomposed into two parts, the *communality* $h_i^2 = \sum_j \lambda_{ij}^2$ and *uniqueness* ψ_{ii} which is thought of as the 'noise' variance.

Fitting the factor analysis model (11.2) is performed by the S function S+ `factanal`. The default method in S-PLUS ('principal factor analysis') dates from the days of limited computational power, and is not intrinsically scale invariant—it should not be used. The preferred method is to maximize the likelihood over Λ and Ψ assuming multivariate normality of the factors (f, u), which depends only on the factor space and is scale-invariant. This likelihood can have multiple local maxima; this possibility is often ignored but `factanal` compares the fit found from several separate starting points. It is possible that the maximum likelihood solution will have some $\widehat{\psi}_{ii} = 0$, so the ith variable lies in the estimated factor space. Opinions differ as to what to do in this case (sometimes known as a *Heywood case*), but often it indicates a lack of data or inadequacy of the factor analysis model. (Bartholomew and Knott, 1999, Section 3.18, discuss possible reasons and actions.)

It is hard to find examples in the literature for which a factor analysis model fits well; many do not give a measure of fit, or have failed to optimize the likelihood well enough and so failed to detect Heywood cases. We consider an example from Smith and Stanley (1983) as quoted by Bartholomew and Knott (1999, pp. 68–72).[14] Six tests were give to 112 individuals, with covariance matrix

```
          general picture  blocks    maze reading    vocab
general    24.641   5.991  33.520   6.023  20.755   29.701
picture     5.991   6.700  18.137   1.782   4.936    7.204
 blocks    33.520  18.137 149.831  19.424  31.430   50.753
   maze     6.023   1.782  19.424  12.711   4.757    9.075
reading    20.755   4.936  31.430   4.757  52.604   66.762
  vocab    29.701   7.204  50.753   9.075  66.762  135.292
```

The tests were of general intelligence, picture completion, block design, mazes, reading comprehension and vocabulary. The S-PLUS default in `factanal` is a single factor, but the fit is not good until we try two. The low uniqueness for reading ability suggests that this is close to a Heywood case, but it definitely is not one.

```
> S: ability.FA <- factanal(covlist = ability.cov, method = "mle")
> R: data(ability.cov)
> R: ability.FA <- factanal(covmat = ability.cov, factors = 1)
> ability.FA
    ....
The chi square statistic is 75.18 on 9 degrees of freedom.
    ....
> (ability.FA <- update(ability.FA, factors = 2))
    ....
The chi square statistic is 6.11 on 4 degrees of freedom.
The p-value is 0.191
    ....
> summary(ability.FA)
Uniquenesses:
 general picture  blocks    maze reading    vocab
 0.45523 0.58933 0.21817 0.76942 0.052463 0.33358
Loadings:
        Factor1 Factor2
general   0.501   0.542
picture   0.158   0.621
 blocks   0.208   0.859
   maze   0.110   0.467
reading   0.957   0.179
  vocab   0.785   0.222
```

[14]Bartholomew & Knott give both covariance and correlation matrices, but these are inconsistent. Neither is in the original paper.

```
> round(loadings(ability.FA) %*% t(loadings(ability.FA)) +
        diag(ability.FA$uniq), 3)

          general picture blocks  maze reading vocab
general    1.000   0.416  0.570 0.308   0.577 0.514
picture    0.416   1.000  0.567 0.308   0.262 0.262
blocks     0.570   0.567  1.000 0.425   0.353 0.355
maze       0.308   0.308  0.425 1.000   0.189 0.190
reading    0.577   0.262  0.353 0.189   1.000 0.791
vocab      0.514   0.262  0.355 0.190   0.791 1.000
```

Remember that the first variable is a composite measure; it seems that the first factor reflects verbal ability, the second spatial reasoning. The main lack of fit is that the correlation 0.193 between picture and maze is fitted as 0.308.

Factor rotations

The usual aim of a rotation is to achieve 'simple structure', that is a pattern of loadings that is easy to interpret with a few large and many small coefficients.

There are many criteria for selecting rotations of the factors and loadings matrix; S-PLUS implements 12. There is an auxiliary function rotate that will rotate the fitted Λ according to one of these criteria, which is called via the rotate argument of factanal. The default varimax criterion is to maximize

$$\sum_{i,j}(d_{ij} - \bar{d}_{.j})^2 \quad \text{where} \quad d_{ij} = \lambda_{ij}^2/\sum_j \lambda_{ij}^2 \qquad (11.4)$$

and $\bar{d}_{.j}$ is the mean of the d_{ij}. Thus the varimax criterion maximizes the sum over factors of the variances of the (normalized) squared loadings. The normalizing factors are the communalities that are invariant under orthogonal rotations.

Following Bartholomew & Knott, we illustrate the oblimin criterion[15] which minimizes the sum over all pairs of factors of the covariance between the squared loadings for those factors.

```
> loadings(rotate(ability.FA, rotation = "oblimin"))
          Factor1 Factor2
general    0.379   0.513
picture            0.640
blocks             0.887
maze               0.483
reading    0.946
vocab      0.757   0.137

Component/Factor Correlations:
         Factor1 Factor2
Factor1  1.000   0.356
Factor2  0.356   1.000
```

---

[15]Not implemented in R at the time of writing.

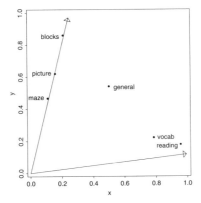

Figure 11.13: The loadings for the intelligence test data after varimax rotation, with the axes for the oblimin rotation shown as arrows.

```
> par(pty = "s")
> L <- loadings(ability.FA)
> eqscplot(L, xlim = c(0,1), ylim = c(0,1))
> identify(L, dimnames(L)[[1]])
> oblirot <- rotate(loadings(ability.FA), rotation = "oblimin")
> naxes <- solve(oblirot$tmat)
> arrows(rep(0, 2), rep(0, 2), naxes[,1], naxes[,2])
```

11.4   Discrete Multivariate Analysis

Most work on visualization and most texts on multivariate analysis implicitly assume continuous measurements. However, large-scale categorical datasets are becoming much more prevalent, often collected through surveys or 'CRM' (customer relationship management: that branch of data mining that collects information on buying habits, for example on shopping baskets) or insurance questionnaires.

There are some useful tools available for exploring categorical data, but it is often essential to use models to understand the data, most often log-linear models. Indeed, 'discrete multivariate analysis' is the title of an early influential book on log-linear models, Bishop *et al.* (1975).

Mosaic plots

There are a few ways to visualize low-dimensional contingency tables. *Mosaic plots* (Hartigan and Kleiner, 1981, 1984; Friendly, 1994; Emerson, 1998; Friendly, 2000) divide the plotting surface recursively according to the proportions of each factor in turn (so the order of the factors matters).

For an example, consider Fisher's (1940) data on colours of eyes and hair of people in Caithness, Scotland:

	fair	red	medium	dark	black
blue	326	38	241	110	3
light	688	116	584	188	4
medium	343	84	909	412	26
dark	98	48	403	681	85

in our dataset `caith`. Figure 11.14 shows mosaic plots for these data and for the housing data we used in Section 7.3, computed by

```
caith1 <- as.matrix(caith)
names(dimnames(caith1)) <- c("eyes", "hair")
mosaicplot(caith1, color = T)
# use xtabs in R
House <- crosstabs(Freq ~ Type + Infl + Cont + Sat, housing)
mosaicplot(House, color = T)
```

Correspondence analysis

Correspondence analysis is applied to two-way tables of counts.

Suppose we have an $r \times c$ table N of counts. Correspondence analysis seeks 'scores' f and g for the rows and columns which are maximally correlated. Clearly the maximum correlation is one, attained by constant scores, so we seek the largest non-trivial solution. Let R and C be matrices of the group indicators of the rows and columns, so $R^T C = N$. Consider the singular value decomposition of their correlation matrix

$$X_{ij} = \frac{n_{ij}/n - (n_{i.}/n)(n_{.j}/n)}{\sqrt{(n_{i.}/n)(n_{.j}/n)}} = \frac{n_{ij} - n\, r_i\, c_j}{n\sqrt{r_i\, c_j}}$$

where $r_i = n_{i.}/n$ and $c_j = n_{.j}/n$ are the proportions in each row and column. Let D_r and D_c be the diagonal matrices of r and c. Correspondence analysis corresponds to selecting the first singular value and left and right singular vectors of X_{ij} and rescaling by $D_r^{-1/2}$ and $D_c^{-1/2}$, respectively. This is done by our function `corresp`:

```
> corresp(caith)
First canonical correlation(s): 0.44637

eyes scores:
    blue      light    medium     dark
 -0.89679 -0.98732 0.075306 1.5743

hair scores:
     fair        red    medium    dark   black
 -1.2187 -0.52258 -0.094147 1.3189 2.4518
```

Can we make use of the subsequent singular values? In what Gower and Hand (1996) call 'classical CA' we consider $A = D_r^{-1/2} U \Lambda$ and $B = D_c^{-1/2} V \Lambda$. Then the first columns of A and B are what we have termed the row and column scores *scaled by* ρ, the first canonical correlation. More generally, we can see

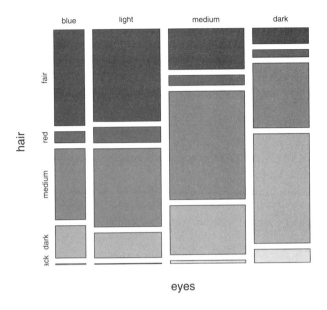

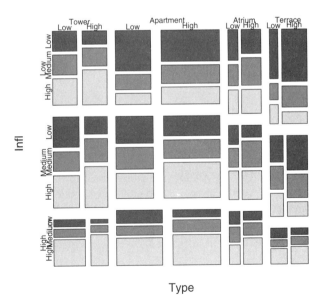

Figure 11.14: Mosaic plots for (top) Fisher's data on people from Caithness and (bottom) Copenhagen housing satisfaction data.

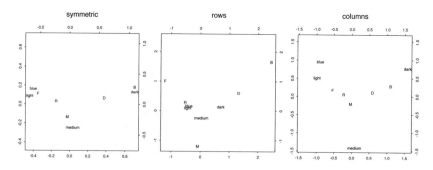

Figure 11.15: Three variants of correspondence analysis plots from Fisher's data on people in Caithness: (left) 'symmetric", (middle) 'row asymmetric' and (right) 'column asymmetric'.

distances between the rows of A as approximating the distances between the row profiles (rows rescaled to unit sum) of the table N, and analogously for the rows of B and the column profiles.

Classical CA plots the first two columns of A and B on the same figure. This is a form of a biplot and is obtained with our software by plotting a correspondence analysis object with nf $\geqslant 2$ or as the default for the method biplot.correspondence. This is sometimes known as a 'symmetric' plot. Other authors (for example, Greenacre, 1992) advocate 'asymmetric' plots. The asymmetric plot for the rows is a plot of the first two columns of A with the column labels plotted at the first two columns of $\Gamma = D_c^{-1/2}V$; the corresponding plot for the columns has columns plotted at B and row labels at $\Phi = D_r^{-1/2}U$. The most direct interpretation for the row plot is that

$$A = D_r^{-1}N\Gamma$$

so A is a plot of the *row profiles* (the rows normalized to sum to one) as convex combinations of the column vertices given by Γ.

By default corresp only retains one-dimensional row and column scores; then plot.corresp plots these scores and indicates the size of the entries in the table by the area of circles. The two-dimensional forms of the plot are shown in Figure 11.15 for Fisher's data on people from Caithness. These were produced by

```
caith2 <- caith
dimnames(caith2)[[2]] <- c("F", "R", "M", "D", "B")
par(mfcol = c(1, 3))
plot(corresp(caith2, nf = 2)); title("symmetric")
plot(corresp(caith2, nf = 2), type = "rows"); title("rows")
plot(corresp(caith2, nf = 2), type = "col"); title("columns")
```

Note that the symmetric plot (left) has the row points from the asymmetric row plot (middle) and the column points from the asymmetric column plot (right) superimposed on the same plot (but with different scales).

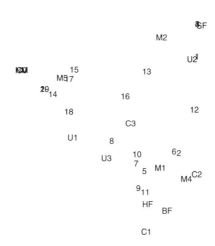

Figure 11.16: Multiple correspondence analysis plot of dataset farms on 20 farms on the Dutch island of Terschelling. Numbers represent the farms and labels levels of moisture (M1, M2, M4 and M5), grassland usage (U1, U2 and U3), manure usage (C0 to C4) and type of grassland management (SF: standard, BF: biological, HF: hobby farming, NM: nature conservation). Levels C0 and NM are coincident (on the extreme left), as are the pairs of farms 3 & 4 and 19 & 20.

Multiple correspondence analysis

Multiple correspondence analysis (MCA) is (confusingly!) a method for visualizing the joint properties of $p \geqslant 2$ categorical variables that does *not* reduce
to correspondence analysis (CA) for $p = 2$, although the methods are closely
related (see, for example, Gower and Hand, 1996, §10.2).

Suppose we have n observations on the p factors with ℓ total levels. Consider G, the $n \times \ell$ indicator matrix whose rows give the levels of each factor for
each observation. Then all the row sums are p. MCA is often (Greenacre, 1992)
defined as CA applied to the table G, that is the singular-value decomposition of
$D_r^{-1/2}(G/\sum_{ij} g_{ij})D_c^{-1/2} = U\Lambda V^T$. Note that $D_r = pI$ since all the row sums
are p, and $\sum_{ij} g_{ij} = np$, so this amounts to the SVD of $p^{-1/2}GD_c^{-1/2}/pn$.[16]

An alternative point of view is that MCA is a principal components analysis
of the data matrix $X = G(pD_c)^{-1/2}$; with PCA it is usual to centre the data, but
it transpires that the largest singular value is one and the corresponding singular
vectors account for the means of the variables. A simple plot for MCA is to plot
the first two principal components of X (which correspond to the second and
third singular vectors of X). This is a form of biplot, but it will not be appropriate
to add axes for the columns of X as the possible values are only $\{0, 1\}$, but it is
usual to add the positions of 1 on each of these axes, and label these by the factor

---

[16]Gower and Hand (1996) omit the divisor pn.

level. (The 'axis' points are plotted at the appropriate row of $(pD_c)^{-1/2}V$.) The point plotted for each observation is the vector sum of the 'axis' points for the levels taken of each of the factors. Gower and Hand seem to prefer (e.g., their Figure 4.2) to rescale the plotted points by p, so they are plotted at the centroid of their levels. This is exactly the asymmetric row plot of the CA of G, apart from an overall scale factor of $p\sqrt{n}$.

We can apply this to the example of Gower and Hand (1996, p. 75) by

```
farms.mca <- mca(farms, abbrev = T)   # Use levels as names
plot(farms.mca, cex = rep(0.7, 2), axes = F)
```

shown in Figure 11.16

Sometimes it is desired to add rows or factors to an MCA plot. Adding rows is easy; the observations are placed at the centroid of the 'axis' points for levels that are observed. Adding factors (so-called *supplementary variables*) is less obvious. The 'axis' points are plotted at the rows of $(pD_c)^{-1/2}V$. Since $U\Lambda V^T = X = G(pD_c)^{-1/2}$, $V = (pD_c)^{-1/2}G^T U\Lambda^{-1}$ and

$$(pD_c)^{-1/2}V = (pD_c)^{-1}G^T U\Lambda^{-1}$$

This tells us that the 'axis' points can be found by taking the appropriate column of G, scaling to total $1/p$ and then taking inner products with the second and third columns of $U\Lambda^{-1}$. This procedure can be applied to supplementary variables and so provides a way to add them to the plot. The `predict` method for class `"mca"` allows rows or supplementary variables to be added to an MCA plot.

Chapter 12

Classification

Classification is an increasingly important application of modern methods in statistics. In the statistical literature the word is used in two distinct senses. The entry (Hartigan, 1982) in the original *Encyclopedia of Statistical Sciences* uses the sense of *cluster analysis* discussed in Section 11.2. Modern usage is leaning to the other meaning (Ripley, 1997) of allocating future cases to one of g prespecified classes. Medical diagnosis is an archetypal classification problem in the modern sense. (The older statistical literature sometimes refers to this as *allocation*.)

In pattern-recognition terminology this chapter is about *supervised* methods. The classical methods of multivariate analysis (Krzanowski, 1988; Mardia, Kent and Bibby, 1979; McLachlan, 1992) have largely been superseded by methods from pattern recognition (Ripley, 1996; Webb, 1999; Duda *et al.*, 2001), but some still have a place.

It is sometimes helpful to distinguish *discriminant analysis* in the sense of describing the differences between the g groups from classification, allocating new observations to the groups. The first provides some measure of explanation; the second can be a 'black box' that makes a decision without any explanation. In many applications no explanation is required (no one cares how machines read postal (zip) codes, only that the envelope is correctly sorted) but in others, especially in medicine, some explanation may be necessary to get the methods adopted.

Classification is related to *data mining*, although some of data mining is exploratory in the sense of Chapter 11. Hand *et al.* (2001) and (especially) Hastie *et al.* (2001) are pertinent introductions.

Some of the methods considered in earlier chapters are widely used for classification, notably classification trees, logistic regression for $g = 2$ groups and multinomial log-linear models (Section 7.3) for $g > 2$ groups.

12.1 Discriminant Analysis

Suppose that we have a set of g classes, and for each case we know the class (assumed correctly). We can then use the class information to help reveal the structure of the data. Let W denote the within-class covariance matrix, that is the covariance matrix of the variables centred on the class mean, and B denote the between-classes covariance matrix, that is, of the predictions by the class means. Let M be the $g \times p$ matrix of class means, and G be the $n \times g$ matrix of class indicator variables (so $g_{ij} = 1$ if and only if case i is assigned to class j). Then the predictions are GM. Let $\bar{x}$ be the means of the variables over the whole sample. Then the sample covariance matrices are

$$W = \frac{(X - GM)^T (X - GM)}{n - g}, \qquad B = \frac{(GM - 1\bar{x})^T (GM - 1\bar{x})}{g - 1} \quad (12.1)$$

Note that B has rank at most $\min(p, g - 1)$.

Fisher (1936) introduced a linear discriminant analysis seeking a linear combination xa of the variables that has a maximal ratio of the separation of the class means to the within-class variance, that is, maximizing the ratio $a^T B a / a^T W a$. To compute this, choose a *sphering* (see page 305) xS of the variables so that they have the identity as their within-group correlation matrix. On the rescaled variables the problem is to maximize $a^T B a$ subject to $\|a\| = 1$, and as we saw for PCA, this is solved by taking a to be the eigenvector of B corresponding to the largest eigenvalue. The linear combination a is unique up to a change of sign (unless there are multiple eigenvalues). The exact multiple of a returned by a program will depend on its definition of the within-class variance matrix. We use the conventional divisor of $n - g$, but divisors of n and $n - 1$ have been used.

As for principal components, we can take further linear components corresponding to the next largest eigenvalues. There will be at most $r = \min(p, g - 1)$ positive eigenvalues. Note that the eigenvalues are the proportions of the between-classes variance explained by the linear combinations, which may help us to choose how many to use. The corresponding transformed variables are called the *linear discriminants* or *canonical variates*. It is often useful to plot the data on the first few linear discriminants (Figure 12.1). Since the within-group covariances should be the identity, we chose an equal-scaled plot. (Using plot(ir.lda) will give this plot without the colours.) The linear discriminants are conventionally centred to have mean zero on dataset.

```
> (ir.lda <- lda(log(ir), ir.species))
Prior probabilities of groups:
      c       s       v
0.33333 0.33333 0.33333

Group means:
    Sepal L. Sepal W. Petal L. Petal W.
c    1.7773   1.0123  1.44293  0.27093
s    1.6082   1.2259  0.37276 -1.48465
v    1.8807   1.0842  1.70943  0.69675

Coefficients of linear discriminants:
              LD1      LD2
Sepal L.   3.7798  4.27690
Sepal W.   3.9405  6.59422
Petal L.  -9.0240  0.30952
Petal W.  -1.5328 -0.13605
```

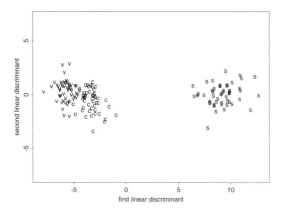

Figure 12.1: The log `iris` data on the first two discriminant axes.

```
Proportion of trace:
   LD1    LD2
 0.9965 0.0035
> ir.ld <- predict(ir.lda, dimen = 2)$x
> eqscplot(ir.ld, type = "n", xlab = "first linear discriminant",
          ylab = "second linear discriminant")
> text(ir.ld, labels = as.character(ir.species[-143]),
          col = 3 + codes(ir.species), cex = 0.8)
```

This shows that 99.65% of the between-group variance is on the first discriminant axis. Using

```
plot(ir.lda, dimen = 1)
plot(ir.lda, type = "density", dimen = 1)
```

will examine the distributions of the groups on the first linear discriminant.

The approach we have illustrated is the conventional one, following Bryan (1951), but it is not the only one. The definition of B at (12.1) weights the groups by their size in the dataset. Rao (1948) used the unweighted covariance matrix of the group means, and our software uses a covariance matrix weighted by the prior probabilities of the classes if these are specified.

Discrimination for normal populations

An alternative approach to discrimination is *via* probability models. Let π_c denote the prior probabilities of the classes, and $p(x \mid c)$ the densities of distributions of the observations for each class. Then the posterior distribution of the classes after observing x is

$$p(c \mid x) = \frac{\pi_c p(x \mid c)}{p(x)} \propto \pi_c p(x \mid c) \qquad (12.2)$$

and it is fairly simple to show that the allocation rule which makes the smallest expected number of errors chooses the class with maximal $p(c \mid x)$; this is known as the *Bayes rule*. (We consider a more general version in Section 12.2.)

Now suppose the distribution for class c is multivariate normal with mean $\boldsymbol{\mu}_c$ and covariance Σ_c. Then the Bayes rule minimizes

$$
\begin{aligned}
Q_c &= -2\,\log p(\boldsymbol{x}\,|\,c) - 2\,\log \pi_c \\
&= (\boldsymbol{x} - \boldsymbol{\mu}_c)\Sigma_c^{-1}(\boldsymbol{x} - \boldsymbol{\mu}_c)^T + \log |\Sigma_c| - 2\,\log \pi_c
\end{aligned}
\tag{12.3}
$$

The first term of (12.3) is the squared *Mahalanobis distance* to the class centre, and can be calculated by the function `mahalanobis`. The difference between the Q_c for two classes is a quadratic function of $\boldsymbol{x}$, so the method is known as *quadratic discriminant analysis* and the boundaries of the decision regions are quadratic surfaces in $\boldsymbol{x}$ space. This is implemented by our function qda.

Further suppose that the classes have a common covariance matrix Σ. Differences in the Q_c are then *linear* functions of $\boldsymbol{x}$, and we can maximize $-Q_c/2$ or

$$
L_c = \boldsymbol{x}\Sigma^{-1}\boldsymbol{\mu}_c^T - \boldsymbol{\mu}_c\Sigma^{-1}\boldsymbol{\mu}_c^T/2 + \log \pi_c
\tag{12.4}
$$

To use (12.3) or (12.4) we have to estimate $\boldsymbol{\mu}_c$ and Σ_c or Σ. The obvious estimates are used, the sample mean and covariance matrix within each class, and W for Σ.

How does this relate to Fisher's linear discrimination? The latter gives new variables, the linear discriminants, with unit within-class sample variance, and the differences between the group means lie entirely in the first r variables. Thus on these variables the Mahalanobis distance (with respect to $\widehat{\Sigma} = W$) is just

$$
\|\boldsymbol{x} - \boldsymbol{\mu}_c\|^2
$$

and only the first r components of the vector depend on c. Similarly, on these variables

$$
L_c = \boldsymbol{x}\boldsymbol{\mu}_c^T - \|\boldsymbol{\mu}_c\|^2/2 + \log \pi_c
$$

and we can work in r dimensions. If there are just two classes, there is a single linear discriminant, and

$$
L_2 - L_1 = \boldsymbol{x}(\boldsymbol{\mu}_2 - \boldsymbol{\mu}_1)^T + \text{const}
$$

This is an affine function of the linear discriminant, which has coefficient $(\boldsymbol{\mu}_2 - \boldsymbol{\mu}_1)^T$ rescaled to unit length.

Note that linear discriminant analysis uses a $p(c\,|\,\boldsymbol{x})$ that is a logistic regression for $g = 2$ and a multinomial log-linear model for $g > 2$. However, it differs from the methods of Chapter 7 in the methods of parameter estimation used. Linear discriminant analysis will be better if the populations really are multivariate normal with equal within-group covariance matrices, but that superiority is fragile, so the methods of Chapter 7 are usually preferred *for classification*.

crabs dataset

Can we construct a rule to predict the sex of a future *Leptograpsus* crab of unknown colour form (species)? We noted that BD is measured differently for males

and females, so it seemed prudent to omit it from the analysis. To start with, we ignore the differences between the forms. Linear discriminant analysis, for what are
highly non-normal populations, finds a variable that is essentially $CL^3 RW^{-2} CW^{-1}$,
a dimensionally neutral quantity. Six errors are made, all for the blue form:

```
> (dcrabs.lda <- lda(crabs$sex ~ FL + RW + CL + CW, lcrabs))
Coefficients of linear discriminants:
        LD1
FL  -2.8896
RW -25.5176
CL  36.3169
CW -11.8280
> table(crabs$sex, predict(dcrabs.lda)$class)
   F  M
F 97  3
M  3 97
```

It does make sense to take the colour forms into account, especially as the
within-group distributions look close to joint normality (look at the Figures 4.13
(page 96) and 11.2 (page 306)). The first two linear discriminants dominate the
between-group variation; Figure 12.2 shows the data on those variables.

```
> (dcrabs.lda4 <- lda(crabs.grp ~ FL + RW + CL + CW, lcrabs))
Proportion of trace:
   LD1    LD2    LD3
0.6422 0.3491 0.0087
> dcrabs.pr4 <- predict(dcrabs.lda4, dimen = 2)
> dcrabs.pr2 <- dcrabs.pr4$post[, c("B", "O")] %*% c(1, 1)
> table(crabs$sex, dcrabs.pr2 > 0.5)
  FALSE TRUE
F    96    4
M     3   97
```

We cannot represent all the decision surfaces exactly on a plot. However,
using the first two linear discriminants as the data will provide a very good approximation; see Figure 12.2.

```
cr.t <- dcrabs.pr4$x[, 1:2]
eqscplot(cr.t, type = "n", xlab = "First LD", ylab = "Second LD")
text(cr.t, labels = as.character(crabs.grp))
perp <- function(x, y) {
    m <- (x+y)/2
    s <- - (x[1] - y[1])/(x[2] - y[2])
    abline(c(m[2] - s*m[1], s))
    invisible()
}
cr.m <- lda(cr.t, crabs$sex)@means   # in R use $means
points(cr.m, pch = 3, mkh = 0.3)
perp(cr.m[1, ], cr.m[2, ])

cr.lda <- lda(cr.t, crabs.grp)
```

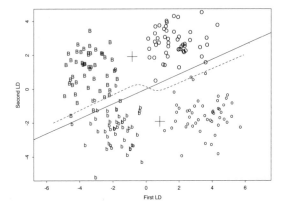

Figure 12.2: Linear discriminants for the `crabs` data. Males are coded as capitals, females as lower case, colours as the initial letter of blue or orange. The crosses are the group means for a linear discriminant for sex (solid line) and the dashed line is the decision boundary for sex based on four groups.

```
x <- seq(-6, 6, 0.25)
y <- seq(-2, 2, 0.25)
Xcon <- matrix(c(rep(x,length(y)), rep(y, each = length(x))),,2)
cr.pr <- predict(cr.lda, Xcon)$post[, c("B", "O")] %*% c(1,1)
contour(x, y, matrix(cr.pr, length(x), length(y)),
        levels = 0.5, labex = 0, add = T, lty= 3)
```

The reader is invited to try quadratic discrimination on this problem. It performs very marginally better than linear discrimination, not surprisingly since the covariances of the groups appear so similar, as can be seen from the result of

```
for(i in c("O", "o",  "B", "b"))
    print(var(lcrabs[crabs.grp == i, ]))
```

Robust estimation of multivariate location and scale

We may wish to consider more robust estimates of W (but not B). Somewhat counter-intuitively, it does not suffice to apply a robust location estimator to each component of a multivariate mean (Rousseeuw and Leroy, 1987, p. 250), and it is easier to consider the estimation of mean and variance simultaneously.

Multivariate variances are very sensitive to outliers. Two methods for robust covariance estimation are available via our function `cov.rob`[1] and the S-PLUS functions `cov.mve` and `cov.mcd` (Rousseeuw, 1984; Rousseeuw and Leroy, 1987) and `covRob` in library section `robust`. Suppose there are n observations of p variables. The *minimum volume ellipsoid* method seeks an ellipsoid containing $h = \lfloor (n + p + 1)/2 \rfloor$ points that is of minimum volume, and the *minimum covariance determinant* method seeks h points whose covariance has minimum determinant (so the conventional confidence ellipsoid for the mean of those points

---

[1] Adopted by R in package `lqs`.

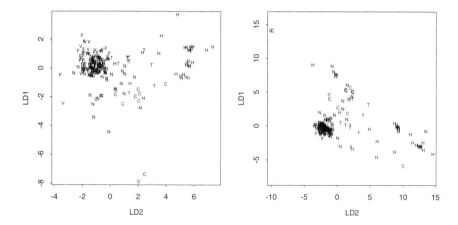

Figure 12.3: The `fgl` data on the first two discriminant axes. The right-hand plot used robust estimation of the common covariance matrix.

has minimum volume). MCD is to be preferred for its higher statistical efficiency. Our function `cov.rob` implements both.

The search for an MVE or MCD provides h points whose mean and variance matrix (adjusted for selection) give an initial estimate. This is refined by selecting those points whose Mahalanobis distance from the initial mean using the initial covariance is not too large (specifically within the 97.5% point under normality), and returning their mean and variance matrix).

An alternative approach is to extend the idea of M-estimation to this setting, fitting a multivariate t_ν distribution for a small number ν of degrees of freedom. This is implemented in our function `cov.trob`; the theory behind the algorithm used is given in Kent, Tyler and Vardi (1994) and Ripley (1996). Normally `cov.trob` is faster than `cov.rob`, but it lacks the latter's extreme resistance. We can use linear discriminant analysis on more than two classes, and illustrate this with the forensic glass dataset `fgl`.

Our function `lda` has an argument `method = "mve"` to use the minimum volume ellipsoid estimate (but without robust estimation of the group centres) or the multivariate t_ν distribution by setting `method = "t"`. This makes a considerable difference for the `fgl` forensic glass data, as Figure 12.3 shows. We use the default $\nu = 5$.

```
fgl.ld <- predict(lda(type ~ ., fgl), dimen = 2)$x
eqscplot(fgl.ld, type = "n", xlab = "LD1", ylab = "LD2")
# either
for(i in seq(along = levels(fgl$type))) {
  set <- fgl$type[-40] == levels(fgl$type)[i]
  points(fgl.ld[set,], pch = 18, cex = 0.6, col = 2 + i)}
key(text = list(levels(fgl$type), col = 3:8))
# or
text(fgl.ld, cex = 0.6,
     labels = c("F", "N", "V", "C", "T", "H")[fgl$type[-40]])
```

```
fgl.rld <- predict(lda(type ~ ., fgl, method = "t"), dimen = 2)$x
eqscplot(fgl.rld, type = "n", xlab = "LD1", ylab = "LD2")
# either
for(i in seq(along = levels(fgl$type))) {
  set <- fgl$type[-40] == levels(fgl$type)[i]
  points(fgl.rld[set,], pch = 18, cex = 0.6, col = 2 + i)}
key(text = list(levels(fgl$type), col = 3:8))
# or
text(fgl.rld, cex = 0.6,
     labels = c("F", "N", "V", "C", "T", "H")[fgl$type[-40]])
```

Try `method = "mve"`, which gives an almost linear plot.

12.2 Classification Theory

In the terminology of pattern recognition the given examples together with their classifications are known as the *training set*, and future cases form the *test set*. Our primary measure of success is the error (or misclassification) rate. Note that we would obtain (possibly seriously) biased estimates by re-classifying the training set, but that the error rate on a test set randomly chosen from the whole population will be an unbiased estimator.

It may be helpful to know the type of errors made. A *confusion matrix* gives the number of cases with true class i classified as of class j. In some problems some errors are considered to be worse than others, so we assign costs L_{ij} to allocating a case of class i to class j. Then we will be interested in the average error cost rather than the error rate.

It is fairly easy to show (Ripley, 1996, p. 19) that the average error cost is minimized by the *Bayes rule*, which is to allocate to the class c minimizing $\sum_i L_{ic} p(i \mid x)$ where $p(i \mid x)$ is the posterior distribution of the classes after observing x. If the costs of all errors are the same, this rule amounts to choosing the class c with the largest posterior probability $p(c \mid x)$. The minimum average cost is known as the *Bayes risk*. We can often estimate a lower bound for it by the method of Ripley (1996, pp. 196–7) (see the example on page 347).

We saw in Section 12.1 how $p(c \mid x)$ can be computed for normal populations, and how estimating the Bayes rule with equal error costs leads to linear and quadratic discriminant analysis. As our functions `predict.lda` and `predict.qda` return posterior probabilities, they can also be used for classification with error costs.

The posterior probabilities $p(c \mid x)$ may also be estimated directly. For just two classes we can model $p(1 \mid x)$ using a logistic regression, fitted by `glm`. For more than two classes we need a multiple logistic model; it may be possible to fit this using a surrogate log-linear Poisson GLM model (Section 7.3), but using the `multinom` function in library section `nnet` will usually be faster and easier.

Classification trees model the $p(c \mid x)$ directly by a special multiple logistic model, one in which the right-hand side is a single factor specifying which leaf the case will be assigned to by the tree. Again, since the posterior probabilities are given by the `predict` method it is easy to estimate the Bayes rule for unequal error costs.

Predictive and 'plug-in' rules

In the last few paragraphs we skated over an important point. To find the Bayes rule we need to know the posterior probabilities $p(c \mid x)$. Since these are unknown we use an explicit or implicit parametric family $p(c \mid x; \theta)$. In the methods considered so far we act as if $p(c \mid x; \hat{\theta})$ were the actual posterior probabilities, where $\hat{\theta}$ is an estimate computed from the training set $\mathcal{T}$, often by maximizing some appropriate likelihood. This is known as the 'plug-in' rule. However, the 'correct' estimate of $p(c \mid x)$ is (Ripley, 1996, §2.4) to use the *predictive* estimates

$$\tilde{p}(c \mid x) = P(c = c \mid X = x, \mathcal{T}) = \int p(c \mid x; \theta) p(\theta \mid \mathcal{T}) \, d\theta \qquad (12.5)$$

If we are very sure of our estimate $\hat{\theta}$ there will be little difference between $p(c \mid x; \hat{\theta})$ and $\tilde{p}(c \mid x)$; otherwise the predictive estimate will normally be less extreme (not as near 0 or 1). The 'plug-in' estimate ignores the uncertainty in the parameter estimate $\hat{\theta}$ which the predictive estimate takes into account.

It is not often possible to perform the integration in (12.5) analytically, but it *is* possible for linear and quadratic discrimination with appropriate 'vague' priors on θ (Aitchison and Dunsmore, 1975; Geisser, 1993; Ripley, 1996). This estimate is implemented by `method = "predictive"` of the `predict` methods for our functions `lda` and `qda`. Often the differences are small, especially for linear discrimination, *provided* there are enough data for a good estimate of the variance matrices. When there are not, Moran and Murphy (1979) argue that considerable improvement can be obtained by using an unbiased estimator of $\log p(x \mid c)$, implemented by the argument `method = "debiased"`.

A simple example: Cushing's syndrome

We illustrate these methods by a small example taken from Aitchison and Dunsmore (1975, Tables 11.1–3) and used for the same purpose by Ripley (1996). The data are on diagnostic tests on patients with Cushing's syndrome, a hypersensitive disorder associated with over-secretion of cortisol by the adrenal gland. This dataset has three recognized types of the syndrome represented as a, b, c. (These encode 'adenoma', 'bilateral hyperplasia' and 'carcinoma', and represent the underlying cause of over-secretion. This can only be determined histopathologically.) The observations are urinary excretion rates (mg/24 h) of the steroid metabolites tetrahydrocortisone and pregnanetriol, and are considered on log scale.

There are six patients of unknown type (marked u), one of whom was later found to be of a fourth type, and another was measured faultily.

Figure 12.4 shows the classifications produced by `lda` and the various options of quadratic discriminant analysis. This was produced by

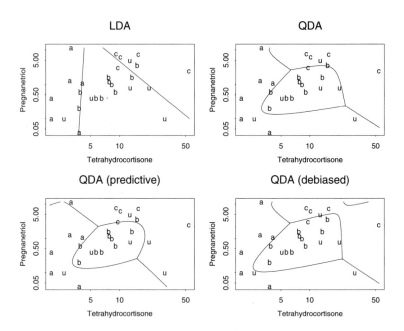

Figure 12.4: Linear and quadratic discriminant analysis applied to the Cushing's syndrome data.

```
cush <- log(as.matrix(Cushings[, -3]))
tp <- Cushings$Type[1:21, drop = T]
cush.lda <- lda(cush[1:21,], tp); predplot(cush.lda, "LDA")
cush.qda <- qda(cush[1:21,], tp); predplot(cush.qda, "QDA")
predplot(cush.qda, "QDA (predictive)", method = "predictive")
predplot(cush.qda, "QDA (debiased)", method = "debiased")
```

(Function `predplot` is given in the scripts.)

We can contrast these with logistic discrimination performed by

```
library(nnet)
Cf <- data.frame(tp = tp,
    Tetrahydrocortisone = log(Cushings[1:21, 1]),
    Pregnanetriol = log(Cushings[1:21, 2]) )
cush.multinom <- multinom(tp ~ Tetrahydrocortisone
    + Pregnanetriol, Cf, maxit = 250)
xp <- seq(0.6, 4.0, length = 100); np <- length(xp)
yp <- seq(-3.25, 2.45, length = 100)
cushT <- expand.grid(Tetrahydrocortisone = xp,
                     Pregnanetriol = yp)
Z <- predict(cush.multinom, cushT, type = "probs")
cushplot(xp, yp, Z)
```

(Function `cushplot` is given in the scripts.) When, as here, the classes have quite different variance matrices, linear and logistic discrimination can give quite different answers (compare Figures 12.4 and 12.5).

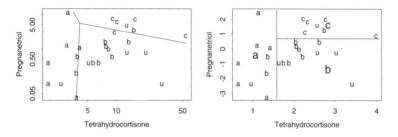

Figure 12.5: Logistic regression and classification trees applied to the Cushing's syndrome data.

For classification trees we can use

```
# R: library(tree)
cush.tr <- tree(tp ~ Tetrahydrocortisone + Pregnanetriol, Cf)
plot(cush[, 1], cush[, 2], type = "n",
    xlab = "Tetrahydrocortisone", ylab = "Pregnanetriol")
for(il in 1:4) {
  set <- Cushings$Type==levels(Cushings$Type)[il]
  text(cush[set, 1], cush[set, 2],
      labels = as.character(Cushings$Type[set]), col = 2 + il) }
par(cex = 1.5); partition.tree(cush.tr, add = T); par(cex = 1)
```

With such a small dataset we make no attempt to refine the size of the tree, shown in Figure 12.5.

Mixture discriminant analysis

Another application of the (plug-in) theory is *mixture discriminant analysis* (Hastie and Tibshirani, 1996) which has an implementation in the library section `mda`. This fits multivariate normal mixture distributions to each class and then applies (12.2).

12.3   Non-Parametric Rules

There are a number of non-parametric classifiers based on non-parametric estimates of the class densities or of the log posterior. Library section `class` implements the k-nearest neighbour classifier and related methods (Devijver and Kittler, 1982; Ripley, 1996) and learning vector quantization (Kohonen, 1990, 1995; Ripley, 1996). These are all based on finding the k nearest examples in some reference set, and taking a majority vote among the classes of these k examples, or, equivalently, estimating the posterior probabilities $p(c \mid x)$ by the proportions of the classes among the k examples.

The methods differ in their choice of reference set. The k-nearest neighbour methods use the whole training set or an edited subset. Learning vector quantization is similar to K-means in selecting points in the space other than the training

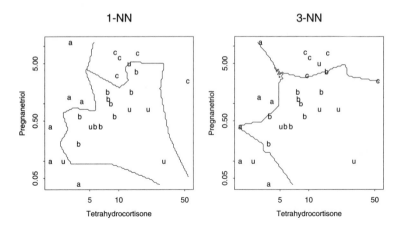

Figure 12.6: k-nearest neighbours applied to the Cushing's syndrome data.

set examples to summarize the training set, but unlike K-means it takes the classes of the examples into account.

These methods almost always measure 'nearest' by Euclidean distance. For the Cushing's syndrome data we use Euclidean distance on the logged covariates, rather arbitrarily scaling them equally.

```
library(class)
Z <- knn(scale(cush[1:21, ], F, c(3.4, 5.7)),
        scale(cushT, F, c(3.4, 5.7)), tp)
cushplot(xp, yp, class.ind(Z))
Z <- knn(scale(cush, F, c(3.4, 5.7)),
        scale(cushT, F, c(3.4, 5.7)), tp, k = 3)
cushplot(xp, yp, class.ind(Z))
```

This dataset is too small to try the editing and LVQ methods in library section class.

12.4 Neural Networks

Neural networks provide a flexible non-linear extension of multiple logistic regression, as we saw in Section 8.10. We can consider them for the Cushing's syndrome example by the following code.[2]

```
library(nnet)
cush <- cush[1:21,]; tpi <- class.ind(tp)
# functions pltnn and plt.bndry given in the scripts
par(mfrow = c(2, 2))
pltnn("Size = 2")
set.seed(1); plt.bndry(size = 2, col = 2)
```

---

[2]The colours are set for a Trellis device, and the random seeds were chosen for a specific S environment.

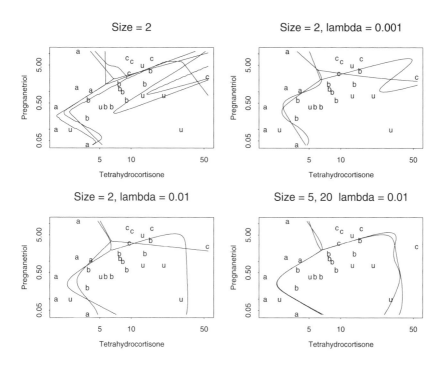

Figure 12.7: Neural networks applied to the Cushing's syndrome data. Each panel shows the fits from two or three local maxima of the (penalized) log-likelihood.

```
set.seed(3); plt.bndry(size = 2, col = 3)
plt.bndry(size = 2, col = 4)

pltnn("Size = 2, lambda = 0.001")
set.seed(1); plt.bndry(size = 2, decay = 0.001, col = 2)
set.seed(2); plt.bndry(size = 2, decay = 0.001, col = 4)

pltnn("Size = 2, lambda = 0.01")
set.seed(1); plt.bndry(size = 2, decay = 0.01, col = 2)
set.seed(2); plt.bndry(size = 2, decay = 0.01, col = 4)

pltnn("Size = 5, 20  lambda = 0.01")
set.seed(2); plt.bndry(size = 5, decay = 0.01, col = 1)
set.seed(2); plt.bndry(size = 20, decay = 0.01, col = 2)
```

The results are shown in Figure 12.7. We see that in all cases there are multiple local maxima of the likelihood, since different runs gave different classifiers.

Once we have a penalty, the choice of the number of hidden units is often not critical (see Figure 12.7). The spirit of the predictive approach is to average the predicted $p(c \mid x)$ over the local maxima. A simple average will often suffice:

```
# functions pltnn and b1 are in the scripts
pltnn("Many local maxima")
```

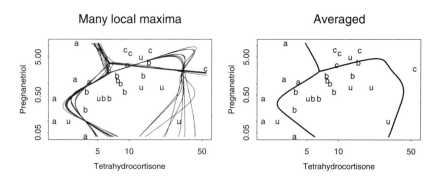

Figure 12.8: Neural networks with three hidden units and $\lambda = 0.01$ applied to the Cushing's syndrome data.

```
Z <- matrix(0, nrow(cushT), ncol(tpi))
for(iter in 1:20) {
    set.seed(iter)
    cush.nn <- nnet(cush, tpi, skip = T, softmax = T, size = 3,
        decay = 0.01, maxit = 1000, trace = F)
    Z <- Z + predict(cush.nn, cushT)
# In R replace @ by $ in next line.
    cat("final value", format(round(cush.nn@value,3)), "\n")
    b1(predict(cush.nn, cushT), col = 2, lwd = 0.5)
}
pltnn("Averaged")
b1(Z, lwd = 3)
```

Note that there are two quite different types of local maxima occurring here, and some local maxima occur several times (up to convergence tolerances). An average does better than either type of classifier.

12.5 Support Vector Machines

Support vector machines (SVMs) are the latest set of methods within this field. They have been promoted enthusiastically, but with little respect to the selection effects of choosing the test problem and the member of the large class of classifiers to present. The original ideas are in Boser *et al.* (1992); Cortes and Vapnik (1995); Vapnik (1995, 1998); the books by Cristianini and Shawe-Taylor (2000) and Hastie *et al.* (2001, §4.5, 12.2, 12.3) present the underlying theory.

The method for $g = 2$ classes is fairly simple to describe. Logistic regression will fit exactly in separable cases where there is a hyperplane that has all class-one points on one side and all class-two points on the other. It would be a coincidence for there to be only one such hyperplane, and fitting a logistic regression will tend to fit a decision surface $p(2 \mid x) = 0.5$ in the middle of the 'gap' between the groups. Support vector methods attempt directly to find a hyperplane in the middle of the gap, that is with maximal margin (the distance from the hyperplane to the nearest point). This is quadratic programming problem that can be solved by standard methods.[3] Such a hyperplane has *support vectors*, data points that are exactly the margin distance away from the hyperplane. It will typically be a very good classifier.

The problem is that usually no separating hyperplane will exist. This difficulty is tackled in two ways. First, we can allow some points to be on the wrong side of their margin (and for some on the wrong side of the hyperplane) subject to a constraint on the total of the 'mis-fit' distances being less than some constant, with Lagrange multiplier $C > 0$. This is still a quadratic programming problem, because of the rather arbitrary use of sum of distances.

Second, the set of variables is expanded greatly by taking non-linear functions of the original set of variables. Thus rather than seeking a classifying hyperplane $f(x) = x^T \beta + \beta_0 = 0$, we seek $f(x) = h(x)^T \beta + \beta_0 = 0$ for a vector of $M \gg p$ functions h_i. Then finding a optimal separating hyperplane is equivalent to solving

$$\min_{\beta_0, \beta} \sum_{i=1}^{n} [1 - y_i f(x_i)]_+ + \frac{1}{2C} \|\beta\|^2$$

where $y_i = \pm 1$ for the two classes. This is yet another penalized fitting problem, not dissimilar (Hastie *et al.*, 2001, p. 380) to a logistic regression with weight decay (which can be fitted by `multinom`). The claimed advantage of SVMs is that because we only have to find the support vectors, the family of functions h can be large, even infinite-dimensional.

There is an implementation of SVMs for R in function `svm` in package `e1071`.[4] The default values do not do well, but after some tuning for the `crabs` data we can get a good discriminant with 21 support vectors. Here `cost` is C and `gamma` is a coefficient of the kernel used to form h.

```
> # R: library(e1071)
> # S: library(libsvm)
> crabs.svm <- svm(crabs$sp ~ ., data = lcrabs, cost = 100,
                   gamma = 1)
> table(true = crabs$sp, predicted = predict(crabs.svm))
     predicted
true   B   O
B    100   0
O      0 100
```

We can try a 10-fold cross-validation by

```
> svm(crabs$sp ~ ., data = lcrabs, cost = 100, gamma = 1,
      cross = 10)
    ....
Total Accuracy: 100
Single Accuracies:
  100 100 100 100 100 100 100 100 100 100
```

---

[3]See Section 16.2 for S software for this problem; however, special-purpose software is often used.
[4]Code by David Meyer based on C++ code by Chih-Chung Chang and Chih-Jen Lin. A port to S-PLUS is available for machines with a C++ compiler.

The extension to $g > 2$ classes is much less elegant, and several ideas have been used. The `svm` function uses one attributed to Knerr *et al.* (1990) in which classifiers are built comparing each pair of classes, and the majority vote amongst the resulting $g(g-1)/2$ classifiers determines the predicted class.

12.6   Forensic Glass Example

The forensic glass dataset `fgl` has 214 points from six classes with nine measurements, and provides a fairly stiff test of classification methods. As we have seen (Figures 4.17 on page 99, 5.4 on page 116, 11.5 on page 309 and 12.3 on page 337) the types of glass do not form compact well-separated groupings, and the marginal distributions are far from normal. There are some small classes (with 9, 13 and 17 examples), so we cannot use quadratic discriminant analysis.

We assess their performance by 10-fold cross-validation, using the same random partition for all the methods. Logistic regression provides a suitable benchmark (as is often the case), and in this example linear discriminant analysis does equally well.

```
set.seed(123); rand <- sample (10, 214, replace = T)
con <- function(...)
{
    print(tab <- table(...)); diag(tab) <- 0
    cat("error rate = ",
        round(100*sum(tab)/length(list(...)[[1]]), 2), "%\n")
    invisible()
}
CVtest <- function(fitfn, predfn, ...)
{
  res <- fgl$type
  for (i in sort(unique(rand))) {
    cat("fold ", i, "\n", sep = "")
    learn <- fitfn(rand != i, ...)
    res[rand == i] <- predfn(learn, rand == i)
  }
  res
}
res.multinom <- CVtest(
    function(x, ...) multinom(type ~ ., fgl[x, ], ...),
    function(obj, x) predict(obj, fgl[x, ], type = "class"),
    maxit = 1000, trace = F)

> con(true = fgl$type, predicted = res.multinom)
    ....
error rate =   37.38 %

> res.lda <- CVtest(
    function(x, ...) lda(type ~ ., fgl[x, ], ...),
    function(obj, x) predict(obj, fgl[x, ])$class )
```

```
> con(true = fgl$type, predicted = res.lda)
    ....
error rate =   37.38 %

> fgl0 <- fgl[ , -10] # drop type
{ res <- fgl$type
  for (i in sort(unique(rand))) {
      cat("fold ", i ,"\n", sep = "")
      sub <- rand == i
      res[sub] <- knn(fgl0[!sub, ], fgl0[sub, ], fgl$type[!sub],
                        k = 1)
  }
  res } -> res.knn1
> con(true = fgl$type, predicted = res.knn1)
      WinF WinNF Veh Con Tabl Head
    ....
error rate =   23.83 %
```

We can use nearest-neighbour methods to estimate the lower bound on the Bayes risk as about 10% (Ripley, 1996, pp. 196–7).

```
> res.lb <- knn(fgl0, fgl0, fgl$type, k = 3, prob = T, use.all = F)
> table(attr(res.lb, "prob"))
  0.333333 0.666667    1
        10       64 140
1/3 * (64/214) = 0.099688
```

We saw in Chapter 9 that we could fit a classification tree of size about six to this dataset. We need to cross-validate over the choice of tree size, which does vary by group from four to seven.

```
library(rpart)
res.rpart <- CVtest(
  function(x, ...) {
    tr <- rpart(type ~ ., fgl[x,], ...)
    cp <- tr$cptable
    r <- cp[, 4] + cp[, 5]
    rmin <- min(seq(along = r)[cp[, 4] < min(r)])
    cp0 <- cp[rmin, 1]
    cat("size chosen was", cp[rmin, 2] + 1, "\n")
    prune(tr, cp = 1.01*cp0)
  },
  function(obj, x)
    predict(obj, fgl[x, ], type = "class"),
  cp = 0.001
)
con(true = fgl$type, predicted = res.rpart)
    ....
error rate =   34.58 %
```

Neural networks

We wrote some general functions for testing neural network models by V-fold cross-validation. First we rescale the dataset so the inputs have range $[0, 1]$.

```
fgl1 <- fgl
fgl1[1:9] <- lapply(fgl[, 1:9], function(x)
                {r <- range(x); (x - r[1])/diff(r)})
```

It is straightforward to fit a fully specified neural network. However, we want to average across several fits and to choose the number of hidden units and the amount of weight decay by an inner cross-validation. To do so we wrote a fairly general function that can easily be used or modified to suit other problems. (See the scripts for the code.)

```
> res.nn2 <- CVnn2(type ~ ., fgl1, skip = T, maxit = 500,
                  nreps = 10)
> con(true = fgl$type, predicted = res.nn2)
    ....
error rate =  28.5 %
```

This fits a neural network 1 000 times, and so is fairly slow (about half an hour on the PC).

This code chooses between neural nets on the basis of their cross-validated error rate. An alternative is to use logarithmic scoring, which is equivalent to finding the deviance on the validation set. Rather than count 0 if the predicted class is correct and 1 otherwise, we count $-\log p(c\,|\,x)$ for the true class c. We can easily code this variant by replacing the line

```
sum(as.numeric(truth) != max.col(res/nreps))
```

by

```
sum(-log(res[cbind(seq(along = truth), as.numeric(truth))]/nreps))
```

in CVnn2.

Support vector machines

```
res.svm <- CVtest(
   function(x, ...) svm(type ~ ., fgl[x, ], ...),
   function(obj, x) predict(obj, fgl[x, ]),
   cost = 100, gamma = 1 )

con(true = fgl$type, predicted = res.svm)
    ....
error rate =  28.04 %
```

The following is faster, but not strictly comparable with the results above, as a different random partition will be used.

```
> svm(type ~ ., data = fgl, cost = 100, gamma = 1, cross = 10)
    ....
Total Accuracy: 71.03
Single Accuracies:
 66.67 61.90 68.18 76.19 77.27 85.71 76.19 72.73 57.14 68.18
```

Learning vector quantization

For LVQ as for k-nearest neighbour methods we have to select a suitable metric. The following experiments used Euclidean distance on the original variables, but the rescaled variables or Mahalanobis distance could also be tried.

```
cd0 <- lvqinit(fgl0, fgl$type, prior = rep(1, 6)/6, k = 3)
cd1 <- olvq1(fgl0, fgl$type, cd0)
con(true = fgl$type, predicted = lvqtest(cd1, fgl0))
```

We set an even prior over the classes as otherwise there are too few representatives of the smaller classes. Our initialization code in `lvqinit` follows Kohonen's in selecting the number of representatives; in this problem 24 points are selected, four from each class.

```
CV.lvq <- function()
{
  res <- fgl$type
  for(i in sort(unique(rand))) {
    cat("doing fold", i, "\n")
    cd0 <- lvqinit(fgl0[rand != i,], fgl$type[rand != i],
                   prior = rep(1, 6)/6, k = 3)
    cd1 <- olvq1(fgl0[rand != i,], fgl$type[rand != i], cd0)
    cd1 <- lvq3(fgl0[rand != i,], fgl$type[rand != i],
                cd1, niter = 10000)
    res[rand == i] <- lvqtest(cd1, fgl0[rand == i, ])
  }
  res
}
con(true = fgl$type, predicted = CV.lvq())
    ....
error rate =   28.5 %
```

The initialization is random, so your results are likely to differ.

12.7   Calibration Plots

One measure that a suitable model for $p(c \mid x)$ has been found is that the predicted probabilities are *well calibrated*; that is, that a fraction of about p of the events we predict with probability p actually occur. Methods for testing calibration of probability forecasts have been developed in connection with weather forecasts (Dawid, 1982, 1986).

For the forensic glass example we are making six probability forecasts for each case, one for each class. To ensure that they are genuine forecasts, we should use the cross-validation procedure. A minor change to the code gives the probability predictions:

```
CVprobs <- function(fitfn, predfn, ...)
{
  res <- matrix(, 214, 6)
```

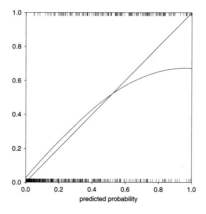

Figure 12.9: Calibration plot for multiple logistic fit to the `fgl` data.

```
for (i in sort(unique(rand))) {
    cat("fold ", i, "\n", sep = "")
    learn <- fitfn(rand != i, ...)
    res[rand == i, ] <- predfn(learn, rand == i)
}
  res
}
probs.multinom <- CVprobs(
    function(x, ...) multinom(type ~ ., fgl[x, ], ...),
    function(obj, x) predict(obj, fgl[x, ], type = "probs"),
    maxit = 1000, trace = F )
```

We can plot these and smooth them by

```
probs.yes <- as.vector(class.ind(fgl$type))
probs <- as.vector(probs.multinom)
par(pty = "s")
plot(c(0, 1), c(0, 1), type = "n", xlab = "predicted probability",
     ylab = "", xaxs = "i", yaxs = "i", las = 1)
rug(probs[probs.yes= = 0], 0.02, side = 1, lwd = 0.5)
rug(probs[probs.yes == 1], 0.02, side = 3, lwd = 0.5)
abline(0, 1)
newp <- seq(0, 1, length = 100)
lines(newp, predict(loess(probs.yes ~ probs, span = 1), newp))
```

A smoothing method with an adaptive bandwidth such as `loess` is needed here, as the distribution of points along the x-axis can be very much more uneven than in this example. The result is shown in Figure 12.9. This plot does show substantial over-confidence in the predictions, especially at probabilities close to one. Indeed, only 22/64 of the events predicted with probability greater than 0.9 occurred. (The underlying cause is the multimodal nature of some of the underlying class distributions.)

Where calibration plots are not straight, the best solution is to find a better model. Sometimes the over-confidence is minor, and mainly attributable to the use of plug-in rather than predictive estimates. Then the plot can be used to adjust the probabilities (which may need further adjustment to sum to one for more than two classes).

Chapter 13

Survival Analysis

Extensive survival analysis facilities written by Terry Therneau (Mayo Foundation) are available in S-PLUS and in the R package `survival`.

Survival analysis is concerned with the distribution of lifetimes, often of humans but also of components and machines. There are two distinct levels of mathematical treatment in the literature. Collett (1994), Cox and Oakes (1984), Hosmer and Lemeshow (1999), Kalbfleisch and Prentice (1980) and Klein and Moeschberger (1997) take a traditional and mathematically non-rigorous approach. The modern mathematical approach based on continuous-parameter martingales is given by Fleming and Harrington (1991) and Andersen *et al.* (1993). (The latter is needed here only to justify some of the distribution theory and for the concept of martingale residuals.) Other aspects closely related to Therneau's software are described in Therneau and Grambsch (2000).

Let T denote a lifetime random variable. It will take values in $(0, \infty)$, and its continuous distribution may be specified by a cumulative distribution function F with a density f. (Mixed distributions can be considered, but many of the formulae used by the software need modification.) For lifetimes it is more usual to work with the *survivor function* $S(t) = 1 - F(t) = P(T > t)$, the *hazard function* $h(t) = \lim_{\Delta t \to 0} P(t \leqslant T < t + \Delta t \mid T \geqslant t)/\Delta t$ and the *cumulative hazard function* $H(t) = \int_0^t h(s)\, \mathrm{d}s$. These are all related; we have

$$h(t) = \frac{f(t)}{S(t)}, \qquad H(t) = -\log S(t)$$

Common parametric distributions for lifetimes are (Kalbfleisch and Prentice, 1980) the exponential, with $S(t) = \exp -\lambda t$ and hazard λ, the Weibull with

$$S(t) = \exp -(\lambda t)^\alpha, \qquad h(t) = \lambda \alpha (\lambda t)^{\alpha - 1}$$

the log-normal, the gamma and the log-logistic which has

$$S(t) = \frac{1}{1 + (\lambda t)^\tau}, \qquad h(t) = \frac{\lambda \tau (\lambda t)^{\tau - 1}}{1 + (\lambda t)^\tau}$$

The major distinguishing feature of survival analysis is *censoring*. An individual case may not be observed on the whole of its lifetime, so that, for example, we may only know that it survived to the end of the trial. More general patterns of censoring are possible, but all lead to data for each case of the form either of a precise lifetime or the information that the lifetime fell in some interval (possibly extending to infinity).

Clearly we must place some restrictions on the censoring mechanism, for if cases were removed from the trial just before death we would be misled. Consider right censoring, in which the case leaves the trial at time C_i, and we know either T_i if $T_i \leqslant C_i$ or that $T_i > C_i$. *Random censoring* assumes that T_i and C_i are independent random variables, and therefore in a strong sense that censoring is uninformative. This includes the special case of *type I* censoring, in which the censoring time is fixed in advance, as well as trials in which the patients enter at random times but the trial is reviewed at a fixed time. It excludes *type II* censoring in which the trial is concluded after a fixed number of failures. Most analyses (including all those based solely on likelihoods) are valid under a weaker assumption that Kalbfleisch and Prentice (1980, §5.2) call *independent* censoring in which the hazard at time t conditional on the whole history of the process only depends on the survival of that individual to time t. (Independent censoring does cover type II censoring.) Conventionally the time recorded is $\min(T_i, C_i)$ together with the indicator variable for observed death $\delta_i = I(T_i \leqslant C_i)$. Then under independent right censoring the likelihood for parameters in the lifetime distribution is

$$L = \prod_{\delta_i=1} f(t_i) \prod_{\delta_i=0} S(t_i) = \prod_{\delta_i=1} h(t_i)S(t_i) \prod_{\delta_i=0} S(t_i) = \prod_{i=1}^{n} h(t_i)^{\delta_i} S(t_i)$$

(13.1)

Usually we are not primarily interested in the lifetime distribution *per se*, but how it varies between groups (usually called *strata* in the survival context) or on measurements on the cases, called *covariates*. In the more complicated problems the hazard will depend on covariates that vary with time, such as blood pressure or changes of treatments.

The function `Surv(times, status)` is used to describe the censored survival data to the S functions, and always appears on the left side of a model formula. In the simplest case of right censoring the variables are $\min(T_i, C_i)$ and δ_i (logical or 0/1 or 1/2). Further forms allow left and interval censoring. The results of printing the object returned by `Surv` are the vector of the information available, either the lifetime or an interval.

We consider three small running examples. Uncensored data on survival times for leukaemia (Feigl and Zelen, 1965; Cox and Oakes, 1984, p. 9) are in data frame `leuk`. This has two covariates, the white blood count `wbc`, and `ag` a test result that returns 'present' or 'absent'. Two-sample data (Gehan, 1965; Cox and Oakes, 1984, p. 7) on remission times for leukaemia are given in data frame `gehan`. This trial has 42 individuals in matched pairs, and no covariates (other than the treatment group).[1] Data frame `motors` contains the results of an accelerated life test experiment with 10 replicates at each of four temperatures reported

---

[1] Andersen *et al.* (1993, p. 22) indicate that this trial had a sequential stopping rule that invalidates most of the methods used here; it should be seen as illustrative only.

by Nelson and Hahn (1972) and Kalbfleisch and Prentice (1980, pp. 4–5). The
times are given in hours, but all but one is a multiple of 12, and only 14 values
occur

17 21 22 56 60 70 73.5 115.5 143.5 147.5833 157.5 202.5 216.5 227

in days, which suggests that observation was not continuous. Thus this is a good
example to test the handling of ties.

13.1   Estimators of Survivor Curves

The estimate of the survivor curve for uncensored data is easy; just take one minus
the empirical distribution function. For the leukaemia data we have

```
plot(survfit(Surv(time) ~ ag, data=leuk), lty = 2:3, col = 2:3)
legend(80, 0.8, c("ag absent", "ag present"), lty = 2:3, col = 2:3)
```

and confidence intervals are obtained easily from the binomial distribution of
$\widehat{S}(t)$. For example, the estimated variance is

$$\widehat{S}(t)[1 - \widehat{S}(t)]/n = r(t)[n - r(t)]/n^3 \qquad (13.2)$$

when $r(t)$ is the number of cases still alive (and hence 'at risk') at time t.

This computation introduces the function $\texttt{survfit}$ and its associated $\texttt{plot}$,
$\texttt{print}$ and $\texttt{summary}$ methods. It takes a model formula, and if there are factors
on the right-hand side, splits the data on those factors, and plots a survivor curve
for each factor combination, here just presence or absence of $\texttt{ag}$. (Although the
factors can be specified additively, the computation effectively uses their interaction.)

For censored data we have to allow for the decline in the number of cases at
risk over time. Let $r(t)$ be the number of cases at risk just before time t, that
is, those that are in the trial and not yet dead. If we consider a set of intervals
$I_i = [t_i, t_{i+1})$ covering $[0, \infty)$, we can estimate the probability p_i of surviving
interval I_i as $[r(t_i) - d_i]/r(t_i)$ where d_i is the number of deaths in interval I_i.
Then the probability of surviving until t_i is

$$P(T > t_i) = S(t_i) \approx \prod_0^{i-1} p_j \approx \prod_0^{i-1} \frac{r(t_i) - d_i}{r(t_i)}$$

Now let us refine the grid of intervals. Non-unity terms in the product will
only appear for intervals in which deaths occur, so the limit becomes

$$\widehat{S}(t) = \prod \frac{r(t_i) - d_i}{r(t_i)}$$

the product being over times at which deaths occur before t (but they could occur
simultaneously). This is the Kaplan–Meier estimator. Note that this becomes
constant after the largest observed t_i, and for this reason the estimate is only
plotted up to the largest t_i. However, the points at the right-hand end of the plot

will be very variable, and it may be better to stop plotting when there are still a few individuals at risk.

We can apply similar reasoning to the cumulative hazard

$$H(t_i) \approx \sum_{j \leqslant i} h(t_j)(t_{j+1} - t_j) \approx \sum_{j \leqslant i} \frac{d_j}{r(t_j)}$$

with limit

$$\widehat{H}(t) = \sum \frac{d_j}{r(t_j)} \tag{13.3}$$

again over times at which deaths occur before t. This is the Nelson estimator of the cumulative hazard, and leads to the Altshuler or Fleming–Harrington estimator of the survivor curve

$$\tilde{S}(t) = \exp -\widehat{H}(t) \tag{13.4}$$

The two estimators are related by the approximation $\exp -x \approx 1 - x$ for small x, so they will be nearly equal for large risk sets. The S functions follow Fleming and Harrington in breaking ties in (13.3), so if there were 3 deaths when the risk set contained 12 people, $3/12$ is replaced by $1/12 + 1/11 + 1/10$.

Similar arguments to those used to derive the two estimators lead to the standard error formula for the Kaplan–Meier estimator

$$\text{var}\left(\widehat{S}(t)\right) = \widehat{S}(t)^2 \sum \frac{d_j}{r(t_j)[r(t_j) - d_j]} \tag{13.5}$$

often called Greenwood's formula after its version for life tables, and

$$\text{var}\left(\widehat{H}(t)\right) = \sum \frac{d_j}{r(t_j)[r(t_j) - d_j]} \tag{13.6}$$

We leave it to the reader to check that Greenwood's formula reduces to (13.2) in the absence of ties and censoring. Note that if censoring can occur, both the Kaplan–Meier and Nelson estimators are biased; the bias results from the inability to give a sensible estimate when the risk set is empty.

Tsiatis (1981) suggested the denominator $r(t_j)^2$ rather than $r(t_j)[r(t_j) - d_j]$ on asymptotic grounds. Both Fleming and Harrington (1991) and Andersen *et al.* (1993) give a rigorous derivation of these formulae (and corrected versions for mixed distributions), as well as calculations of bias and limit theorems that justify asymptotic normality. Klein (1991) discussed the bias and small-sample behaviour of the variance estimators; his conclusions for $\widehat{H}(t)$ are that the bias is negligible and the Tsiatis form of the standard error is accurate (for practical use) provided the expected size of the risk set at t is at least five. For the Kaplan–Meier estimator Greenwood's formula is preferred, and is accurate enough (but biased downwards) again provided the expected size of the risk set is at least five.

We can use these formulae to indicate confidence intervals based on asymptotic normality, but we must decide on what scale to compute them. By default the function `survfit` computes confidence intervals on the log survivor (or cumulative hazard) scale, but linear and complementary log-log scales are also available (via the `conf.type` argument). These choices give

$$\widehat{S}(t) \exp\left[\pm k_\alpha \text{ s.e.}(\widehat{H}(t))\right]$$

$$\widehat{S}(t) \left[1 \pm k_\alpha \text{ s.e.}(\widehat{H}(t))\right]$$

$$\exp\left\{-\widehat{H}(t) \exp\left[\pm k_\alpha \frac{\text{s.e.}(\widehat{H}(t))}{\widehat{H}(t)}\right]\right\}$$

the last having the advantage of taking values in $(0, 1)$. Bie, Borgan and Liestøl (1987) and Borgan and Liestøl (1990) considered these and an arc-sine transformation; their results indicate that the complementary log-log interval is quite satisfactory for sample sizes as small as 25.

We do not distinguish clearly between log-survivor curves and cumulative hazards, which differ only by sign, yet the natural estimator of the first is the Kaplan–Meier estimator on log scale, and for the second it is the Nelson estimator. This is particularly true for confidence intervals, which we would expect to transform just by a change of sign. Fortunately, practical differences only emerge for very small risk sets, and are then swamped by the very large variability of the estimators.

The function `survfit` also handles censored data, and uses the Kaplan–Meier estimator by default. We try it on the `gehan` data:

```
> attach(gehan)
> Surv(time, cens)
 [1]  1  10  22   7   3  32+ 12  23   8  22  17   6   2  16
[15] 11  34+  8  32+ 12  25+  2  11+  5  20+  4  19+ 15   6
[29]  8  17+ 23  35+  5   6  11  13   4   9+  1   6+  8  10+
> plot(log(time) ~ pair)
> gehan.surv <- survfit(Surv(time, cens) ~ treat, data = gehan,
       conf.type = "log-log")
> summary(gehan.surv)
   ....
> plot(gehan.surv, conf.int = T, lty = 3:2, log = T,
       xlab = "time of remission (weeks)", ylab = "survival")
> lines(gehan.surv, lty = 3:2, lwd = 2, cex = 2)
> legend(25, 0.1 , c("control", "6-MP"), lty = 2:3, lwd = 2)
```

which calculates and plots (as shown in Figure 13.1) the product-limit estimators for the two groups, giving standard errors calculated using Greenwood's formula. (Confidence intervals are plotted automatically if there is only one group.) Other options are available, including `error = "tsiatis"` and `type = "fleming-harrington"` (which can be abbreviated to the first character). Note that the `plot` method has a `log` argument that plots $\widehat{S}(t)$ on log scale, effectively showing the negative cumulative hazard.

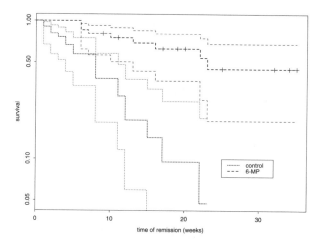

Figure 13.1: Survivor curves (on log scale) for the two groups of the gehan data. The crosses (on the 6-MP curve) represent censoring times. The thicker lines are the estimates, the thinner lines pointwise 95% confidence intervals.

Testing survivor curves

We can test for differences between the groups in the gehan example by

```
> survdiff(Surv(time, cens) ~ treat, data = gehan)
               N Observed Expected (O-E)^2/E (O-E)^2/V
treat=6-MP 21         9     19.3       5.46      16.8
treat=control 21     21     10.7       9.77      16.8

 Chisq= 16.8   on 1 degrees of freedom, p= 4.17e-05
```

This is one of a family of tests with parameter ρ defined by Fleming and Harrington (1981) and Harrington and Fleming (1982). The default $\rho = 0$ corresponds to the *log-rank test*. Suppose t_j are the observed death times. If we condition on the risk set and the number of deaths D_j at time t_j, the mean of the number of deaths D_{jk} in group k is clearly $E_{jk} = D_k r_k(t_j)/r(t_j)$ under the null hypothesis (where $r_k(t_j)$ is the number from group k at risk at time j). The statistic used is $(O_k - E_k) = \sum_j \widehat{S}(t_j-)^\rho [D_{jk} - E_{jk}]$,[2] and from this we compute a statistic $(O - E)^T V^{-1}(O - E)$ with an approximately chi-squared distribution. There are a number of different approximations to the variance matrix V, the one used being the weighted sum over death times of the variance matrices of $D_{jk} - E_{jk}$ computed from the hypergeometric distribution. The sum of $(O_k - E_k)^2/E_k$ provides a conservative approximation to the chi-squared statistic. The final column is $(O - E)^2$ divided by the diagonal of V; the final line gives the overall statistic computed from the full quadratic form.

---

[2] $\widehat{S}(t-)$ is the Kaplan–Meier estimate of survival just prior to t, ignoring the grouping.

The value rho = 1 corresponds approximately to the Peto–Peto modification (Peto and Peto, 1972) of the Wilcoxon test, and is more sensitive to early differences in the survivor curves.

A warning: tests of differences between groups are often used inappropriately. The gehan dataset has no other covariates, but where there are covariates the differences between the groups may reflect or be masked by differences in the covariates. Thus for the leuk dataset

```
> survdiff(Surv(time) ~ ag, data = leuk)
              N Observed Expected (O-E)^2/E (O-E)^2/V
ag=absent  16       16      9.3      4.83      8.45
ag=present 17       17     23.7      1.90      8.45

Chisq= 8.4  on 1 degrees of freedom, p= 0.00365
```

is inappropriate as there are differences in distribution of wbc between the two groups. A model is needed to adjust for the covariates (see page 368).

13.2 Parametric Models

Parametric models for survival data have fallen out of fashion with the advent of less parametric approaches such as the Cox proportional hazard models considered in the next section, but they remain a very useful tool, particularly in exploratory work (as usually they can be fitted very much faster than the Cox models).

The simplest parametric model is the exponential distribution with hazard $\lambda_i > 0$. The natural way to relate this to a covariate vector x for the case (including a constant if required) and to satisfy the positivity constraint is to take

$$\log \lambda_i = \beta^T x_i, \qquad \lambda_i = e^{\beta^T x_i}$$

For the Weibull distribution the hazard function is

$$h(t) = \lambda^\alpha \alpha t^{\alpha-1} = \alpha t^{\alpha-1} \exp(\alpha \beta^T x) \tag{13.7}$$

if we again make λ an exponential function of the covariates, and so we have the first appearance of the *proportional hazards* model

$$h(t) = h_0(t) \exp \beta^T x \tag{13.8}$$

which we consider again later. This identification suggests re-parametrizing the Weibull by replacing λ^α by λ, but as this just rescales the coefficients we can move easily from one parametrization to the other.

The Weibull is also a member of the class of *accelerated life* models, which have survival time T such that $T \exp \beta^T x$ has a fixed distribution; that is, time is speeded up by the factor $\exp \beta^T x$ for an individual with covariate x. This corresponds to replacing t in the survivor function and hazard by $t \exp \beta^T x$, and for models such as the exponential, Weibull and log-logistic with parametric

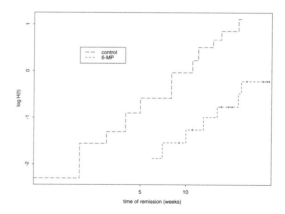

Figure 13.2: A log-log plot of cumulative hazard for the gehan dataset.

dependence on λt, this corresponds to taking $\lambda = \exp \boldsymbol{\beta}^T \boldsymbol{x}$. For all accelerated-life models we will have

$$\log T = \log T_0 - \boldsymbol{\beta}^T \boldsymbol{x} \tag{13.9}$$

for a random variable T_0 whose distribution does not depend on $\boldsymbol{x}$, so these are naturally considered as regression models.

For the Weibull the cumulative hazard is linear on a log-log plot, which provides a useful diagnostic aid. For example, for the gehan data

```
> plot(gehan.surv, lty = 3:4, col = 2:3, fun = "cloglog",
      xlab = "time of remission (weeks)", ylab = "log H(t)")
> legend(2, 0.5, c("control","6-MP"), lty = 4:3, col = 3:2)
```

we see excellent agreement with the proportional hazards hypothesis and with a Weibull baseline (Figure 13.2).

The function survReg[3] fits parametric survival models of the form

$$\ell(T) \sim \boldsymbol{\beta}^T \boldsymbol{x} + \sigma \, \epsilon \tag{13.10}$$

where $\ell(\ )$ is usually a log transformation. The dist argument specifies the distribution of ϵ and $\ell(\ )$, and σ is known as the *scale*. The distribution can be weibull (the default) exponential, rayleigh, lognormal or loglogistic, all with a log transformation, or extreme, logistic, gaussian or t with an identity transformation.

The default for distribution corresponds to the model

$$\log T \sim \boldsymbol{\beta}^T \boldsymbol{x} + \sigma \log E$$

for a standard exponential E whereas our Weibull parametrization corresponds to

$$\log T \sim -\log \lambda + \frac{1}{\alpha} \log E$$

---

[3] survreg in R.

Thus `survReg` uses a log-linear Weibull model for $-\log \lambda$ and the scale factor σ estimates $1/\alpha$. The `exponential` distribution comes from fixing $\sigma = \alpha = 1$.

We consider exponential analyses, followed by Weibull and log-logistic regression analyses.

```
> options(contrasts = c("contr.treatment", "contr.poly"))
> survReg(Surv(time) ~ ag*log(wbc), data = leuk,
          dist = "exponential")
    . . . .
Coefficients:
 (Intercept)      ag log(wbc) ag:log(wbc)
      4.3433   4.135 -0.15402    -0.32781

Scale fixed at 1

Loglik(model)= -145.7   Loglik(intercept only)= -155.5
        Chisq= 19.58 on 3 degrees of freedom, p= 0.00021
> summary(survReg(Surv(time) ~ ag + log(wbc), data = leuk,
                  dist = "exponential"))
              Value Std. Error     z        p
(Intercept)   5.815      1.263  4.60 4.15e-06
         ag   1.018      0.364  2.80 5.14e-03
   log(wbc) -0.304      0.124 -2.45 1.44e-02


> summary(survReg(Surv(time) ~ ag + log(wbc), data = leuk))
# Weibull is the default
    . . . .
              Value Std. Error     z        p
(Intercept)  5.8524      1.323  4.425 9.66e-06
         ag  1.0206      0.378  2.699 6.95e-03
   log(wbc) -0.3103      0.131 -2.363 1.81e-02
 Log(scale)  0.0399      0.139  0.287 7.74e-01

Scale= 1.04

Weibull distribution
Loglik(model)= -146.5   Loglik(intercept only)= -153.6
        Chisq= 14.18 on 2 degrees of freedom, p= 0.00084
    . . . .
> summary(survReg(Surv(time) ~ ag + log(wbc), data = leuk,
                  dist = "loglogistic"))
              Value Std. Error     z        p
(Intercept)  8.027      1.701  4.72 2.37e-06
         ag  1.155      0.431  2.68 7.30e-03
   log(wbc) -0.609      0.176 -3.47 5.21e-04
 Log(scale) -0.374      0.145 -2.58 9.74e-03

Scale= 0.688

Log logistic distribution
```

```
Loglik(model)= -146.6   Loglik(intercept only)= -155.4
      Chisq= 17.58 on 2 degrees of freedom, p= 0.00015
```

The Weibull analysis shows no support for non-exponential shape. For later reference, in the proportional hazards parametrization (13.8) the estimate of the coefficients is $\widehat{\beta} = -(5.85, 1.02, -0.310)^T/1.04 = (-5.63, -0.981, 0.298)^T$. The log-logistic distribution, which is an accelerated life model but not a proportional hazards model (in our parametrization), gives a considerably more significant coefficient for log(wbc). Its usual scale parameter τ (as defined on page 353) is estimated as $1/0.688 \approx 1.45$.

We can test for a difference in groups within the Weibull model by the Wald test (the 'z' value for ag) or we can perform a likelihood ratio test by the anova method.

```
> anova(survReg(Surv(time) ~ log(wbc), data = leuk),
        survReg(Surv(time) ~ ag + log(wbc), data = leuk))
    ....
            Terms Resid. Df  -2*LL Test Df Deviance Pr(Chi)
1       log(wbc)        30 299.35
2 ag + log(wbc)        29 293.00  +ag  1   6.3572 0.01169
```

The likelihood ratio test statistic is somewhat less significant than the result given by survdiff.

An extension is to allow different scale parameters σ for each group, by adding a strata argument to the formula. For example,

```
> summary(survReg(Surv(time) ~ strata(ag) + log(wbc), data=leuk))
    ....
              Value Std. Error    z        p
(Intercept)   7.499      1.475  5.085 3.68e-07
   log(wbc)  -0.422      0.149 -2.834 4.59e-03
  ag=absent   0.152      0.221  0.688 4.92e-01
 ag=present   0.142      0.216  0.658 5.11e-01

Scale:
 ag=absent ag=present
      1.16       1.15

Weibull distribution
Loglik(model)= -149.7   Loglik(intercept only)= -153.2
    ....
```

If the accelerated-life model holds, $T \exp(-\beta^T x)$ has the same distribution for all subjects, being standard Weibull, log-logistic and so on. Thus we can get some insight into what the common distribution should be by studying the distribution of $(T_i \exp(-\widehat{\beta}^T x_i))$. Another way to look at this is that the residuals from the regression are $\log T_i - \widehat{\beta}^T x_i$ which we have transformed back to the scale of time. For the leuk data we could use, for example,

```
leuk.wei <- survReg(Surv(time) ~ ag + log(wbc), data = leuk)
ntimes <- leuk$time * exp(-leuk.wei$linear.predictors)
plot(survfit(Surv(ntimes)), log = T)
```

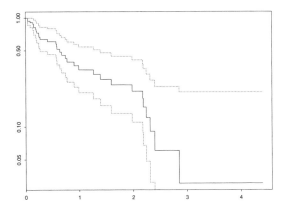

Figure 13.3: A log plot of $S(t)$ (equivalently, a linear plot of $-H(t)$) for the `leuk` dataset with pointwise confidence intervals.

The result (Figure 13.3) is plausibly linear, confirming the suitability of an exponential model. If we wished to test for a general Weibull distribution, we should plot $\log(-\log \widehat{S}(t))$ against $\log t$. (This is provided by the `fun="cloglog"` argument to `plot.survfit`)

Moving on to the `gehan` dataset, which includes right censoring, we find

```
> survReg(Surv(time, cens) ~ factor(pair) + treat, data = gehan,
            dist = "exponential")
   ....
Loglik(model)= -101.6    Loglik(intercept only)= -116.8
        Chisq= 30.27 on 21 degrees of freedom, p= 0.087
> summary(survReg(Surv(time, cens) ~ treat, data = gehan,
                dist = "exponential"))
            Value Std. Error     z        p
(Intercept)  3.69       0.333 11.06 2.00e-28
      treat -1.53       0.398 -3.83 1.27e-04
Scale fixed at 1

Exponential distribution
Loglik(model)= -108.5    Loglik(intercept only)= -116.8
        Chisq= 16.49 on 1 degrees of freedom, p= 4.9e-05
> summary(survReg(Surv(time, cens) ~ treat, data = gehan))
            Value Std. Error     z        p
(Intercept)  3.516      0.252 13.96 2.61e-44
      treat -1.267      0.311 -4.08 4.51e-05
 Log(scale) -0.312      0.147 -2.12 3.43e-02

Scale= 0.732

Weibull distribution
Loglik(model)= -106.6    Loglik(intercept only)= -116.4
        Chisq= 19.65 on 1 degrees of freedom, p= 9.3e-06
```

There is no evidence of close matching of pairs. The difference in log hazard between treatments is $-(-1.267)/0.732 = 1.73$ with a standard error of $0.42 = 0.311/0.732$.

Finally, we consider the `motors` data, which are analysed by Kalbfleisch and Prentice (1980, §3.8.1). According to Nelson and Hahn (1972), the data were collected to assess survival at $130°$C, for which they found a median of 34 400 hours and a 10 percentile of 17 300 hours.

```
> plot(survfit(Surv(time, cens) ~ factor(temp), data = motors),
        conf.int = F)
> motor.wei <- survReg(Surv(time, cens) ~ temp, data = motors)
> summary(motor.wei)
              Value Std. Error      z          p
(Intercept) 16.3185    0.62296   26.2  3.03e-151
       temp -0.0453    0.00319  -14.2   6.74e-46
 Log(scale) -1.0956    0.21480   -5.1   3.38e-07

Scale= 0.334

Weibull distribution
Loglik(model)= -147.4   Loglik(intercept only)= -169.5
        Chisq= 44.32 on 1 degrees of freedom, p= 2.8e-11
        ....
> unlist(predict(motor.wei, data.frame(temp=130), se.fit = T))
    fit se.fit
  33813 7506.3
```

The `predict` method by default predicts the centre of the distribution. We can obtain predictions for quantiles by

```
> predict(motor.wei, data.frame(temp=130), type = "quantile",
        p = c(0.5, 0.1))
[1] 29914 15935
```

We can also use `predict` to find standard errors, but we prefer to compute confidence intervals on log-time scale by

```
> t1 <- predict(motor.wei, data.frame(temp=130),
                type = "uquantile", p = 0.5, se = T)
> exp(c(LL=t1$fit - 2*t1$se, UL=t1$fit + 2*t1$se))
    LL    UL
 19517 45849
> t1 <- predict(motor.wei, data.frame(temp=130),
                type = "uquantile", p = 0.1, se = T)
> exp(c(LL=t1$fit - 2*t1$se, UL=t1$fit + 2*t1$se))
    LL    UL
 10258 24752
```

Nelson & Hahn worked with $z = 1000/(\text{temp} + 273.2)$. We leave the reader to try this; it gives slightly larger quantiles.

Function `censorReg`

S+   S-PLUS has a function `censorReg` for parametric survival analysis by Bill
Meeker; this has a very substantial overlap with `survReg` but is more general
in that it allows *truncation* as well as *censoring*. Either or both of censoring and
truncation occur when subjects are only observed for part of the time axis. An
observation T_i is right-censored if it is known only that $T_i > U_i$ for a censoring time U_i, and left-censored if it is known only that $T_i \leqslant L_i$. (Both left- and
right-censoring can occur in a study, but not for the same individual.) Interval
censoring is usually taken to refer to subjects known to have an event in $(L_i, U_i]$,
but with the time of the event otherwise unknown. Truncation is similar but subtly
different. For left and right truncation, subjects with events before L_i or after U_i
are not included in the study, and interval truncation refers to both left and right
truncation. (Notice the inconsistency with interval censoring.)

Confusingly, `censorReg` uses `"logexponential"` and `"lograyleigh"`
for what are known to `survReg` as the `"exponential"` and `"rayleigh"` distributions and are accelerated-life models for those distributions.

Let us consider a simple example using `gehan`. We can fit a Weibull model
by

```
> options(contrasts = c("contr.treatment", "contr.poly"))
> summary(censorReg(censor(time, cens) ~ treat, data = gehan))
    . . . .
Coefficients:
  Est. Std.Err. 95% LCL 95% UCL z-value  p-value
  3.52    0.252    3.02   4.009   13.96 2.61e-44
 -1.27    0.311   -1.88  -0.658   -4.08 4.51e-05

Extreme value distribution: Dispersion (scale) = 0.73219
Observations: 42 Total; 12 Censored
-2*Log-Likelihood: 213
```

which agrees with our results on page 363.

The potential advantages of `censorReg` come from its wider range of options. As noted previously, it allows truncation, by specifying a call to `censor`
with a `truncation` argument. Distributions can be fitted with a *threshold*, that
is, a parameter $\gamma > 0$ such that the failure-time model is fitted to $T - \gamma$ (and
hence no failures can occur before time γ).

There is a `plot` method for `censorReg` that produces up to seven figures.

A `strata` argument in a `censorReg` model has a completely different
meaning: it fits separate models at each level of the stratifying factor, unlike
`survReg` which has common regression coefficients across strata.

13.3   Cox Proportional Hazards Model

Cox (1972) introduced a less parametric approach to proportional hazards. There
is a baseline hazard function $h_0(t)$ that is modified multiplicatively by covariates

(including group indicators), so the hazard function for any individual case is

$$h(t) = h_0(t) \exp \boldsymbol{\beta}^T \boldsymbol{x}$$

and the interest is mainly in the proportional factors rather than the baseline hazard. Note that the cumulative hazards will also be proportional, so we can examine the hypothesis by plotting survivor curves for sub-groups on log scale. Later we allow the covariates to depend on time.

The parameter vector $\boldsymbol{\beta}$ is estimated by maximizing a *partial likelihood*. Suppose one death occurred at time t_j. Then conditional on this event the probability that case i died is

$$\frac{h_0(t) \exp \boldsymbol{\beta}^T \boldsymbol{x}_i}{\sum_l I(T_l \geqslant t) h_0(t) \exp \boldsymbol{\beta}^T \boldsymbol{x}_l} = \frac{\exp \boldsymbol{\beta}^T \boldsymbol{x}_i}{\sum_l I(T_l \geqslant t) \exp \boldsymbol{\beta}^T \boldsymbol{x}_l} \qquad (13.11)$$

which does not depend on the baseline hazard. The partial likelihood for $\boldsymbol{\beta}$ is the product of such terms over all observed deaths, and usually contains most of the information about $\boldsymbol{\beta}$ (the remainder being in the observed times of death). However, we need a further condition on the censoring (Fleming and Harrington, 1991, pp. 138–9) that it is independent and *uninformative* for this to be so; the latter means that the likelihood for censored observations in $[t, t + \Delta t)$ does not depend on $\boldsymbol{\beta}$.

The correct treatment of ties causes conceptual difficulties as they are an event of probability zero for continuous distributions. Formally (13.11) may be corrected to include all possible combinations of deaths. As this increases the computational load, it is common to employ the Breslow approximation[4] in which each death is always considered to precede all other events at that time. Let $\tau_i = I(T_i \geqslant t) \exp \boldsymbol{\beta}^T \boldsymbol{x}_i$, and suppose there are d deaths out of m possible events at time t. Breslow's approximation uses the term

$$\prod_{i=1}^{d} \frac{\tau_i}{\sum_1^m \tau_j}$$

in the partial likelihood at time t. Other options are Efron's approximation

$$\prod_{i=1}^{d} \frac{\tau_i}{\sum_1^m \tau_j - \frac{i}{d} \sum_1^d \tau_j}$$

and the 'exact' partial likelihood

$$\prod_{i=1}^{d} \tau_i \bigg/ \sum \prod_{k=1}^{d} \tau_{j_k}$$

where the sum is over subsets of $1, \ldots, m$ of size d. One of these terms is selected by the method argument of the function coxph, with default efron.

---

[4]First proposed by Peto (1972).

The baseline cumulative hazard $H_0(t)$ is estimated by rescaling the contributions to the number at risk by $\exp \widehat{\beta}^T x$ in (13.3). Thus in that formula $r(t) = \sum I(T_i \geqslant t) \exp \widehat{\beta}^T x_i$.

The Cox model is easily extended to allow different baseline hazard functions for different groups, and this is automatically done if they are declared as strata. For our leukaemia example we have:

```
> leuk.cox <- coxph(Surv(time) ~ ag + log(wbc), data = leuk)
> summary(leuk.cox)
    . . . .
          coef exp(coef) se(coef)     z      p
     ag -1.069    0.343    0.429 -2.49 0.0130
log(wbc)  0.368    1.444    0.136  2.70 0.0069

         exp(coef) exp(-coef) lower .95 upper .95
     ag     0.343      2.913     0.148     0.796
log(wbc)    1.444      0.692     1.106     1.886

Rsquare= 0.377   (max possible= 0.994 )
Likelihood ratio test= 15.6  on 2 df,   p=0.000401
Wald test              = 15.1  on 2 df,   p=0.000537
Score (logrank) test = 16.5  on 2 df,   p=0.000263

> update(leuk.cox, ~ . -ag)
    . . . .
Likelihood ratio test=9.19  on 1 df, p=0.00243  n= 33

> (leuk.coxs <- coxph(Surv(time) ~ strata(ag) + log(wbc),
                      data = leuk))
    . . . .
          coef exp(coef) se(coef)    z      p
log(wbc) 0.391      1.48    0.143 2.74 0.0062
    . . . .
Likelihood ratio test=7.78  on 1 df, p=0.00529  n= 33

> (leuk.coxs1 <- update(leuk.coxs, . ~ . + ag:log(wbc)))
    . . . .
              coef exp(coef) se(coef)     z    p
   log(wbc) 0.183      1.20    0.188 0.978 0.33
ag:log(wbc) 0.456      1.58    0.285 1.598 0.11
    . . . .
> plot(survfit(Surv(time) ~ ag), lty = 2:3, log = T)
> lines(survfit(leuk.coxs), lty = 2:3, lwd = 3)
> legend(80, 0.8, c("ag absent", "ag present"), lty = 2:3)
```

The 'likelihood ratio test' is actually based on (log) partial likelihoods, not the full likelihood, but has similar asymptotic properties. The tests show that there is a significant effect of wbc on survival, but also that there is a significant difference between the two ag groups (although as Figure 13.4 shows, this is less than before adjustment for the effect of wbc).

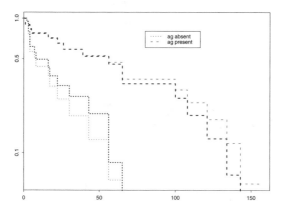

Figure 13.4: Log-survivor curves for the `leuk` dataset. The thick lines are from a Cox model with two strata, the thin lines Kaplan–Meier estimates that ignore the blood counts.

Note how `survfit` can take the result of a fit of a proportional hazard model. In the first fit the hazards in the two groups differ only by a factor whereas later they are allowed to have separate baseline hazards (which look very close to proportional). There is marginal evidence for a difference in slope within the two strata. Note how straight the log-survivor functions are in Figure 13.4, confirming the good fit of the exponential model for these data. The Kaplan–Meier survivor curves refer to the populations; those from the `coxph` fit refer to a patient in the stratum with an average `log(wbc)` for the whole dataset. This example shows why it is inappropriate just to test (using `survdiff`) the difference between the two groups; part of the difference is attributable to the lower `wbc` in the `ag` absent group.

The test statistics refer to the whole set of covariates. The likelihood ratio test statistic is the change in deviance on fitting the covariates over just the baseline hazard (by strata); the score test is the expansion at the baseline, and so does not need the parameters to be estimated (although this has been done). The R^2 measure quoted by `summary.coxph` is taken from Nagelkerke (1991).

The general proportional hazards model gives estimated (non-intercept) coefficients $\hat{\beta} = (-1.07, 0.37)^T$, compared to the Weibull fit of $(-0.98, 0.30)^T$ (on page 362). The log-logistic had coefficients $(-1.16, 0.61)^T$ which under the approximations of Solomon (1984) would be scaled by $\tau/2$ to give $(-0.79, 0.42)^T$ for a Cox proportional-hazards fit if the log-logistic regression model (an accelerated-life model) were the true model.

We next consider the Gehan data. We saw before that the pairing has a negligible effect for the exponential model. Here the effect is a little larger, with $P \approx 8\%$. The Gehan data have a large number of (mainly pairwise) ties, so we use the 'exact' partial likelihood.

```
> coxph(Surv(time, cens) ~ treat, data = gehan, method = "exact")
      coef exp(coef) se(coef)    z       p
treat 1.63      5.09    0.433 3.76 0.00017
```

```
Likelihood ratio test=16.2  on 1 df, p=5.54e-05  n= 42

# The next fit is slow
> coxph(Surv(time, cens) ~ treat + factor(pair), data = gehan,
      method = "exact")
   ....
Likelihood ratio test=45.5  on 21 df, p=0.00148  n= 42
   ....
> 1 - pchisq(45.5 - 16.2, 20)
[1] 0.082018
```

Finally we consider the `motors` data. The exact fit is much the slowest, as it has large groups of ties.

```
> (motor.cox <- coxph(Surv(time, cens) ~ temp, motors))
   ....
      coef exp(coef) se(coef)    z       p
temp 0.0918     1.1   0.0274 3.36 0.00079
   ....
> coxph(Surv(time, cens) ~ temp, motors, method = "breslow")
   ....
      coef exp(coef) se(coef)   z       p
temp 0.0905    1.09   0.0274 3.3 0.00098
   ....
> coxph(Surv(time, cens) ~ temp, motors, method = "exact")
   ....
      coef exp(coef) se(coef)    z       p
temp 0.0947     1.1   0.0274 3.45 0.00056
   ....
> plot( survfit(motor.cox, newdata = data.frame(temp=200),
               conf.type = "log-log") )
> summary( survfit(motor.cox, newdata = data.frame(temp=130)) )
time n.risk n.event survival  std.err lower 95% CI upper 95% CI
 408    40      4    1.000 0.000254        0.999            1
 504    36      3    1.000 0.000499        0.999            1
1344    28      2    0.999 0.001910        0.995            1
1440    26      1    0.998 0.002698        0.993            1
1764    20      1    0.996 0.005327        0.986            1
2772    19      1    0.994 0.007922        0.978            1
3444    18      1    0.991 0.010676        0.971            1
3542    17      1    0.988 0.013670        0.962            1
3780    16      1    0.985 0.016980        0.952            1
4860    15      1    0.981 0.020697        0.941            1
5196    14      1    0.977 0.024947        0.929            1
```

The function `survfit` has a special method for `coxph` objects that plots the mean and confidence interval of the survivor curve for an average individual (with average values of the covariates). As we see, this can be overridden by giving new data, as shown in Figure 13.5. The non-parametric method is unable to extrapolate to $130°C$ as none of the test examples survived long enough to estimate the baseline hazard beyond the last failure at 5 196 hours.

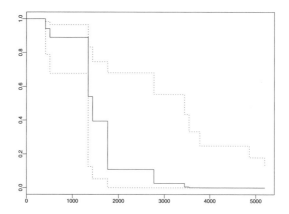

Figure 13.5: The survivor curve for a motor at $200°\mathrm{C}$ estimated from a Cox proportional hazards model (solid line) with pointwise 95% confidence intervals (dotted lines).

Residuals

The concept of a residual is a difficult one for binary data, especially as here the event may not be observed because of censoring. A straightforward possibility is to take

$$r_i = \delta_i - \widehat{H}(t_i)$$

which is known as the *martingale residual* after a derivation from the mathematical theory given by Fleming and Harrington (1991, §4.5). They show that it is appropriate for checking the functional form of the proportional hazards model, for if

$$h(t) = h_0(t)\phi(x^*)\exp\beta^T x$$

for an (unknown) function of a covariate x^* then

$$E[R \mid X^*] \approx [\phi(X^*) - \overline{\phi}]\sum \delta_i/n$$

and this can be estimated by smoothing a plot of the martingale residuals versus x^*, for example, using `lowess` or the function `scatter.smooth` based on `loess`. (The term $\overline{\phi}$ is a complexly weighted mean.) The covariate x^* can be one not included in the model, or one of the terms to check for non-linear effects.

The martingale residuals are the default output of `residuals` on a `coxph` fit.

The martingale residuals can have a very skewed distribution, as their maximum value is 1, but they can be arbitrarily negative. The *deviance residuals* are a transformation

$$\mathrm{sign}\,(r_i)\sqrt{2[-r_i - \delta_i \log(\delta_i - r_i)]}$$

which reduces the skewness, and for a parametric survival model when squared and summed give (approximately) the deviance. Deviance residuals are best used in plots that will indicate cases not fitted well by the model.

The *Schoenfeld residuals* (Schoenfeld, 1982) are defined at death times as $x_i - \overline{x}(t_i)$ where $\overline{x}(s)$ is the mean weighted by $\exp \widehat{\beta}^T x$ of the x over only the cases still in the risk set at time s. These residuals form a matrix with one row for each case that died and a column for each covariate. The scaled Schoenfeld residuals (type = "scaledsch") are the I^{-1} matrix multiplying the Schoenfeld residuals, where I is the (partial) information matrix at the fitted parameters in the Cox model.

The *score residuals* are the terms of efficient score for the partial likelihood, this being a sum over cases of

$$L_i = \left[x_i - \overline{x}(t_i) \right] \delta_i - \int_0^{t_i} \left[x_i(s) - \overline{x}(s) \right] \widehat{h}(s) \, \mathrm{d}s$$

Thus the score residuals form an $n \times p$ matrix. They can be used to examine leverage of individual cases by computing (approximately) the change in $\widehat{\beta}$ if the observation were dropped; type = "dfbeta" gives this, whereas type = "dfbetas" scales by the standard errors for the components of $\widehat{\beta}$.

Tests of proportionality of hazards

Once a type of departure from the base model is discovered or suspected, the proportional hazards formulation is usually flexible enough to allow an extended model to be formulated and the significance of the departure tested within the extended model. Nevertheless, some approximations can be useful, and are provided by the function cox.zph for departures of the type

$$\beta(t) = \beta + \theta g(t)$$

for some postulated smooth function g. Grambsch and Therneau (1994) show that the scaled Schoenfeld residuals for case i have, approximately, mean $g(t_i)\theta$ and a computable variance matrix.

The function cox.zph has both print and plot methods. The printed output gives an estimate of the correlation between $g(t_i)$ and the scaled Schoenfeld residuals and a chi-squared test of $\theta = 0$ for each covariate, and an overall chi-squared test. The plot method gives a plot for each covariate, of the scaled Schoenfeld residuals against $g(t)$ with a spline smooth and pointwise confidence bands for the smooth. (Figure 13.8 on page 375 is an example.)

The function g has to be specified. The default in cox.zph is $1 - \widehat{S}(t)$ for the Kaplan–Meier estimator, with options for the ranks of the death times, $g \equiv 1$ and $g = \log$ as well as a user-specified function. (The x-axis of the plots is labelled by the death times, not $\{g(t_i)\}$.)

13.4 Further Examples

VA lung cancer data

S-PLUS supplies[5] the dataset cancer.vet on a Veterans Administration lung cancer trial used by Kalbfleisch and Prentice (1980), but as it has no on-line help,

---

[5]For R it is supplied in package MASS .

it is not obvious what it is! It is a matrix of 137 cases with right-censored survival
time and the covariates

treatment	standard or test
celltype	one of four cell types
Karnofsky score	of performance on scale 0–100, with high values for relatively well patients
diagnosis	time since diagnosis in months at entry to trial
age	in years
therapy	logical for prior therapy

As there are several covariates, we use the Cox model to establish baseline hazards.

```
> # R: data(VA) # is all that is required.
> # S: VA.temp <- as.data.frame(cancer.vet)
> # S: dimnames(VA.temp)[[2]] <- c("treat", "cell", "stime",
    "status", "Karn", "diag.time","age","therapy")
> # S: attach(VA.temp)
> # S: VA <- data.frame(stime, status, treat = factor(treat), age,
    Karn, diag.time, cell = factor(cell), prior = factor(therapy))
> # S: detach(VA.temp)
> (VA.cox <- coxph(Surv(stime, status) ~ treat + age  + Karn +
                 diag.time + cell + prior, data = VA))
                  coef exp(coef) se(coef)        z        p
   treat  2.95e-01     1.343  0.20755  1.41945 1.6e-01
     age -8.71e-03     0.991  0.00930 -0.93612 3.5e-01
    Karn -3.28e-02     0.968  0.00551 -5.95801 2.6e-09
diag.time 8.18e-05     1.000  0.00914  0.00895 9.9e-01
   cell2  8.62e-01     2.367  0.27528  3.12970 1.7e-03
   cell3  1.20e+00     3.307  0.30092  3.97474 7.0e-05
   cell4  4.01e-01     1.494  0.28269  1.41955 1.6e-01
   prior  7.16e-02     1.074  0.23231  0.30817 7.6e-01

Likelihood ratio test=62.1  on 8 df, p=1.8e-10  n= 137

> (VA.coxs <- coxph(Surv(stime, status) ~ treat + age + Karn +
    diag.time + strata(cell) + prior, data = VA))
              coef exp(coef) se(coef)       z       p
   treat  0.28590     1.331  0.21001  1.361 1.7e-01
     age -0.01182     0.988  0.00985 -1.201 2.3e-01
    Karn -0.03826     0.962  0.00593 -6.450 1.1e-10
diag.time -0.00344     0.997  0.00907 -0.379 7.0e-01
   prior  0.16907     1.184  0.23567  0.717 4.7e-01

Likelihood ratio test=44.3  on 5 df, p=2.04e-08  n= 137

> plot(survfit(VA.coxs), log = T, lty = 1:4, col = 2:5)
> legend(locator(1), c("squamous", "small", "adeno", "large"),
    lty = 1:4, col = 2:5)
> plot(survfit(VA.coxs), fun = "cloglog", lty = 1:4, col = 2:5)
```

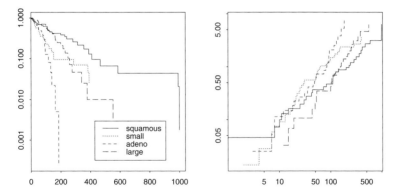

Figure 13.6: Cumulative hazard functions for the cell types in the VA lung cancer trial. The left-hand plot is labelled by survival probability on log scale. The right-hand plot is on log-log scale.

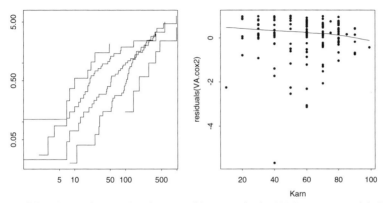

Figure 13.7: Diagnostic plots for the Karnofsky score in the VA lung cancer trial. Left: log-log cumulative hazard plot for five groups. Right: martingale residuals versus Karnofsky score, with a smoothed fit.

```
> cKarn <- factor(cut(VA$Karn, 5))
> VA.cox1 <- coxph(Surv(stime, status) ~ strata(cKarn) + cell,
               data = VA)
> plot(survfit(VA.cox1), fun="cloglog")
> VA.cox2 <- coxph(Surv(stime, status) ~ Karn + strata(cell),
               data = VA)
> scatter.smooth(VA$Karn, residuals(VA.cox2))
```

Figures 13.6 and 13.7 show some support for proportional hazards among the cell types (except perhaps squamous), and suggest a Weibull or even exponential distribution.

```
> VA.wei <- survReg(Surv(stime, status) ~ treat + age + Karn +
      diag.time + cell + prior, data = VA)
```

```
> summary(VA.wei, cor = F)
    ....
               Value Std. Error        z        p
(Intercept)  3.262014    0.66253   4.9236 8.50e-07
      treat -0.228523    0.18684  -1.2231 2.21e-01
        age  0.006099    0.00855   0.7131 4.76e-01
       Karn  0.030068    0.00483   6.2281 4.72e-10
  diag.time -0.000469    0.00836  -0.0561 9.55e-01
      cell2 -0.826185    0.24631  -3.3542 7.96e-04
      cell3 -1.132725    0.25760  -4.3973 1.10e-05
      cell4 -0.397681    0.25475  -1.5611 1.19e-01
      prior -0.043898    0.21228  -0.2068 8.36e-01
 Log(scale) -0.074599    0.06631  -1.1250 2.61e-01

Scale= 0.928

Weibull distribution
Loglik(model)= -715.6   Loglik(intercept only)= -748.1
        Chisq= 65.08 on 8 degrees of freedom, p= 4.7e-11

> VA.exp <- survReg(Surv(stime, status) ~ Karn + cell,
                    data = VA, dist = "exponential")
> summary(VA.exp, cor = F)
               Value Std. Error        z        p
(Intercept)  3.4222     0.35463   9.65 4.92e-22
       Karn  0.0297     0.00486   6.11 9.97e-10
      cell2 -0.7102     0.24061  -2.95 3.16e-03
      cell3 -1.0933     0.26863  -4.07 4.70e-05
      cell4 -0.3113     0.26635  -1.17 2.43e-01

Scale fixed at 1

Exponential distribution
Loglik(model)= -717   Loglik(intercept only)= -751.2
        Chisq= 68.5 on 4 degrees of freedom, p= 4.7e-14
```

Note that `scale` does not differ significantly from one, so an exponential distribution is an appropriate summary.

```
> cox.zph(VA.coxs)
               rho  chisq        p
    treat -0.0607  0.545  0.46024
      age  0.1734  4.634  0.03134
     Karn  0.2568  9.146  0.00249
diag.time  0.1542  2.891  0.08909
    prior -0.1574  3.476  0.06226
   GLOBAL      NA 13.488  0.01921
> par(mfrow = c(3, 2)); plot(cox.zph(VA.coxs))
```

Closer investigation does show some suggestion of time-varying coefficients in the Cox model. The plot is Figure 13.8. Note that some coefficients that are not

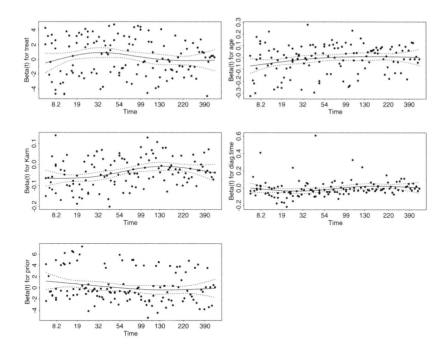

Figure 13.8: Diagnostics plots from `cox.zph` of the constancy of the coefficients in the proportional hazards model `VA.coxs`. Each plot is of a component of the Schoenfeld residual against a non-linear scale of time. A spline smoother is shown, together with ±2 standard deviations.

significant in the basic model show evidence of varying with time. This suggests that the model with just `Karn` and `cell` may be too simple, and that we need to consider interactions. We automate the search of interactions using `stepAIC`, which has methods for both `coxph` and `survReg` fits. With hindsight, we centre the data.

```
> VA$Karnc <- VA$Karn - 50
> VA.coxc <- update(VA.cox, ~ . - Karn + Karnc)
> VA.cox2 <- stepAIC(VA.coxc, ~ .^2)
> VA.cox2$anova
Initial Model:
Surv(stime, status) ~ treat + age + diag.time + cell + prior +
        Karnc

Final Model:
Surv(stime, status) ~ treat + diag.time + cell + prior + Karnc +
        prior:Karnc + diag.time:cell + treat:prior + treat:Karnc

            Step Df Deviance Resid. Df Resid. Dev    AIC
1                                  129      948.79 964.79
2   + prior:Karnc  1    9.013       128      939.78 957.78
```

```
3 + diag.time:cell 3    11.272         125    928.51 952.51
4                - age 1     0.415      126    928.92 950.92
5        + treat:prior 1    2.303       125    926.62 950.62
6        + treat:Karnc 1    2.904       124    923.72 949.72
```

(The 'deviances' here are minus twice log partial likelihoods.) Applying stepAIC to VA.wei leads to the same sequence of steps. As the variables diag.time and Karn are not factors, this will be easier to interpret using nesting:

```
> (VA.cox3 <- update(VA.cox2, ~ treat/Karnc + prior*Karnc
+    + treat:prior + cell/diag.time))
                      coef exp(coef) se(coef)      z       p
       treat     0.8065     2.240  0.27081   2.978 2.9e-03
       prior     0.9191     2.507  0.31568   2.912 3.6e-03
       Karnc    -0.0107     0.989  0.00949  -1.129 2.6e-01
       cell2     1.7068     5.511  0.37233   4.584 4.6e-06
       cell3     1.5633     4.775  0.44205   3.536 4.1e-04
       cell4     0.7476     2.112  0.48136   1.553 1.2e-01
Karnc %in% treat -0.0187   0.981  0.01101  -1.695 9.0e-02
  prior:Karnc   -0.0481     0.953  0.01281  -3.752 1.8e-04
  treat:prior   -0.7264     0.484  0.41833  -1.736 8.3e-02
cell1diag.time   0.0532     1.055  0.01595   3.333 8.6e-04
cell2diag.time  -0.0245     0.976  0.01293  -1.896 5.8e-02
cell3diag.time   0.0161     1.016  0.04137   0.388 7.0e-01
cell4diag.time   0.0150     1.015  0.04033   0.373 7.1e-01
```

Thus the hazard increases with time since diagnosis in squamous cells, only, and the effect of the Karnofsky score is only pronounced in the group with prior therapy. We tried replacing diag.time with a polynomial, with negligible benefit. Using cox.zph shows a very significant change with time in the coefficients of the treat*Karn interaction.

```
> cox.zph(VA.cox3)
                     rho      chisq        p
       treat     0.18012   6.10371 0.013490
       prior     0.07197   0.76091 0.383044
       Karnc     0.27220  14.46103 0.000143
       cell2     0.09053   1.31766 0.251013
       cell3     0.06247   0.54793 0.459164
       cell4     0.00528   0.00343 0.953318
Karnc %in% treat -0.20606  7.80427 0.005212
  prior:Karnc   -0.04017   0.26806 0.604637
  treat:prior   -0.13061   2.33270 0.126682
cell1diag.time   0.11067   1.62464 0.202446
cell2diag.time  -0.01680   0.04414 0.833596
cell3diag.time   0.09713   1.10082 0.294086
cell4diag.time   0.16912   3.16738 0.075123
       GLOBAL        NA   25.52734 0.019661


> par(mfrow = c(2, 2))
> plot(cox.zph(VA.cox3), var = c(1, 3, 7)) ## not shown
```

Stanford heart transplants

This set of data is analysed by Kalbfleisch and Prentice (1980, §5.5.3). (The data given in Kalbfleisch & Prentice are rounded, but the full data are supplied as data frame `heart`.) It is on survival from early heart transplant operations at Stanford. The new feature is that patients may change treatment during the study, moving from the control group to the treatment group at transplantation, so some of the covariates such as waiting time for a transplant are time-dependent (in the simplest possible way). Patients who received a transplant are treated as two cases, before and after the operation, so cases in the transplant group are in general both right-censored and left-truncated. This is handled by `Surv` by supplying entry and exit times. For example, patient 4 has the rows

```
start  stop event       age          year  surgery transplant
 0.0   36.0   0  -7.73716632   0.49007529        0          0
36.0   39.0   1  -7.73716632   0.49007529        0          1
```

which show that he waited 36 days for a transplant and then died after 3 days. The proportional hazards model is fitted from this set of cases, but some summaries need to take account of the splitting of patients.

The covariates are age (in years minus 48), year (after 1 October 1967) and an indicator for previous surgery. Rather than use the six models considered by Kalbfleisch & Prentice, we do our own model selection.

```
> R: data(heart)
> coxph(Surv(start, stop, event) ~ transplant*
      (age + surgery + year), data = heart)
    ....
Likelihood ratio test=18.9  on 7 df, p=0.00852  n= 172
> coxph(Surv(start, stop, event) ~ transplant*(age + year) +
      surgery, data = heart)
    ....
Likelihood ratio test=18.4  on 6 df, p=0.0053  n= 172
> (stan <- coxph(Surv(start, stop, event) ~ transplant*year +
      age + surgery, data = heart))
    ....
                 coef exp(coef) se(coef)      z     p
    transplant -0.6213    0.537   0.5311  -1.17 0.240
          year -0.2526    0.777   0.1049  -2.41 0.016
           age  0.0299    1.030   0.0137   2.18 0.029
       surgery -0.6641    0.515   0.3681  -1.80 0.071
transplant:year 0.1974    1.218   0.1395   1.42 0.160

Likelihood ratio test=17.1  on 5 df, p=0.00424  n= 172
> stan1 <- coxph(Surv(start, stop, event) ~ strata(transplant) +
      year + year:transplant + age + surgery, heart)
> plot(survfit(stan1), conf.int = T, log = T, lty = c(1, 3),
       col = 2:3)
> legend(locator(1), c("before", "after"), lty = c(1, 3),
        col = 2:3)
```

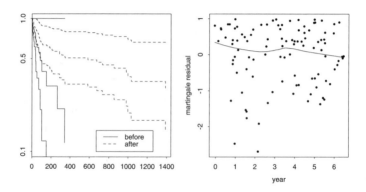

Figure 13.9: Plots for the Stanford heart transplant study. Left: log survivor curves and confidence limits for the two groups. Right: martingale residuals against calendar time.

```
> attach(heart)
> plot(year[transplant==0], residuals(stan1, collapse = id),
       xlab = "year", ylab = "martingale residual")
> lines(lowess(year[transplant == 0],
                residuals(stan1, collapse = id)))
> sresid <- resid(stan1, type = "dfbeta", collapse = id)
> detach()
> -100 * sresid %*% diag(1/stan1$coef)
```

This analysis suggests that survival rates over the study improved *prior* to transplantation, which Kalbfleisch & Prentice suggest could be due to changes in recruitment. The diagnostic plots of Figure 13.9 show nothing amiss. The `collapse` argument is needed as those patients who received transplants are treated as two cases, and we need the residual per patient.

Now consider predicting the survival of future patient aged 50 on 1 October 1971 with prior surgery, transplanted after six months.

```
# Survivor curve for the "average" subject
> summary(survfit(stan))
#  follow-up for two years
> stan2 <- data.frame(start = c(0, 183), stop= c(183, 2*365),
       event = c(0, 0), year = c(4, 4), age = c(50, 50) - 48,
       surgery = c(1, 1), transplant = c(0, 1))
> summary(survfit(stan, stan2, individual = T,
                conf.type = "log-log"))
```

time	n.risk	n.event	survival	std.err	lower 95% CI	upper 95% CI
....						
165	43	1	0.654	0.11509	0.384	0.828
186	41	1	0.643	0.11602	0.374	0.820
188	40	1	0.632	0.11697	0.364	0.812
207	39	1	0.621	0.11790	0.353	0.804
219	38	1	0.610	0.11885	0.343	0.796
263	37	1	0.599	0.11978	0.332	0.788

285	35	2	0.575 0.11524	0.325	0.762
308	33	1	0.564 0.11618	0.314	0.753
334	32	1	0.552 0.11712	0.302	0.744
340	31	1	0.540 0.11799	0.291	0.735
343	29	1	0.527 0.11883	0.279	0.725
584	21	1	0.511 0.12018	0.263	0.713
675	17	1	0.492 0.12171	0.245	0.699

The argument `individual = T` is needed to avoid averaging the two cases (which are the same individual).

Australian AIDS survival

The data on the survival of AIDS patients within Australia are of unusually high quality within that field, and jointly with Dr Patty Solomon we have studied survival up to 1992.[6] There are a large number of difficulties in defining survival from AIDS (acquired immunodeficiency syndrome), in part because as a syndrome its diagnosis is not clear-cut and has almost certainly changed with time. (To avoid any possible confusion, we are studying survival from AIDS and not the HIV infection which is generally accepted as the cause of AIDS.)

The major covariates available were the reported transmission category, and the state or territory within Australia. The AIDS epidemic had started in New South Wales and then spread, so the states have different profiles of cases in calendar time. A factor that was expected to be important in survival is the widespread availability of zidovudine (AZT) to AIDS patients from mid-1987 which has enhanced survival, and the use of zidovudine for HIV-infected patients from mid-1990, which it was thought might delay the onset of AIDS without necessarily postponing death further.

The transmission categories were:

hs	male homosexual or bisexual contact
hsid	as **hs** and also intravenous drug user
id	female or heterosexual male intravenous drug user
het	heterosexual contact
haem	haemophilia or coagulation disorder
blood	receipt of blood, blood components or tissue
mother	mother with or at risk of HIV infection
other	other or unknown

The data file gave data on all patients whose AIDS status was diagnosed prior to January 1992, with their status then. Since there is a delay in notification of death, some deaths in late 1991 would not have been reported and we adjusted the endpoint of the study to 1 July 1991. A total of 2 843 patients were included, of whom about 1 770 had died by the end date. The file contained an ID number, the dates of first diagnosis, birth and death (if applicable), as well as the state and the coded transmission category. We combined the states ACT and NSW (as

---

[6]We are grateful to the Australian National Centre in HIV Epidemiology and Clinical Research for making these data available to us.

Australian Capital Territory is a small enclave within New South Wales), and to maintain confidentiality the dates have been jittered and the smallest states combined. Only the transformed file Aids2 is included in our library.

As there are a number of patients who are diagnosed at (strictly, after) death, there are a number of zero survivals. The software used to have problems with these, so all deaths were shifted by 0.9 days to occur after other events the same day. To transform Aids2 to a form suitable for time-dependent-covariate analysis we used

```
time.depend.covar <- function(data) {
  id <- row.names(data); n <- length(id)
  events <- c(0, 10043, 11139, 12053) # julian days
  crit1 <- matrix(events[1:3], n, 3 ,byrow = T)
  crit2 <- matrix(events[2:4], n, 3, byrow = T)
  diag <- matrix(data$diag,n,3); death <- matrix(data$death,n,3)
  incid <- (diag < crit2) & (death >= crit1); incid <- t(incid)
  indr <- col(incid)[incid]; indc <- row(incid)[incid]
  ind <- cbind(indr, indc); idno <- id[indr]
  state <- data$state[indr]; T.categ <- data$T.categ[indr]
  age <- data$age[indr]; sex <- data$sex[indr]
  late <- indc - 1
  start <- t(pmax(crit1 - diag, 0))[incid]
  stop <- t(pmin(crit2, death + 0.9) - diag)[incid]
  status <- matrix(as.numeric(data$status),n,3)-1 # 0/1
  status[death > crit2] <- 0; status <- status[ind]
  levels(state) <- c("NSW", "Other", "QLD", "VIC")
  levels(T.categ) <- c("hs", "hsid", "id", "het", "haem",
                       "blood", "mother", "other")
  levels(sex) <- c("F", "M")
  data.frame(idno, zid=factor(late), start, stop, status,
             state, T.categ, age, sex)
}
Aids3 <- time.depend.covar(Aids2)
```

The factor zid indicates whether the patient is likely to have received zidovudine at all, and if so whether it might have been administered during HIV infection.

Our analysis was based on a proportional hazards model that allowed a proportional change in hazard from 1 July 1987 to 30 June 1990 and another from 1 July 1990; the results show a halving of hazard from 1 July 1987 but a nonsignificant change in 1990.

```
> attach(Aids3)
> aids.cox <- coxph(Surv(start, stop, status)
    ~ zid + state + T.categ + sex + age, data = Aids3)
> summary(aids.cox)
```

	coef	exp(coef)	se(coef)	z	p
zid1	-0.69087	0.501	0.06578	-10.5034	0.0e+00
zid2	-0.78274	0.457	0.07550	-10.3675	0.0e+00
stateOther	-0.07246	0.930	0.08964	-0.8083	4.2e-01

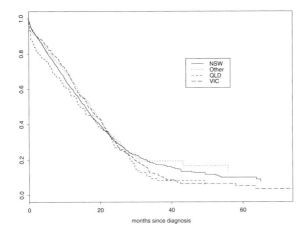

Figure 13.10: Survival of AIDS patients in Australia by state.

stateQLD	0.18315	1.201	0.08752	2.0927	3.6e-02
stateVIC	0.00464	1.005	0.06134	0.0756	9.4e-01
T.categhsid	-0.09937	0.905	0.15208	-0.6534	5.1e-01
T.categid	-0.37979	0.684	0.24613	-1.5431	1.2e-01
T.categhet	-0.66592	0.514	0.26457	-2.5170	1.2e-02
T.categhaem	0.38113	1.464	0.18827	2.0243	4.3e-02
T.categblood	0.16856	1.184	0.13763	1.2248	2.2e-01
T.categmother	0.44448	1.560	0.58901	0.7546	4.5e-01
T.categother	0.13156	1.141	0.16380	0.8032	4.2e-01
sex	0.02421	1.025	0.17557	0.1379	8.9e-01
age	0.01374	1.014	0.00249	5.5060	3.7e-08

```
. . . .
Likelihood ratio test= 185   on 14 df,    p=0
```

The effect of sex is nonsignificant, and so dropped in further analyses. There is
no detected difference in survival during 1990.

Note that Queensland has a significantly elevated hazard relative to New South
Wales (which has over 60% of the cases), and that the intravenous drug users
have a longer survival, whereas those infected via blood or blood products have a
shorter survival, relative to the first category who form 87% of the cases. We can
use stratified Cox models to examine these effects (Figures 13.10 and 13.11).

```
> aids1.cox <- coxph(Surv(start, stop, status)
    ~ zid + strata(state) + T.categ + age, data = Aids3)
> (aids1.surv <- survfit(aids1.cox))
               n events mean se(mean) median 0.95LCL 0.95UCL
  state=NSW 1780    1116  639     17.6    481     450     509
state=Other  249     142  658     42.2    525     453     618
  state=QLD  226     149  519     33.5    439     360     568
  state=VIC  588     355  610     26.3    508     476     574
> plot(aids1.surv, mark.time = F, lty = 1:4, col = 2:5,
      xscale = 365.25/12, xlab = "months since diagnosis")
```

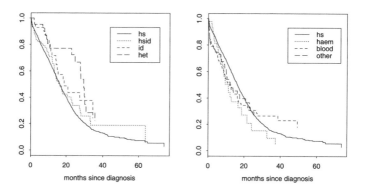

Figure 13.11: Survival of AIDS patients in Australia by transmission category.

```
> legend(locator(1), levels(state), lty = 1:4, col = 2:5)

> aids2.cox <- coxph(Surv(start, stop, status)
    ~ zid + state + strata(T.categ) + age, data = Aids3)
> (aids2.surv <- survfit(aids2.cox))
                 n events mean se(mean) median 0.95LCL 0.95UCL
    T.categ=hs 2465   1533  633     15.6    492   473.9     515
  T.categ=hsid   72     45  723     86.7    493   396.9     716
    T.categ=id   48     19  653     54.3    568   447.9      NA
   T.categ=het   40     17  775     57.3    897   842.9      NA
  T.categ=haem   46     29  431     53.9    337   252.9     657
 T.categ=blood   94     76  583     86.1    358   267.9     507
T.categ=mother    7      3  395     92.6    655    15.9      NA
 T.categ=other   70     40  421     40.7    369   300.9     712

> par(mfrow = c(1, 2))
> plot(aids2.surv[1:4], mark.time = F, lty = 1:4, col = 2:5,
    xscale = 365.25/12, xlab = "months since diagnosis")
> legend(locator(1), levels(T.categ)[1:4], lty = 1:4, col = 2:5)

> plot(aids2.surv[c(1, 5, 6, 8)], mark.time = F, lty = 1:4,
    col = 2:5, xscale = 365.25/12, xlab = "months since diagnosis")
> legend(locator(1), levels(T.categ)[c(1, 5, 6, 8)],
         lty = 1:4, col = 2:5)
```

We now consider the possible non-linear dependence of log-hazard on age.
First we consider the martingale residual plot.

```
cases <- diff(c(0,idno)) != 0
aids.res <- residuals(aids.cox, collapse = idno)
scatter.smooth(age[cases], aids.res, xlab = "age",
               ylab = "martingale residual")
```

This shows a slight rise in residual with age over 60, but no obvious effect. The
next step is to augment a linear term in age by a step function, with breaks chosen

from prior experience. We set the base level to be the 31–40 age group by using
`relevel`, which re-orders the factor levels.

```
age2 <- cut(age, c(-1, 15, 30, 40, 50, 60, 100))
c.age <- factor(as.numeric(age2), labels = c("0-15", "16-30",
   "31-40", "41-50", "51-60", "61+"))
table(c.age)
 0-15 16-30 31-40 41-50 51-60 61+
   39  1022  1583   987   269  85
c.age <- relevel(c.age, "31-40")

summary(coxph(Surv(start, stop, status) ~ zid + state
   + T.categ + age + c.age, data = Aids3))
   ....
                 coef   exp(coef) se(coef)       z       p
   ....
         age  0.009218     1.009  0.00818   1.1266 0.2600
  c.age0-15  0.499093     1.647  0.36411   1.3707 0.1700
 c.age16-30 -0.019631     0.981  0.09592  -0.2047 0.8400
 c.age41-50 -0.004818     0.995  0.09714  -0.0496 0.9600
 c.age51-60  0.198136     1.219  0.18199   1.0887 0.2800
   c.age61+  0.413690     1.512  0.30821   1.3422 0.1800
   ....
Likelihood ratio test= 193  on 18 df,  p=0
   ....
detach()
```

which is not a significant improvement in fit. Beyond this we could fit a smooth
function of `age` via splines, but to save computational time we deferred this to
the parametric analysis, which we now consider. From the survivor curves the
obvious model is the Weibull. Since this is both a proportional hazards model and
an accelerated-life model, we can include the effect of the introduction of zidovudine by assuming a doubling of survival after July 1987. With 'time' computed
on this basis we find

```
make.aidsp <- function(){
    cutoff <- 10043
    btime <- pmin(cutoff, Aids2$death) - pmin(cutoff, Aids2$diag)
    atime <- pmax(cutoff, Aids2$death) - pmax(cutoff, Aids2$diag)
    survtime <- btime + 0.5*atime
    status <- as.numeric(Aids2$status)
    data.frame(survtime, status = status - 1, state = Aids2$state,
       T.categ = Aids2$T.categ, age = Aids2$age, sex = Aids2$sex)
}

Aidsp <- make.aidsp()
aids.wei <- survReg(Surv(survtime + 0.9, status) ~   state
                    + T.categ + sex + age, data = Aidsp)
summary(aids.wei, cor = F)
   ....
Coefficients:
```

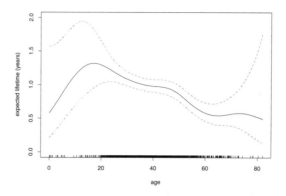

Figure 13.12: Predicted survival versus age of a NSW hs patient (solid line), with point-wise 95% confidence intervals (dashed lines) and a rug of all observed ages.

	Value	Std. Error	z	p
(Intercept)	6.41825	0.2098	30.5970	1.34e-205
stateOther	0.09387	0.0931	1.0079	3.13e-01
stateQLD	-0.18213	0.0913	-1.9956	4.60e-02
stateVIC	-0.00750	0.0637	-0.1177	9.06e-01
T.categhsid	0.09363	0.1582	0.5918	5.54e-01
T.categid	0.40132	0.2552	1.5727	1.16e-01
T.categhet	0.67689	0.2744	2.4667	1.36e-02
T.categhaem	-0.34090	0.1956	-1.7429	8.14e-02
T.categblood	-0.17336	0.1429	-1.2131	2.25e-01
T.categmother	-0.40186	0.6123	-0.6563	5.12e-01
T.categother	-0.11279	0.1696	-0.6649	5.06e-01
sex	-0.00426	0.1827	-0.0233	9.81e-01
age	-0.01374	0.0026	-5.2862	1.25e-07
Log(scale)	0.03969	0.0193	2.0572	3.97e-02

```
Scale= 1.04
```

Note that we continue to avoid zero survival. This shows good agreement with the parameters for the Cox model. The parameter α (the reciprocal of the scale) is close to one. For practical purposes the exponential is a good fit, and the parameters are little changed.

We also considered parametric non-linear functions of age by using a spline function. We use the P-splines of Eilers and Marx (1996) as this is implemented in both survReg and coxph; it can be seen as a convenient approximation to smoothing splines. For useful confidence intervals we include the constant term in the predictions, which are for a NSW hs patient. Note that for valid prediction with pspline the range of the new data must exactly match that of the old data.

```
> survReg(Surv(survtime + 0.9, status) ~ state + T.categ
  + age, data = Aidsp)
  ....
Scale= 1.0405
```

```
Loglik(model)= -12111   Loglik(intercept only)= -12140

> (aids.ps <- survReg(Surv(survtime + 0.9, status) ~  state
    + T.categ + pspline(age, df=6), data = Aidsp))
    ....
                                coef se(coef)     se2 Chisq    DF
              (Intercept)    4.83189 0.82449  0.60594 34.34 1.00
    ....
pspline(age, df = 6), lin -0.01362 0.00251  0.00251 29.45 1.00
pspline(age, df = 6), non                           9.82 5.04
                                  p
    ....
pspline(age, df = 6), lin 5.8e-08
pspline(age, df = 6), non 8.3e-02
    ....
> zz <- predict(aids.ps, data.frame(
    state = factor(rep("NSW", 83), levels = levels(Aidsp$state)),
    T.categ = factor(rep("hs", 83), levels = levels(Aidsp$T.categ)),
    age = 0:82), se = T, type = "linear")
> plot(0:82, exp(zz$fit)/365.25, type = "l", ylim = c(0, 2),
    xlab = "age", ylab = "expected lifetime (years)")
> lines(0:82, exp(zz$fit+1.96*zz$se.fit)/365.25, lty = 3, col = 2)
> lines(0:82, exp(zz$fit-1.96*zz$se.fit)/365.25, lty = 3, col = 2)
> rug(Aidsp$age + runif(length(Aidsp$age), -0.5, 0.5),
      ticksize = 0.015)
```

The results (Figure 13.12) suggest that a non-linear in age term is not worthwhile, although there are too few young people to be sure. We predict log-time to get confidence intervals on that scale.

Chapter 14

Time Series Analysis

There are now many books on time series. Our philosophy and notation are close to those of the applied book by Diggle (1990) (from which some of our examples are taken). Brockwell and Davis (1991) and Priestley (1981) provide more theoretical treatments, and Bloomfield (2000) and Priestley are particularly thorough on spectral analysis. Brockwell and Davis (1996) and Shumway and Stoffer (2000) provide readable introductions to time series theory and practice.

Functions for time series have been included in S for some years, and further time-series support was one of the earliest enhancements of S-PLUS. In S-PLUS regularly spaced time series are of class "rts", and are created by the function rts. (R uses class "ts", and most of its time-series functions are in package ts.) S-PLUS has a further set of time series classes[1] aimed at event- and calendar-based series. These supersede the older classes cts for series of dates and its for irregularly spaced series, but like them are only useful for manipulating and plotting such time series; no new analysis functions are provided.

Our first running example is lh, a series of 48 observations at 10-minute intervals on luteinizing hormone levels for a human female taken from Diggle (1990). Printing it in S-PLUS gives

```
> lh
 1: 2.4 2.4 2.4 2.2 2.1 1.5 2.3 2.3 2.5 2.0 1.9 1.7 2.2 1.8
15: 3.2 3.2 2.7 2.2 2.2 1.9 1.9 1.8 2.7 3.0 2.3 2.0 2.0 2.9
29: 2.9 2.7 2.7 2.3 2.6 2.4 1.8 1.7 1.5 1.4 2.1 3.3 3.5 3.5
43: 3.1 2.6 2.1 3.4 3.0
 start deltat frequency
     1      1         1
```

which shows the attribute vector tspar of the class "rts", which is used for plotting and other computations. The components are the start, the label for the first observation, deltat (Δt), the increment between observations and frequency, the reciprocal of deltat. Note that the final index can be deduced from the attributes and length. Any of start, deltat, frequency and end can be specified in the call to rts, provided they are specified consistently.

Our second example is a seasonal series. Our dataset deaths gives monthly deaths in the UK from a set of common lung diseases for the years 1974 to 1979,

---

[1]See the help on timeSeries and signalSeries.

from Diggle (1990). This prints as

```
> deaths
      Jan  Feb  Mar  Apr  May  Jun  Jul  Aug  Sep  Oct  Nov  Dec
1974: 3035 2552 2704 2554 2014 1655 1721 1524 1596 2074 2199 2512
1975: 2933 2889 2938 2497 1870 1726 1607 1545 1396 1787 2076 2837
  ....
start    deltat frequency
 1974 0.083333        12
```

Notice how it is laid out by years. Quarterly data are also treated specially.

There is a series of functions to extract aspects of the time base:

```
> tspar(deaths)      # tsp(deaths) in R
start    deltat frequency
 1974 0.083333        12
> start(deaths)
[1] 1974
> end(deaths)
[1] 1979.9
> frequency(deaths)
 frequency
        12
> cycle(deaths)
      Jan Feb Mar Apr May Jun Jul Aug Sep Oct Nov Dec
1974:   1   2   3   4   5   6   7   8   9  10  11  12
1975:   1   2   3   4   5   6   7   8   9  10  11  12
  ....
start    deltat frequency
 1974 0.083333        12
```

Time series can be plotted by `plot`, but the functions `ts.plot`, `ts.lines` and `ts.points` are provided for time-series objects. All can plot several related series together. For example, the `deaths` series is the sum of two series `mdeaths` and `fdeaths` for males and females. Figure 14.1 was created by

```
> par(mfrow = c(2, 2))
> ts.plot(lh)
> ts.plot(deaths, mdeaths, fdeaths,
          lty = c(1, 3, 4), xlab = "year", ylab = "deaths")
```

The functions `ts.union` and `ts.intersect` bind together multiple time series. The time axes are aligned and only observations at times that appear in all the series are retained with `ts.intersect`; with `ts.union` the combined series covers the whole range of the components, possibly as `NA` values.

The function `window` extracts a sub-series of a single or multiple time series, by specifying `start` and/or `end`.

The function `lag` shifts the time axis of a series back by k positions, default one. Thus `lag(deaths, k = 3)` is the series of deaths shifted one quarter into the past. This can cause confusion, as most people think of lags as shifting time and not the series; that is, the current value of a series lagged by one year is last year's, not next year's.

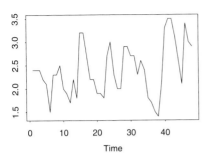 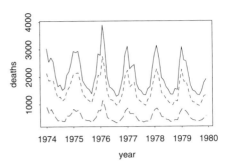

Figure 14.1: Plots by `ts.plot` of `lh` and the three series on deaths by lung diseases. In the right-hand plot the dashed series is for males, the long dashed series for females and the solid line for the total.

The function `diff` takes the difference between a series and its lagged values, and so returns a series of length $n - k$ with values lost from the beginning (if $k > 0$) or end. (The argument `lag` specifies k and defaults to one. Note that the lag is used in the usual sense here, so `diff(deaths, lag = 3)` is equal to `deaths - lag(deaths, k = -3)`!) The function `diff` has an argument `differences` which causes the operation to be iterated. For later use, we denote the dth difference of series X_t by $\nabla^d X_t$, and the dth difference at lag s by $\nabla_s^d X_t$.

The (generic) function `aggregate` can be used to change the frequency of the time base. For example, to obtain quarterly sums or annual means of `deaths`:

```
> aggregate(deaths, 4, sum)
         1    2    3    4
1974: 8291 6223 4841 6785
    ....
> aggregate(deaths, 1, mean)
1974: 2178.3 2175.1 2143.2 1935.8 1995.9 1911.5
```

Each of the functions `lag`, `diff` and `aggregate` can also be applied to multiple time series objects formed by `ts.union` or `ts.intersect`.

14.1 Second-Order Summaries

The theory for time series is based on the assumption of second-order stationarity after removing any trends (which will include seasonal trends). Thus second moments are particularly important in the practical analysis of time series. We assume that the series X_t runs throughout time, but is observed only for $t = 1, \ldots, n$. We use the notations X_t and $X(t)$ interchangeably. The series has a mean μ, often taken to be zero, and the covariance and correlation

$$\gamma_t = \text{cov}\,(X_{t+\tau}, X_\tau), \qquad \rho_t = \text{corr}\,(X_{t+\tau}, X_\tau)$$

do not depend on τ. The covariance is estimated for $t > 0$ from the $n - t$ observed pairs $(X_{1+t}, X_1), \ldots, (X_n, X_{n-t})$. If we just take the standard correlation or covariance of these pairs we use different estimates of the mean and variance for each of the subseries $X_{1+t}, \ldots, X_n$ and $X_1, \ldots, X_{n-t}$, whereas under our assumption of second-order stationarity these have the same mean and variance. This suggests the estimators

$$c_t = \frac{1}{n} \sum_{s=\max(1,-t)}^{\min(n-t,n)} [X_{s+t} - \overline{X}][X_s - \overline{X}], \qquad r_t = \frac{c_t}{c_0}$$

Note that we use divisor n even though there are $n - |t|$ terms. This is to ensure that the sequence (c_t) is the covariance sequence of some second-order stationary time series.[2] Note that all of γ, ρ, c, r are symmetric functions ($\gamma_{-t} = \gamma_t$ and so on).

The function acf computes and by default plots the sequences (c_t) and (r_t), known as the *autocovariance* and *autocorrelation* functions. The argument type controls which is used, and defaults to the correlation.

Our definitions are easily extended to several time series observed over the same interval. Let

$$\gamma_{ij}(t) = \mathrm{cov}\,(X_i(t + \tau), X_j(\tau))$$

$$c_{ij}(t) = \frac{1}{n} \sum_{s=\max(1,-t)}^{\min(n-t,n)} [X_i(s+t) - \overline{X_i}][X_j(s) - \overline{X_j}]$$

which are not symmetric in t for $i \neq j$. These forms are used by acf for multiple time series:

```
acf(lh)
acf(lh, type = "covariance")
acf(deaths)
acf(ts.union(mdeaths, fdeaths))
```

The type may be abbreviated in any unique way, for example cov. The output is shown in Figures 14.2 and 14.3. Note that approximate 95% confidence limits are shown for the autocorrelation plots; these are for an independent series for which $\rho_t = I(t = 0)$. As with a time series *a priori* one is expecting autocorrelation, these limits must be viewed with caution. In particular, if any ρ_t is non-zero, all the limits are invalid.

Note that for a series with a non-unit frequency such as deaths the lags are expressed in the basic time unit, here years. The function acf chooses the number of lags to plot unless this is specified by the argument lag.max. Plotting can be suppressed by setting argument plot = F. The function returns a list that can be plotted subsequently by acf.plot (S-PLUS) or plot.acf (R).

The plots of the deaths series show the pattern typical of seasonal series, and the autocorrelations do not damp down for large lags. Note how one of the cross-series is only plotted for negative lags. We have $c_{ji}(t) = c_{ij}(-t)$, so the cross

---

[2]That is, the covariance sequence is positive-definite.

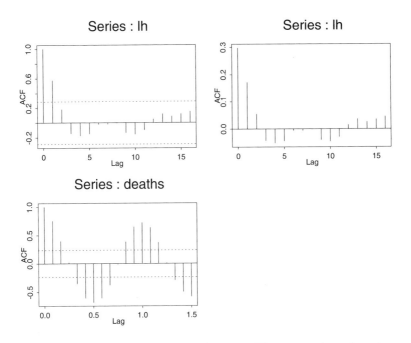

Figure 14.2: `acf` plots for the series `lh` and `deaths`. The top row shows the autocorrelation (left) and autocovariance (right).

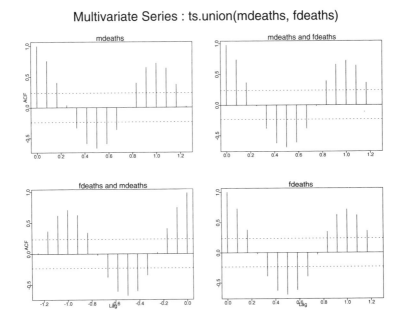

Figure 14.3: Autocorrelation plots for the multiple time series of male and female deaths.

terms are needed for all lags, whereas the terms for a single series are symmetric about 0. The labels are confusing; the plot in row 2 column 1 shows c_{12} for negative lags, a reflection of the plot of c_{21} for positive lags.

Spectral analysis

The spectral approach to second-order properties is better able to separate short-term and seasonal effects, and also has a sampling theory that is easier to use for non-independent series.

We only give a brief treatment; extensive accounts are given by Bloomfield (2000) and Priestley (1981). Be warned that accounts differ in their choices of where to put the constants in spectral analysis; we have tried to follow S-PLUS as far as possible.

The covariance sequence of a second-order stationary time series can always be expressed as

$$\gamma_t = \frac{1}{2\pi} \int_{-\pi}^{\pi} e^{i\omega t} \, dF(\omega)$$

for the *spectrum* F, a finite measure on $(-\pi, \pi]$. Under mild conditions that exclude purely periodic components of the series, the measure has a density known as the *spectral density* f, so

$$\gamma_t = \frac{1}{2\pi} \int_{-\pi}^{\pi} e^{i\omega t} f(\omega) \, d\omega = \int_{-1/2}^{1/2} e^{2\pi i \omega_f t} f(2\pi \omega_f) \, d\omega_f \qquad (14.1)$$

where in the first form the frequency ω is in units of radians/time and in the second form ω_f is in units of cycles/time, and in both cases time is measured in units of Δt. If the time series object has a `frequency` greater than one and time is measured in the base units, the spectral density will be divided by `frequency`.

The Fourier integral can be inverted to give

$$f(\omega) = \sum_{-\infty}^{\infty} \gamma_t e^{-i\omega t} = \gamma_0 \left[1 + 2 \sum_{1}^{\infty} \rho_t \cos(\omega t) \right] \qquad (14.2)$$

By the symmetry of γ_t, $f(-\omega) = f(\omega)$, and we need only consider f on $(0, \pi)$. Equations (14.1) and (14.2) are the first place the differing constants appear. Bloomfield and Brockwell & Davis omit the factor $1/2\pi$ in (14.1) which therefore appears in (14.2).

The basic tool in estimating the spectral density is the *periodogram*. For a frequency ω we effectively compute the squared correlation between the series and the sine/cosine waves of frequency ω by

$$I(\omega) = \left| \sum_{t=1}^{n} e^{-i\omega t} X_t \right|^2 \Big/ n = \frac{1}{n} \left[\left\{ \sum_{t=1}^{n} X_t \sin(\omega t) \right\}^2 + \left\{ \sum_{t=1}^{n} X_t \cos(\omega t) \right\}^2 \right]$$

$$(14.3)$$

Frequency 0 corresponds to the mean, which is normally removed. The frequency π corresponds to a cosine series of alternating ± 1 with no sine series. S-PLUS appears to divide by the `frequency` to match its view of the spectral density.

The periodogram is related to the autocovariance function by

$$I(\omega) = \sum_{-\infty}^{\infty} c_t e^{-i\omega t} = c_0 \left[1 + 2 \sum_{1}^{\infty} r_t \cos(\omega t) \right]$$

$$c_t = \frac{1}{2\pi} \int_{-\pi}^{\pi} e^{i\omega t} I(\omega) \, d\omega$$

and so conveys the same information. However, each form makes some of that information easier to interpret.

Asymptotic theory shows that $I(\omega) \sim f(\omega)E$ where E has a standard exponential distribution, except for $\omega = 0$ and $\omega = \pi$. Thus if $I(\omega)$ is plotted on log scale, the variation about the spectral density is the same for all $\omega \in (0, \pi)$ and is given by a Gumbel distribution (as that is the distribution of $\log E$). Furthermore, $I(\omega_1)$ and $I(\omega_2)$ will be asymptotically independent at distinct frequencies. Indeed if ω_k is a *Fourier frequency* of the form $\omega_k = 2\pi k/n$, then the periodogram at two Fourier frequencies will be approximately independent for large n. Thus although the periodogram itself does not provide a consistent estimator of the spectral density, if we assume that the latter is smooth, we can average over adjacent independently distributed periodogram ordinates and obtain a much less variable estimate of $f(\omega)$. A kernel smoother is used of the form

$$\hat{f}(\omega) = \frac{1}{h} \int K\left(\frac{\lambda - \omega}{h}\right) I(\lambda) \, d\lambda$$

$$\approx \frac{2\pi}{nh} \sum_k K\left(\frac{\omega_k - \omega}{h}\right) I(\omega_k) = \sum_k g_k I(\omega_k)$$

for a probability density K. The parameter h controls the degree of smoothing. To see its effect we approximate the mean and variance of $\hat{f}(\omega)$:

$$\operatorname{var}\left(\hat{f}(\omega)\right) \approx \sum_k g_k^2 f(\omega_k)^2 \approx f(\omega)^2 \sum_k g_k^2 \approx \frac{2\pi}{nh} f(\omega)^2 \int K(x)^2 \, dx$$

$$E\left(\hat{f}(\omega)\right) \approx \sum_k g_k f(\omega_k) \approx f(\omega) + \frac{f''(\omega)}{2} \sum_k g_k (\omega_k - \omega)^2$$

$$\operatorname{bias}\left(\hat{f}(\omega)\right) \approx \frac{f''(\omega)}{2} h^2 \int x^2 K(x) \, dx$$

so as h increases the variance decreases but the bias increases. We see that the ratio of the variance to the squared mean is approximately $g^2 = \sum_k g_k^2$. If $\hat{f}(\omega)$ had a distribution proportional to χ_ν^2, this ratio would be $2/\nu$, so $2/g^2$ is referred to as the equivalent degrees of freedom. Bloomfield and S-PLUS refer to $\sqrt{2 \operatorname{bias}\left(\hat{f}(\omega)\right)/f''(\omega)}$ as the *bandwidth*, which is proportional to h.

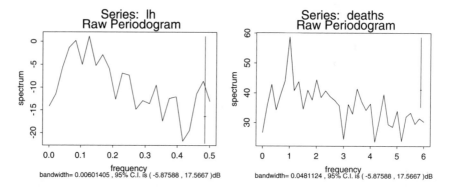

Figure 14.4: Periodogram plots for `lh` and `deaths`.

To understand these quantities, consider a simple moving average over $2m + 1$ Fourier frequencies centred on a Fourier frequency ω. Then the variance is $f(\omega)^2/(2m+1)$ and the equivalent degrees of freedom are $2(2m+1)$, as we would expect on averaging $2m+1$ exponential (or χ_2^2) variates. The bandwidth is approximately

$$\frac{(2m+1)2\pi}{n}\,\frac{1}{\sqrt{12}}$$

and the first factor is the width of the window in frequency space. (Since S-PLUS works in cycles rather than radians, the bandwidth is about $(2m + 1)/n\sqrt{12}$ frequency on its scale.) The bandwidth is thus a measure of the size of the smoothing window, but rather smaller than the effective width.

The workhorse function for spectral analysis is `spectrum`, which with its default options computes and plots the periodogram on log scale. The function `spectrum` calls `spec.pgram` to do most of the work. (Note: `spectrum` by default removes a linear trend from the series before estimating the spectral density.) For our examples we can use:

```
par(mfrow = c(2, 2))
spectrum(lh)
spectrum(deaths)
```

with the result shown in Figure 14.4.

Note how elaborately labelled[3] the figures are. The plots are on log scale, in units of *decibels*; that is, the plot is of $10\log_{10} I(\omega)$. The function `spec.pgram` returns the bandwidth and (equivalent) degrees of freedom as components `bandwidth` and `df`.

The function `spectrum` also produces smoothed plots, using repeated smoothing with modified Daniell smoothers (Bloomfield, 2000, p. 157), which are moving averages giving half weight to the end values of the span. Trial-and-error is needed to choose the spans (Figures 14.5 and 14.6):

---

[3]In S-PLUS; less so in R, even if `options(ts.S.compat = TRUE)` has been set.

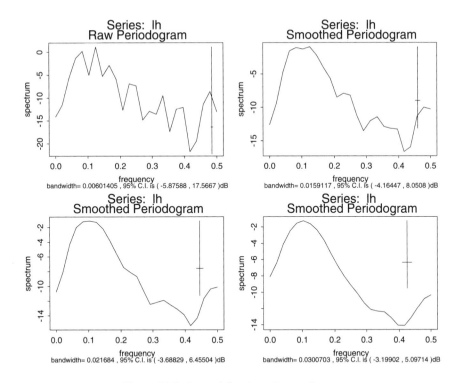

Figure 14.5: Spectral density estimates for `lh`.

```
par(mfrow = c(2, 2))
spectrum(lh)
spectrum(lh, spans = 3)
spectrum(lh, spans = c(3, 3))
spectrum(lh, spans = c(3, 5))

spectrum(deaths)
spectrum(deaths, spans = c(3, 3))
spectrum(deaths, spans = c(3, 5))
spectrum(deaths, spans = c(5, 7))
```

The spans should be odd integers, and it helps to produce a smooth plot if they
are different and at least two are used. The width of the centre mark on the 95%
confidence interval indicator indicates the bandwidth.

The periodogram has other uses. If there are periodic components in the series
the distribution theory given previously does not apply, but there will be peaks in
the plotted periodogram. Smoothing will reduce those peaks, but they can be seen
quite clearly by plotting the *cumulative periodogram*

$$U(\omega) = \sum_{0 < \omega_k \leqslant \omega} I(\omega_k) \, \Big/ \, \sum_{1}^{\lfloor n/2 \rfloor} I(\omega_k)$$

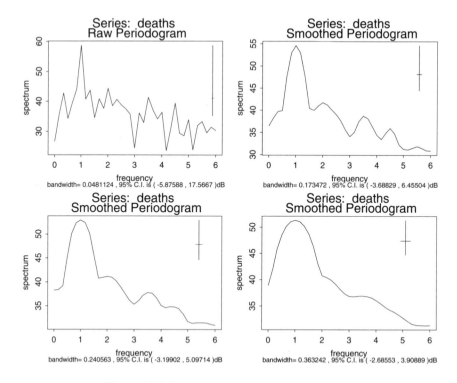

Figure 14.6: Spectral density estimates for `deaths`.

against ω. The cumulative periodogram is also very useful as a test of whether a particular spectral density is appropriate, as if we replace $I(\omega)$ by $I(\omega)/f(\omega)$, $U(\omega)$ should be a straight line. Furthermore, asymptotically, the maximum deviation from that straight line has a distribution given by that of the Kolmogorov–Smirnov statistic, with a 95% limit approximately $1.358/[\sqrt{m}+0.11+0.12/\sqrt{m}]$ where $m = \lfloor n/2 \rfloor$ is the number of Fourier frequencies included. This is particularly useful for a residual series with f constant in testing if the series is uncorrelated.

We wrote a function `cpgram` to plot the cumulative periodogram: the results for our examples are shown in Figure 14.7, with 95% confidence bands.

```
cpgram(lh)
cpgram(deaths)
```

The distribution theory can be made more accurate, and the peaks made sharper, by *tapering* the de-meaned series (Bloomfield, 2000). The magnitude of the first α and last α of the series is tapered down towards zero by a cosine bell; that is, X_t is replaced by

$$
X'_t = \begin{cases}
(1 - \cos\frac{\pi(t-0.5)}{\alpha n})X_t & t \leqslant \alpha n \\
X_t & \alpha n < t < (1-\alpha)n \\
(1 - \cos\frac{\pi(n-t+0.5)}{\alpha n})X_t & t \geqslant (1-\alpha)n
\end{cases}
$$

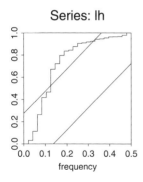
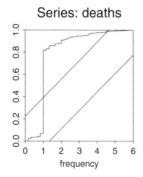

Figure 14.7: Cumulative periodogram plots for `lh` and `deaths`.

The proportion α is controlled by the parameter `taper` of `spec.pgram`, and defaults to 10%. It should rarely need to be altered. (The taper function `spec.taper` can be called directly if needed.) Tapering does increase the variance of the periodogram and hence the spectral density estimate, by about 12% for the default taper, but it will decrease the bias near peaks very markedly, if those peaks are not at Fourier frequencies.

14.2 ARIMA Models

In the late 1960s Box and Jenkins advocated a methodology for time series based on finite-parameter models for the second-order properties, so this approach is often named after them. Let ϵ_t denote a series of uncorrelated random variables with mean zero and variance σ^2. A moving average process of order q (MA(q)) is defined by

$$X_t = \sum_0^q \beta_j \epsilon_{t-j} \qquad (14.4)$$

an autoregressive process of order p (AR(p)) is defined by

$$X_t = \sum_1^p \alpha_i X_{t-i} + \epsilon_t \qquad (14.5)$$

and an ARMA(p, q) process is defined by

$$X_t = \sum_1^p \alpha_i X_{t-i} + \sum_0^q \beta_j \epsilon_{t-j} \qquad (14.6)$$

Note that we do not need both σ^2 and β_0 for a MA(q) process, and we take $\beta_0 = 1$. Some authors put the regression terms of (14.5) and (14.6) on the left-hand side and reverse the sign of α_i. Any of these processes can be given mean μ by adding μ to each observation.

An ARIMA(p, d, q) process (where the I stands for integrated) is a process whose dth difference $\nabla^d X$ is an ARMA(p, q) process.

Equation (14.4) will always define a second-order stationary time series, but (14.5) and (14.6) need not. They need the condition that all the (complex) roots of the polynomial

$$\phi_\alpha(z) = 1 - \alpha_1 z - \cdots - \alpha_p z^p$$

lie outside the unit disc. (The function `polyroot` can be used to check this.) However, there are in general 2^q sets of coefficients in (14.4) that give the same second-order properties, and it is conventional to take the set with roots of

$$\phi_\beta(z) = 1 + \beta_1 z + \cdots + \beta_q z^q$$

on or outside the unit disc. Let B be the backshift or lag operator defined by $BX_t = X_{t-1}$. Then we conventionally write an ARMA process as

$$\phi_\alpha(B)X = \phi_\beta(B)\epsilon \tag{14.7}$$

The function `arima.sim` simulates an ARIMA process. Simple usage is of the form

```
ts.sim <- arima.sim(list(order = c(1,1,0), ar = 0.7), n = 200)
```

which generates a series whose first differences follow an AR(1) process.

Model identification

A lot of attention has been paid to *identifying* ARMA models, that is choosing plausible values of p and q by looking at the second-order properties. Much of the literature is reviewed by de Gooijer *et al.* (1985). Nowadays it is computationally feasible to fit all plausible models and choose on the basis of their goodness of fit, but some simple diagnostics are still useful. For an MA(q) process we have

$$\gamma_k = \sigma^2 \sum_{i=0}^{q-|k|} \beta_i \beta_{i+|k|}$$

which is zero for $|k| > q$, and this may be discernible from plots of the ACF. For an AR(p) process the population autocovariances are generally all non-zero, but they satisfy the Yule–Walker equations

$$\rho_k = \sum_{1}^{p} \alpha_i \rho_{k-i}, \qquad k > 0 \tag{14.8}$$

This motivates the *partial autocorrelation function*. The partial correlation between X_s and X_{s+t} is the correlation after regression on $X_{s+1}, \ldots, X_{s+t-1}$, and is zero for $t > p$ for an AR(p) process. The PACF can be estimated by solving the Yule–Walker equations (14.8) with $p = t$ and ρ replaced by r, and is given by the `type = "partial"` option of `acf`:

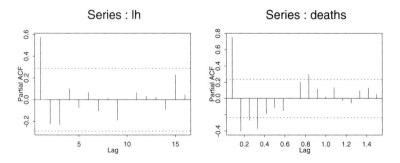

Figure 14.8: Partial autocorrelation plots for the series `lh` and `deaths`.

```
acf(lh, type = "partial")
acf(deaths, type = "partial")
```

as shown in Figure 14.8. These are short series, so no definitive pattern emerges, but `lh` might be fitted well by an AR(1) or perhaps an AR(3) process.

Function `ARMAacf` will compute the theoretical ACF or PACF for any ARMA process.

Model fitting

Selection among ARMA processes can be done by Akaike's information criterion (AIC) which penalizes the deviance by twice the number of parameters; the model with the smallest AIC is chosen. (All likelihoods considered assume a Gaussian distribution for the time series.) Fitting can be done by the functions `ar` or `arima`. The output here is from R.

```
> (lh.ar1 <- ar(lh, aic = F, order.max = 1))
Coefficients:
    1
0.576

Order selected 1  sigma^2 estimated as  0.208
> cpgram(lh.ar1$resid, main = "AR(1) fit to lh")
> (lh.ar <- ar(lh, order.max = 9))
Coefficients:
     1      2       3
 0.653  -0.064  -0.227

Order selected 3  sigma^2 estimated as  0.196
> lh.ar$aic
 [1] 18.30668  0.99567  0.53802  0.00000  1.49036  3.21280
 [7]  4.99323  6.46950  8.46258  8.74120
> cpgram(lh.ar$resid, main = "AR(3) fit to lh")
```

This first fits an AR(1) process and obtains, after removing the mean,

$$X_t = 0.576 X_{t-1} + \epsilon_t$$

AR(1) fit to lh

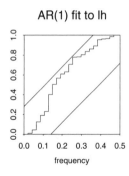

AR(3) fit to lh

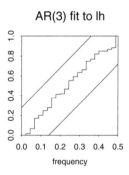

Figure 14.9: Cumulative periodogram plots for residuals of AR models fitted to lh.

with $\hat{\sigma}^2 = 0.208$. It then uses AIC to choose the order among AR processes, selects $p = 3$ and fits

$$X_t = 0.653X_{t-1} - 0.064X_{t-2} - 0.227X_{t-3} + \epsilon_t$$

with $\hat{\sigma}^2 = 0.196$ and AIC is reduced by 0.996. (For ar the component aic is the excess over the best fitting model, and it starts from $p = 0$.) The function ar by default fits the model by solving the Yule–Walker equations (14.8) with ρ replaced by r. An alternative is to use method = "burg"; R also has methods "ols" and "mle".

The diagnostic plots are shown in Figure 14.9. The cumulative periodograms of the residuals show that the AR(1) process has not removed all the correlation.

We can also use the function arima, in library section MASS for S-PLUS and in package ts for R. This includes a mean in the model (for $d = 0$) and maximizes the full likelihood, whereas S-PLUS has a function arima.mle which maximizes a likelihood conditional on $p + d$ starting values for a non-seasonal series (or n.cond if this is larger) but does not include a mean.

```
> (lh.arima1 <- arima(lh, order = c(1,0,0)))
Coefficients:
         ar1  intercept
       0.574      2.413
s.e.   0.116      0.147

sigma^2 = 0.197:  log likelihood = -29.38,  aic = 64.76
> tsdiag(lh.arima1)
> (lh.arima3 <- arima(lh, order = c(3,0,0)))
Coefficients:
         ar1      ar2      ar3  intercept
       0.645   -0.063   -0.220      2.393
s.e.   0.139    0.167    0.142      0.096

sigma^2 = 0.179:  log likelihood = -27.09,  aic = 64.18
> tsdiag(lh.arima3)
> (lh.arima11 <- arima(lh, order = c(1,0,1)))
```

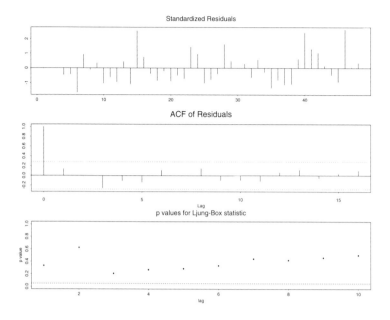

Figure 14.10: Diagnostic plots for AR(1) model fitted to `lh`.

```
Coefficients:
         ar1    ma1 intercept
      0.452 0.198     2.410
s.e.  0.136 0.178     0.136

sigma^2 = 0.192:  log likelihood = -28.76,  aic = 65.52
> tsdiag(lh.arima11)
```

The diagnostic plots are shown in Figure 14.10. The bottom panel shows the P values for the Ljung and Box (1978) *portmanteau test*

$$Q_K = n(n+2) \sum_{1}^{K} (n-k)^{-1} c_k^2 \tag{14.9}$$

applied to the residuals, for a range of values of K. Here the maximum K is set by the parameter `gof.lag` (which defaults to 10). Note that although an AR(3) model fits better according to AIC, a formal likelihood ratio test of twice the difference, 4.58, using a χ_2^2 distribution is not significant.

Function `arima` can also include differencing and so fit an ARIMA model (the middle integer in `order` is d).

Forecasting

Forecasting is straightforward using the `predict` method:

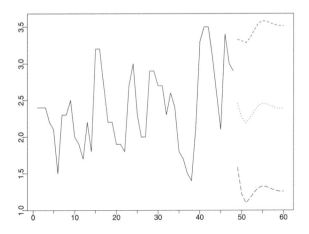

Figure 14.11: Forecasts for 12 periods (of 10mins) ahead for the series `lh`. The dashed curves are approximate pointwise 95% confidence intervals.

```
lh.fore <- predict(lh.arima3, 12)
ts.plot(lh, lh.fore$pred, lh.fore$pred + 2*lh.fore$se,
        lh.fore$pred - 2*lh.fore$se)
```

(see Figure 14.11) but the standard errors do not include the effect of estimating the mean and the parameters of the ARIMA model.

Spectral densities via AR processes

The spectral density of an ARMA process has a simple form; it is given by

$$f(\omega) = \sigma^2 \left| \frac{1 + \sum_s \beta_s e^{-is\omega}}{1 - \sum_t \alpha_t e^{-it\omega}} \right|^2 \qquad (14.10)$$

and so we can estimate the spectral density by substituting parameter estimates in (14.10). It is most usual to fit high-order AR models, both because they can be fitted rapidly, and since they can produce peaks in the spectral density estimate by small values of $|1 - \sum \alpha_t e^{-it\omega}|$ (which correspond to nearly non-stationary fitted models since there must be roots of $\phi_\alpha(z)$ near $e^{-i\omega}$).

This procedure is implemented by function `spectrum` with `method = "ar"`, which calls `spec.ar`. Although popular because it often produces visually pleasing spectral density estimates, it is not recommended (for example, Thomson, 1990).

Regression terms

The `arima` function can also handle regressions with ARIMA residual processes, that is, models of the form

$$X_t = \sum \gamma_i Z_t^{(i)} + \eta_t, \qquad \phi_\alpha(B) \Delta^d \eta_t = \phi_\beta(B)\epsilon \qquad (14.11)$$

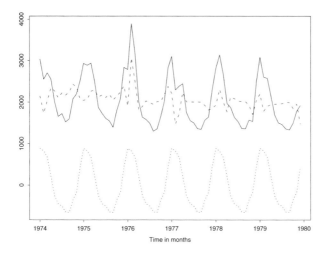

Figure 14.12: stl decomposition for the deaths series (solid line). The dotted series is the seasonal component, the dashed series the remainder.

for one or more external time series $Z^{(i)}$. Again, the variability of the parameter estimates $\widehat{\gamma}$ is not taken into account in the computed prediction standard errors.

Some examples are given in Section 14.5.

14.3 Seasonality

For a seasonal series there are two possible approaches. One is to decompose the series, usually into a trend, a seasonal component and a residual, and to apply non-seasonal methods to the residual component. The other is to model all the aspects simultaneously.

Decompositions

The function stl is based on Cleveland *et al.* (1990). It is complex, and the details differ between S environments, so the on-line documentation and the reference should be consulted.

We can extract a strictly periodic component plus a remainder (Figure 14.12).

```
deaths.stl <- stl(deaths, "periodic")
dsd <- deaths.stl$rem
ts.plot(deaths, deaths.stl$sea, dsd)   # R version in scripts
```

We now complete the analysis of the deaths series by analysing the non-seasonal component. The results are shown in Figure 14.13.

```
> ts.plot(dsd); acf(dsd); acf(dsd, type = "partial")
> spectrum(dsd, span = c(3, 3)); cpgram(dsd)
> dsd.ar <- ar(dsd)
```

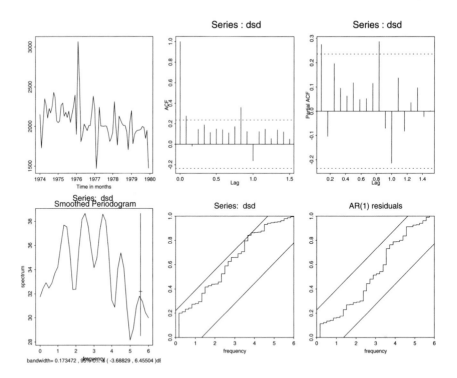

Figure 14.13: Diagnostics for an AR(1) fit to the remainder of an `stl` decomposition of the `deaths` series.

```
> dsd.ar$order
[1] 1
> dsd.ar$aic
 [1]  3.64856  0.00000  1.22644  0.40857  1.75586  3.46936
    ....
> dsd.ar$ar
    ....
[1,] 0.27469
> cpgram(dsd.ar$resid, main = "AR(1) residuals")
```

The large jump in the cumulative periodogram at the lowest (non-zero) Fourier frequency is caused by the downward trend in the series. The spectrum has dips at the integers since we have removed the seasonal component and hence all of the components at that frequency and its multiples. (The dip at frequency 1 is obscured by the peak at the Fourier frequency to its left; see the cumulative periodogram.) The plot of the remainder series shows exceptional values for February–March 1976 and 1977. As there are only six cycles, the seasonal pattern is difficult to establish at all precisely.

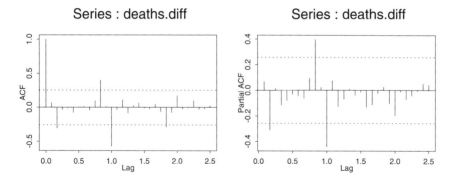

Figure 14.14: Autocorrelation and partial autocorrelation plots for the seasonally differenced `deaths` series. The negative values at lag 12 suggest over-differencing.

Seasonal ARIMA models

The function `diff` allows us to compute differences at lags greater than one, so for a monthly series the difference at lag 12 is the difference from his time last year. Let s denote the period, often 12. We can then consider ARIMA models for the sub-series sampled s apart, for example, for all Januaries. This corresponds to replacing B by B^s in the definition (14.7). Thus an ARIMA $(P, D, Q)_s$ process is a seasonal version of an ARIMA process. However, we may include both seasonal and non-seasonal terms, obtaining a process of the form

$$\Phi_{AR}(B)\Phi_{SAR}(B^s)Y = \Phi_{MA}(B)\Phi_{SMA}(B^s)\epsilon, \qquad Y = (I-B)^d(I-B^s)^D X$$

If we expand this, we see that it is an ARMA $(p + sP, q + sQ)$ model for Y_t, but parametrized in a special way with large numbers of zero coefficients. It can still be fitted as an ARMA process, and `arima` can handle models specified in this form, by specifying the argument `seasonal` with the period set. (Examples follow.)

Perhaps the most commonly used seasonal ARIMA model is the 'airline model', ARIMA$((0, 1, 1) \times (0, 1, 1)_{12})$.

To identify a suitable model for the `nottem` series we first look at the seasonally differenced series. Figure 14.14 suggests that this may be over-differencing, but that the non-seasonal term should be an AR(2).

```
> deaths.diff <- diff(deaths, 12)
> acf(deaths.diff, 30); acf(deaths.diff, 30, type = "partial")
> ar(deaths.diff)
$order:
[1] 12

    ....
$aic:
 [1]  7.8143  9.5471  5.4082  7.3929  8.5839 10.1979 12.1388
 [8] 14.0201 15.7926 17.2504  8.9905 10.9557  0.0000  1.6472
[15]  2.6845  4.4097  6.4047  8.3152
```

```
# this suggests the seasonal effect is still present.
> (deaths.arima1 <- arima(deaths, order = c(2,0,0),
      seasonal = list(order = c(0,1,0), period = 12)) )
Coefficients:
         ar1      ar2
       0.118   -0.300
s.e.   0.126    0.125

sigma^2 = 118960:  log likelihood = -435.83,  aic = 877.66
> tsdiag(deaths.arima1, gof.lag = 30)
# suggests need a seasonal AR term
> (deaths.arima2 <- arima(deaths, order = c(2,0,0),
      list(order = c(1,0,0), period = 12)) )
Coefficients:
         ar1      ar2    sar1   intercept
       0.801   -0.231   0.361     2062.45
s.e.   0.446    0.252   0.426      133.90

sigma^2 = 116053:  log likelihood = -523.16,  aic = 1056.3
> tsdiag(deaths.arima2, gof.lag = 30)
> cpgram(resid(deaths.arima2))
> (deaths.arima3 <- arima(deaths, order = c(2,0,0),
      list(order = c(1,1,0), period = 12)) )
Coefficients:
         ar1      ar2    sar1
       0.293   -0.271  -0.571
s.e.   0.137    0.141   0.103

sigma^2 = 77145:  log likelihood = -425.22,  aic = 858.43
> tsdiag(deaths.arima3, gof.lag = 30)
```

The AR-fitting suggests a model of order 12 (of up to 16) which indicates that seasonal effects are still present. The diagnostics from the ARIMA($(2,0,0) \times (0,1,0)_{12}$) model suggest problems at lag 12. We tried replacing the differencing by a seasonal AR term. (The AICs are not comparable, as a differenced model is not an explanation of all the observations.) The diagnostics suggested that there was still seasonal structure in the residuals, so we next tried including both differencing and a seasonal AR term, for which the diagnostics plots look good.

14.4 Nottingham Temperature Data

We now consider a substantial example. The data are mean monthly air temperatures ($°F$) at Nottingham Castle for the months January 1920–December 1939, from 'Meteorology of Nottingham', in *City Engineer and Surveyor*. They also occur in Anderson (1976). We use the years 1920–1936 to forecast the years 1937–1939 and compare with the recorded temperatures. The data are series `nottem` in `MASS`.

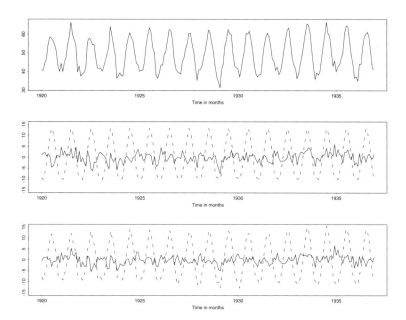

Figure 14.15: Plots of the first 17 years of the `nottem` dataset. Top is the data, middle the `stl` decomposition with a seasonal periodic component and bottom the `stl` decomposition with a 'local' seasonal component.

```
# R versions in the scripts
nott <- window(nottem, end = c(1936, 12))
ts.plot(nott)
nott.stl <- stl(nott, "period")
ts.plot(nott.stl$rem-49, nott.stl$sea,
        ylim  =  c(-15, 15), lty = c(1, 3))
nott.stl <- stl(nott, 5)
ts.plot(nott.stl$rem-49, nott.stl$sea,
        ylim  =  c(-15, 15), lty = c(1, 3))
boxplot(split(nott, cycle(nott)), names = month.abb)
```

Figures 14.15 and 14.16 show clearly that February 1929 is an outlier. It *is* correct—it was an exceptionally cold month in England. The `stl` plots show that the seasonal pattern is fairly stable over time. Since the value for February 1929 will distort the fitting process, we altered it to a low value for February of 35°. We first model the remainder series:

```
> nott[110] <- 35
> nott.stl <- stl(nott, "period")
> nott1 <- nott.stl$rem - mean(nott.stl$rem)
> acf(nott1)
> acf(nott1, type = "partial")
> cpgram(nott1)
> ar(nott1)$aic
 [1] 13.67432  0.00000  0.11133  2.07849  3.40381  5.40125
```

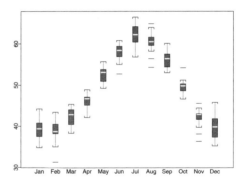

Figure 14.16: Monthly boxplots of the first 17 years of the `nottem` dataset.

```
    ....
> plot(0:23, ar(nott1)$aic, xlab = "order", ylab = "AIC",
      main = "AIC for AR(p)")
> (nott1.ar1 <- arima(nott1, order = c(1,0,0)))
Coefficients:
       ar1 intercept
     0.272     0.005
s.e. 0.067     0.207

sigma^2 = 4.65:  log likelihood = -446.31,  aic = 898.61
> nott1.fore <- predict(nott1.ar1, 36)
> nott1.fore$pred <- nott1.fore$pred + mean(nott.stl$rem) +
                     as.vector(nott.stl$sea[1:36])
> ts.plot(window(nottem, 1937), nott1.fore$pred,
      nott1.fore$pred+2*nott1.fore$se,
      nott1.fore$pred-2*nott1.fore$se, lty = c(3, 1, 2, 2))
> title("via Seasonal Decomposition")
```

(see Figures 14.17 and 14.18), all of which suggest an AR(1) model. (Remember that a seasonal term has been removed, so we expect negative correlation at lag 12.) The confidence intervals in Figure 14.18 for this method ignore the variability of the seasonal terms. We can easily make a rough adjustment. Each seasonal term is approximately the mean of 17 approximately independent observations (since 0.272^{12} is negligible). Those observations have variance about $4.65/(1 - 0.272^2) = 5.02$ about the seasonal term, so the seasonal term has standard error about $\sqrt{5.02/17} = 0.54$, compared to the 2.25 for the forecast. The effect of estimating the seasonal terms is in this case negligible. (Note that the forecast errors are correlated with errors in the seasonal terms.)

We now move to the Box–Jenkins methodology of using differencing:

```
> acf(diff(nott,12), 30)
> acf(diff(nott,12), 30, type = "partial")
> cpgram(diff(nott, 12))
> (nott.arima1 <- arima(nott, order = c(1,0,0),
      list(order = c(2,1,0), period = 12)) )
```

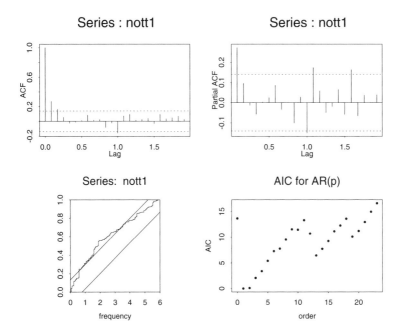

Figure 14.17: Summaries for the remainder series of the nottem dataset.

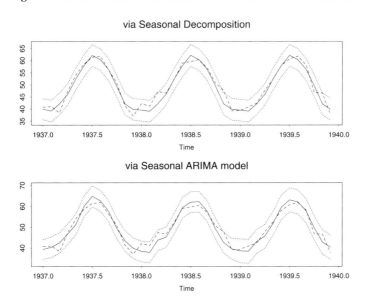

Figure 14.18: Forecasts (solid), true values (dashed) and approximate 95% confidence intervals for the nottem series. The upper plot is via a seasonal decomposition, the lower plot via a seasonal ARIMA model.

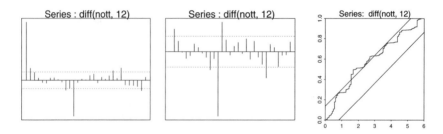

Figure 14.19: Seasonal ARIMA modelling of the `nottem` series. The ACF, partial ACF and cumulative periodogram of the yearly differences.

```
Coefficients:
         ar1     sar1     sar2
       0.324  -0.907   -0.323
s.e.   0.069   0.071    0.075

sigma^2 = 5.45:  log likelihood = -440.34,  aic = 888.68
> tsdiag(nott.arima1, gof.lag = 30)
> (nott.arima2 <- arima(nott, order = c(0,0,2),
      list(order = c(0,1,2), period = 12)) )
Coefficients:
          ma1    ma2    sma1    sma2
        0.261  0.160  -1.011   0.245
s.e.    0.071  0.072   0.070   0.068

sigma^2 = 5.06:  log likelihood = -435.26,  aic = 880.52
> tsdiag(nott.arima2, gof.lag = 30)
> (nott.arima3 <- arima(nott, order = c(1,0,0),
      list(order = c(0,1,2), period = 12)) )
Coefficients:
          ar1    sma1    sma2
        0.293  -1.012   0.245
s.e.    0.069   0.070   0.067

sigma^2 = 5.09:  log likelihood = -435.93,  aic = 879.86
> tsdiag(nott.arima3, gof.lag = 30)
> nott.fore <- predict(nott.arima3, 36)
> ts.plot(window(nottem, 1937), nott.fore$pred,
      nott.fore$pred+2*nott.fore$se,
      nott.fore$pred-2*nott.fore$se, lty = c(3, 1, 2, 2))
> title("via Seasonal ARIMA model")
```

The autocorrelation plots (Figure 14.19) suggest a model with seasonal terms, and either AR or MA models. There is not much to choose between the three models we tried. The predictions are shown in Figure 14.18.

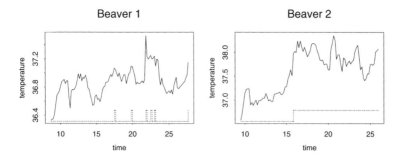

Figure 14.20: Plots of temperature (solid) and activity (dashed) for two beavers. The time is shown in hours since midnight of the first day of observation.

14.5   Regression with Autocorrelated Errors

We touched briefly on the use of regression terms with the functions `arima` on page 402. In this section we consider other ways to use S to study regression with autocorrelated errors. They are most pertinent when the regression rather than time-series prediction is of primary interest.

Our main example is taken from Reynolds (1994). She describes a small part of a study of the long-term temperature dynamics of beaver (*Castor canadensis*) in north-central Wisconsin. Body temperature was measured by telemetry every 10 minutes for four females, but data from one period of less than a day for each of two animals is used there (and here). Columns indicate the day (December 12–13, 1990 and November 3–4, 1990 for the two examples), time (hhmm on a 24-hour clock), temperature ($^\circ$C) and a binary index of activity (0 = animal inside retreat; 1 = animal outside retreat).

Figure 14.20 shows the two series. The first series has a missing observation (at 22:20), and this and the pattern of activity suggest that it is easier to start with beaver 2.

```
beav1 <- beav1; beav2 <- beav2
attach(beav1)
beav1$hours <- 24*(day-346) + trunc(time/100) + (time%%100)/60
detach(); attach(beav2)
beav2$hours <- 24*(day-307) + trunc(time/100) + (time%%100)/60
detach()
par(mfrow = c(2, 2))
plot(beav1$hours, beav1$temp, type = "l", xlab = "time",
     ylab = "temperature", main = "Beaver 1")
usr <- par("usr"); usr[3:4] <- c(-0.2, 8); par(usr = usr)
lines(beav1$hours, beav1$activ, type = "s", lty = 2)
plot(beav2$hours, beav2$temp, type = "l", xlab = "time",
     ylab = "temperature", main = "Beaver 2")
usr <- par("usr"); usr[3:4] <- c(-0.2, 8); par(usr = usr)
lines(beav2$hours, beav2$activ, type = "s", lty = 2)
```

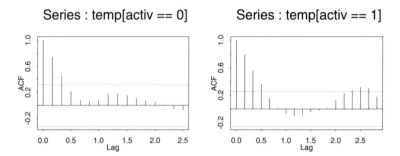

Figure 14.21: ACF of the beaver 2 temperature series before and after activity begins.

Beaver 2

Looking at the series before and after activity begins suggests a moderate amount of autocorrelation, confirmed by the plots in Figure 14.21.

```
attach(beav2)
temp2 <- rts(temp, start = 8+2/3, frequency = 6, units = "hours")
activ2 <- rts(activ, start = 8+2/3, frequency = 6, units = "hours")
acf(temp2[activ2 == 0]); acf(temp2[activ2 == 1]) # also look at PACFs
ar(temp2[activ2 == 0]); ar(temp2[activ2 == 1])
detach(); rm(temp2, activ2)
```

Fitting an $AR(p)$ model to each part of the series selects $AR(1)$ models with coefficients 0.74 and 0.79, so a common $AR(1)$ model for the residual series looks plausible.

We can use the function gls to fit a regression with $AR(1)$ errors.[4]

```
> # R: library(nlme)
> beav2.gls <- gls(temp ~ activ, data = beav2,
                    corr = corAR1(0.8), method = "ML")
> summary(beav2.gls)
    ....
Correlation Structure: AR(1)
 Parameter estimate(s):
     Phi
 0.87318

Coefficients:
             Value Std.Error t-value p-value
(Intercept) 37.192   0.11313  328.75  <.0001
      activ  0.614   0.10873    5.65  <.0001
```

There are some end effects due to the sharp initial rise in temperature:

```
> summary(update(beav2.gls, subset = 6:100))
    ....
Correlation Structure: AR(1)
```

---

[4] gls can use general ARMA error models specified via the corARMA function.

```
Parameter estimate(s):
    Phi
0.83803
Fixed effects: temp ~ activ
            Value Std.Error t-value p-value
(Intercept) 37.250  0.096340  386.65  <.0001
      activ  0.603  0.099319    6.07  <.0001
```

and the REML estimates of the standard errors are somewhat larger.

We can also use arima:

```
> arima(beav2$temp, c(1, 0, 0), xreg = beav2$activ)
Coefficients:
        ar1  intercept  beav2$activ
      0.873     37.192        0.614
s.e.  0.068      0.119        0.138

sigma^2 = 0.0152:  log likelihood = 66.78,  aic = -125.55
```

This computes standard errors from the observed information matrix and not from the asymptotic formula; the difference is usually small.

Beaver 1

Applying the same ideas to beaver 2, we can select an initial covariance model based on the observations before the first activity at 17:30. The autocorrelations again suggest an $AR(1)$ model, whose coefficient is fitted as 0.82. We included as regressors the activity now and 10, 20 and 30 minutes ago.

```
attach(beav1)
temp1 <- rts(c(temp[1:82], NA, temp[83:114]), start = 9.5,
            frequency = 6, units = "hours")
activ1 <- rts(c(activ[1:82], NA, activ[83:114]), start = 9.5,
            frequency = 6, units = "hours")
acf(temp1[1:53]) # and also type = "partial"
ar(temp1[1:53])

act <- c(rep(0, 10), activ1)
beav1b <- data.frame(Time = time(temp1), temp = as.vector(temp1),
            act = act[11:125], act1 = act[10:124],
            act2 = act[9:123], act3 = act[8:122])
detach(); rm(temp1, activ1)
summary(gls(temp ~ act + act1 + act2 + act3,
            data = beav1b, na.action = na.omit,
            corr = corCAR1(0.82^6, ~Time), method = "ML"))

Correlation Structure: Continuous AR(1)
 Formula:  ~ Time
 Parameter estimate(s):
     Phi
0.45557
```

```
Coefficients:
            Value Std.Error t-value p-value
(Intercept) 36.808  0.060973  603.67  <.0001
       act   0.246  0.038868    6.33  <.0001
      act1   0.159  0.051961    3.06  0.0028
      act2   0.150  0.053095    2.83  0.0055
      act3   0.097  0.042581    2.27  0.0252
```

Because there is a missing value, this is a continuous-time AR process with correlation in units per hour, corresponding to 0.877 for a 10-minute lag. We can also use `arima`:

```
> arima(beav1b$temp, c(1, 0, 0), xreg = beav1b[, 3:6])
Coefficients:
        ar1  intercept    act   act1   act2   act3
      0.877     36.808  0.246  0.159  0.150  0.097
s.e.  0.052      0.060  0.038  0.052  0.053  0.042

sigma^2 = 0.0068:  log likelihood = 118.21,  aic = -222.41
```

Our analysis shows that there is a difference in temperature between activity and inactivity, and that temperature may build up gradually with activity (which seems physiologically reasonable). A *caveat* is that the apparent outlier at 21:50, at the start of a period of activity, contributes considerably to this conclusion.

14.6 Models for Financial Series

Most financial time series have characteristics which make ARIMA models unsuitable. Figure 14.22 shows the closing prices of the S&P 500 index on trading days throughout the 1990s. Also shown are the 'daily' returns, $100 \log(X_t/X_{t-1})$. Note that the returns after non-trading days refer to two or more days. A characteristic of such series is the occasional large excursions seen in the returns, and general long-tailed behaviour.

Models for such series fall into two classes, *conditional heteroscedasticity* models and *stochastic volatility* models; both are concisely surveyed by Shephard (1996). Campbell *et al.* (1997, §12.2) give an econometric perspective.

Conditional heteroscedasticity models start with Figure 14.23. Whereas the returns are uncorrelated, their squares have long-range correlations. So one way to think of the process is as white noise with variance depending on the past variances (and perhaps past values). We can have

$$y_t = \sigma_t \epsilon_t, \qquad \sigma_t^2 = \alpha_0 + \alpha_1 y_{t-1}^2 + \cdots + \alpha_p y_{t-p}^2$$

the ARCH model of Engle (1982). This was extended to GARCH(p, q) models,

$$y_t = \sigma_t \epsilon_t, \qquad \sigma_t^2 = \alpha_0 + \sum_{i=1}^{p} \alpha_i y_{t-i}^2 + \sum_{j=1}^{q} \beta_j \sigma_{t-j}^2$$

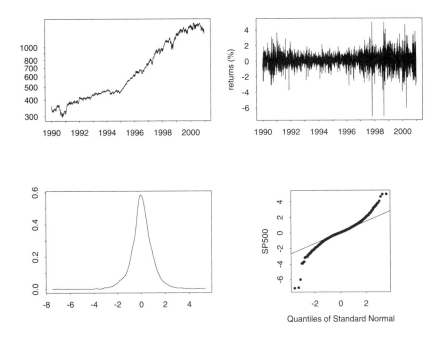

Figure 14.22: Closing prices (upper left) and daily returns (upper right) of the S&P 500 stock index, with density and normal probability plots.

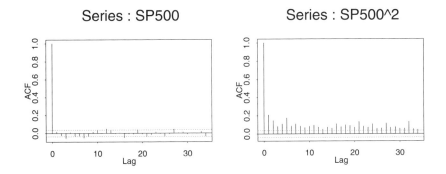

Figure 14.23: ACFs of the daily returns and their squares.

by Bollerslev (1986) and Taylor (1986). Stationarity is assured if $\alpha_0 > 0$ and $\sum \alpha_i + \sum \beta_j \leqslant 1$.

Stochastic volatility models are similar to the case $p = 0$ with added noise; the simplest version of Taylor (1986) is

$$y_t = \exp(h_t/2)\epsilon_t, \qquad h_t = \gamma_0 + \gamma_1 h_{t-1} + \eta_t$$

Taking logs of the squares shows

$$z_t = \log y_t^2 = h_t + \log \epsilon_t^2$$

so the z_t process is a Gaussian $AR(1)$ process observed with added noise, and $y_t = \pm \exp z_t/2$ where the sign is chosen randomly at each time point.

There are many extensions of GARCH models surveyed by Shephard (1996). One of the most useful is the EGARCH class of Nelson (1991) which has

$$y_t = \sigma_t \epsilon_t, \qquad h_t = \log \sigma_t^2 = \alpha_0 + \sum_{i=1}^{p} \alpha_i g(y_{t-i}) + \sum_{j=1}^{q} \beta_j h_{t-j}$$

where

$$g(x) = |x| + wx$$

and so the EGARCH model responds asymmetrically to disturbances (as stock markets are observed to do).

The S+GARCH module for S-PLUS fits a wide range of GARCH models and extensions. We have not yet mentioned the distribution of ϵ_t which is usually taken to be normal but the `garch` function allows to be taken as t or generalized Gaussian (replace the square in the density by the νth power of the absolute value).

We can use `garch` on our S&P 500 example.

```
> module(garch)
> summary(garch(SP500 ~ 1, ~garch(1, 1)))
    ....

Estimated Coefficients:
          Value Std.Error t value  Pr(>|t|)
       C 0.05978  0.01465    4.08  2.30e-05
       A 0.00526  0.00121    4.37  6.57e-06
 ARCH(1) 0.05671  0.00494   11.49  0.00e+00
GARCH(1) 0.93936  0.00530  177.32  0.00e+00
    ....

Normality Test:
 Jarque-Bera P-value
        763       0
    ....
```

The Jarque–Bera test (Cromwell *et al.*, 1994, 20–22) is of normality of the residuals. This suggests we use a model with non-normal ϵ_t.

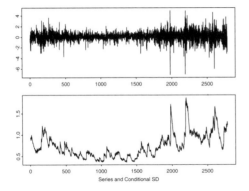
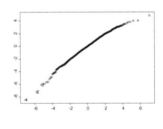

Series and Conditional SD

Figure 14.24: The S&P500 returns, the estimates of the σ_t^2 and a QQ-plot of the residuals from a GARCH fit.

```
> fit <- garch(SP500 ~ 1, ~garch(1, 1), cond.dist = "t")
> summary(fit)
   . . . .
Conditional Distribution:  t
 with estimated parameter 6.1266 and standard error 0.67594

Estimated Coefficients:
             Value Std.Error t value  Pr(>|t|)
        C 0.06068   0.01339    4.53  3.04e-06
        A 0.00201   0.00101    2.00  2.29e-02
  ARCH(1) 0.03144   0.00491    6.40  8.92e-11
 GARCH(1) 0.95182   0.00716  132.91  0.00e+00
   . . . .
> plot(fit)
```

Finally, we consider an EGARCH model. Argument `leverage` allows $w \neq 0$ in $g(x)$.

```
> summary(garch(SP500 ~ 1, ~egarch(1, 1), cond.dist = "t",
           leverage = T))
   . . . .
Conditional Distribution:  t
 with estimated parameter 6.7977 and standard error 0.79991

Estimated Coefficients:
             Value Std.Error t value  Pr(>|t|)
        C  0.0454   0.01324    3.43  3.12e-04
        A -0.0914   0.01279   -7.15  5.65e-13
  ARCH(1)  0.0934   0.01337    6.98  1.78e-12
 GARCH(1)  0.9900   0.00288  343.20  0.00e+00
   LEV(1) -0.6334   0.13439   -4.71  1.28e-06
   . . . .
```

There is significant support for asymmetry.

R has a `garch` function in package `tseries` by Adrian Trapletti. That expects a zero-mean series, so we subtract the median (as a very robust estimate).

```
> library(tseries)
> summary(garch(x = SP500 - median(SP500), order = c(1, 1)))
    Estimate  Std. Error  t value  Pr(>|t|)
a0   0.00457     0.00109     4.19   2.8e-05
a1   0.05193     0.00453    11.46   < 2e-16
b1   0.94467     0.00486   194.51   < 2e-16

Diagnostic Tests:
        Jarque Bera Test

data:  Residuals
X-squared = 777.06, df = 2, p-value = < 2.2e-16
    . . . .
```

Chapter 15

Spatial Statistics

Spatial statistics is a recent and graphical subject that is ideally suited to implementation in S; S-PLUS itself includes one spatial interpolation method, `akima`, and `loess` which can be used for two-dimensional smoothing, but the specialist methods of spatial statistics have been added and are given in our library section `spatial`. The main references for spatial statistics are Ripley (1981, 1988), Diggle (1983), Upton and Fingleton (1985) and Cressie (1991). Not surprisingly, our notation is closest to that of Ripley (1981).

The S-PLUS module[1] S+SPATIALSTATS (Kaluzny and Vega, 1997) provides more comprehensive (and more polished) facilities for spatial statistics than those provided in our library section `spatial`. Details of how to work through our examples in that module may be found in the on-line complements[2] to this book.

More recently other contributed software has become available. There are geostatistical packages called `geoR`/`geoS`[3] and `sgeostat`,[4] and point-process packages `splancs`[5] and `spatstat`.[6]

15.1 Spatial Interpolation and Smoothing

We provide three examples of datasets for spatial interpolation. The dataset `topo` contains 52 measurements of topographic height (in feet) within a square of side 310 feet (labelled in 50 feet units). The dataset `npr1` contains permeability measurements (a measure of the ease of oil flow in the rock) and porosity (the volumetric proportion of the rock which is pore space) measurements from an oil reserve in the USA.

Suppose we are given n observations $Z(x_i)$ and we wish to map the process $Z(x)$ within a region D. (The sample points x_i are usually, but not always, within D.) Although our treatment is quite general, our S code assumes D to be a two-dimensional region, which covers the majority of examples. There are

---

[1]S-PLUS modules are additional-cost products; contact your S-PLUS distributor for details.

[2]See page 461 for where to obtain these.

[3]http://www.maths.lancs.ac.uk/~ribeiro/geoR.html and on CRAN.

[4]http://www.gis.iastate.edu/SGeoStat/homepage.html; R port on CRAN

[5]http://www.maths.lancs.ac.uk/~rowlings/Splancs/; R port on CRAN.

[6]http://www.maths.uwa.edu.au/~adrian/spatstat.html and on CRAN.

however applications to the terrestrial sphere and in three dimensions in mineral and oil applications.

Trend surfaces

One of the earliest methods was fitting *trend surfaces*, polynomial regression surfaces of the form

$$f((x,y)) = \sum_{r+s\leqslant p} a_{rs} x^r y^s \tag{15.1}$$

where the parameter p is the order of the surface. There are $P = (p+1)(p+2)/2$ coefficients. Originally (15.1) was fitted by least squares, and could for example be fitted using `lm` with `poly` which will give polynomials in one or more variables. However, there will be difficulties in prediction, and this is rather inefficient in applications such as ours in which the number of points at which prediction is needed may far exceed n. Our function `surf.ls` implicitly rescales x and y to $[-1, 1]$, which ensures that the first few polynomials are far from collinear. We show some low-order trend surfaces for the `topo` dataset in Figure 15.1, generated by:

```
library(spatial)
par(mfrow = c(2, 2), pty = "s")
topo.ls <- surf.ls(2, topo)
trsurf <- trmat(topo.ls, 0, 6.5, 0, 6.5, 30)
eqscplot(trsurf, , xlab = "", ylab = "", type = "n")
contour(trsurf, levels = seq(600, 1000, 25), add = T)
points(topo)
title("Degree = 2")
topo.ls <- surf.ls(3, topo)
    . . . .
topo.ls <- surf.ls(4, topo)
    . . . .
topo.ls <- surf.ls(6, topo)
    . . . .
```

Notice how `eqscplot` is used to generate geometrically accurate plots.

Figure 15.1 shows trend surfaces for the `topo` dataset. The highest degree, 6, has 28 coefficients fitted from 52 points. The higher-order surfaces begin to show the difficulties of fitting by polynomials in two or more dimensions, when inevitably extrapolation is needed at the edges.

There are several other ways to show trend surfaces. Figure 15.2 uses Trellis to show a greyscale plot from `levelplot` and a perspective plot from `wireframe`. They were generated by

```
topo.ls <- surf.ls(4, topo)
trs <- trsurf <- trmat(topo.ls, 0, 6.5, 0, 6.5, 30)
trs[c("x", "y")] <- expand.grid(x = trs$x, y = tr$y)
plt1 <- levelplot(z ~ x * y, trs, aspect = 1,
    at = seq(650, 1000, 10), xlab = "", ylab = "")
```

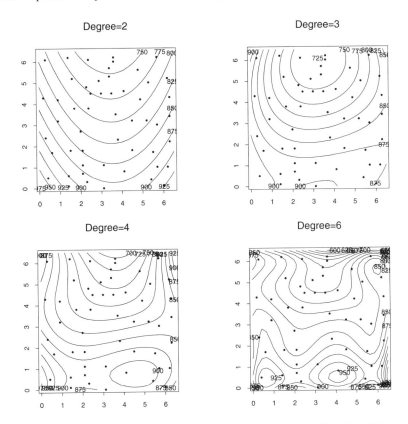

Figure 15.1: Trend surfaces for the topo dataset, of degrees 2, 3, 4 and 6.

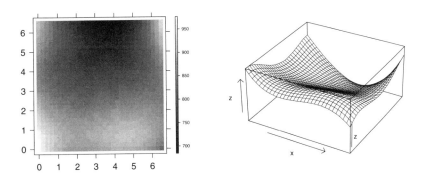

Figure 15.2: The quartic trend surfaces for the topo dataset.

```
plt2 <- wireframe(z ~ x * y, trs, aspect = c(1, 0.5),
            screen = list(z = -30, x = -60))
print(plt1, position = c(0, 0, 0.5, 1), more = T)
print(plt2, position = c(0.45, 0, 1, 1))
```

Users of S-PLUS under Windows can use the rotatable 3D-plots in the GUI graphics, for example by

```
tr <- data.frame(x = trs$x, y = trs$y, z = as.vector(trs$z))
guiPlot(PlotType = "32 Color Surface", Dataset = "tr")
guiModify("Graph3D", Name = guiGetGraphName(),
          xSizeRatio = 2.2, ySizeRatio = 2.2)
```

where the final commands sets a square box. For R under Windows we can get a rotatable 3D-plot by (see page 69 for package rgl)

```
library(rgl)
persp3d(trsurf)
```

One difficulty with fitting trend surfaces is that in most applications the observations are not regularly spaced, and sometimes they are most dense where the surface is high (for example, in mineral prospecting). This makes it important to take the spatial correlation of the errors into consideration. We thus suppose that

$$Z(x) = f(x)^T \beta + \epsilon(x)$$

for a parametrized trend term such as (15.1) and a zero-mean spatial stochastic process $\epsilon(x)$ of errors. We assume that $\epsilon(x)$ possesses second moments, and has covariance matrix

$$C(x, y) = \mathrm{cov}\,(\epsilon(x), \epsilon(y))$$

(this assumption is relaxed slightly later). Then the natural way to estimate β is by *generalized least squares,* that is, to minimize

$$[Z(x_i) - f(x_i)^T \beta]^T [C(x_i, x_j)]^{-1} [Z(x_i) - f(x_i)^T \beta]$$

We need some simplified notation. Let $Z = F\beta + \epsilon$ where

$$F = \begin{bmatrix} f(x_1)^T \\ \vdots \\ f(x_n)^T \end{bmatrix}, \qquad Z = \begin{bmatrix} Z(x_1) \\ \vdots \\ Z(x_n) \end{bmatrix}, \qquad \epsilon = \begin{bmatrix} \epsilon(x_1) \\ \vdots \\ \epsilon(x_n) \end{bmatrix}$$

and let $K = [C(x_i, x_j)]$. We assume that K is of full rank. Then the problem is to minimize

$$[Z - F\beta]^T K^{-1} [Z - F\beta] \tag{15.2}$$

The Choleski decomposition (Golub and Van Loan, 1989; Nash, 1990) finds a lower-triangular matrix L such that $K = LL^T$. (The S function chol is unusual in working with $U = L^T$.) Then minimizing (15.2) is equivalent to

$$\min_{\beta} \|L^{-1}[Z - F\beta]\|^2$$

which reduces the problem to one of ordinary least squares. To solve this we use the QR decomposition (Golub and Van Loan, 1989) of $L^{-1}F$ as

$$QL^{-1}F = \begin{bmatrix} R \\ 0 \end{bmatrix}$$

for an orthogonal matrix Q and upper-triangular $P \times P$ matrix R. Write

$$QL^{-1}\mathbf{Z} = \begin{bmatrix} \mathbf{Y}_1 \\ \mathbf{Y}_2 \end{bmatrix}$$

as the upper P and lower $n - P$ rows. Then $\widehat{\beta}$ solves

$$R\widehat{\beta} = \mathbf{Y}_1$$

which is easy to compute as R is triangular.

Trend surfaces for the `topo` data fitted by generalized least squares are shown later (Figure 15.5), where we discuss the choice of the covariance function C.

Local trend surfaces

We have commented on the difficulties of using polynomials as global surfaces. There are two ways to make their effect local. The first is to fit a polynomial surface for each predicted point, using only the nearby data points. The function `loess` is of this class, and provides a wide range of options. By default it fits a quadratic surface by weighted least squares, the weights ensuring that 'local' data points are most influential. We only give details for the span parameter α less than one. Let $q = \lfloor \alpha n \rfloor$, and let δ denote the Euclidean distance to the qth nearest point to x. Then the weights are

$$w_i = \left[1 - \left(\frac{d(\mathbf{x}, \mathbf{x}_i)}{\delta} \right)^3 \right]_+^3$$

for the observation at $\mathbf{x}_i$. ($[\quad]_+$ denotes the positive part.) Full details of `loess` are given by Cleveland, Grosse and Shyu (1992). For our example we have (Figure 15.3):

```
par(mfcol = c(2,2), pty = "s")
topo.loess <- loess(z ~ x * y, topo, degree = 2, span = 0.25,
    normalize = F)
topo.mar <- list(x = seq(0, 6.5, 0.1), y = seq(0, 6.5, 0.1))
topo.lo <- predict(topo.loess, expand.grid(topo.mar), se = T)
eqscplot(topo.mar, xlab = "fit", ylab = "", type = "n")
contour(topo.mar$x, topo.mar$y, topo.lo$fit,
    levels = seq(700, 1000, 25), add = T)
points(topo)
eqscplot(trsurf, , xlab = "standard error", ylab = "", type = "n")
contour(topo.mar$x,topo.mar$y,topo.lo$se.fit,
    levels = seq(5, 25, 5), add = T)
points(topo)
title("Loess degree = 2")
topo.loess <- loess(z ~ x * y, topo, degree = 1, span = 0.25,
    normalize = F, xlab = "", ylab = "")
    . . . .
```

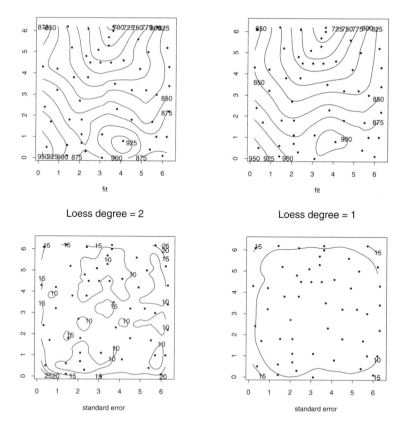

Figure 15.3: loess surfaces and prediction standard errors for the topo dataset.

We turn normalization off to use Euclidean distance on unscaled variables. Note that the predictions from loess are confined to the range of the data in each of the x and y directions even though we requested them to cover the square; this is a side effect of the algorithms used. The standard-error calculations are slow; loess is much faster without them.

Although loess allows a wide range of smoothing via its parameter span, it is designed for exploratory work and has no way to choose the smoothness except to 'look good'.

The Dirichlet tessellation[7] of a set of points is the set of *tiles*, each of which is associated with a data point, and is the set of points nearer to that data point than any other. There is an associated triangulation, the Delaunay triangulation, in which data points are connected by an edge of the triangulation if and only if their Dirichlet tiles share an edge. (Algorithms and examples are given in Ripley, 1981, §4.3.) There is S code in library section delaunay available from statlib (see page 464), and in the R packages deldir and tripack on CRAN.) Akima's (1978) fitting method fits a fifth-order trend surface within each triangle of the

---

[7]Also known as Voronoi or Thiessen polygons.

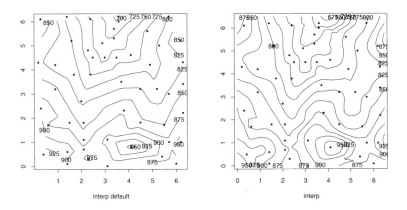

Figure 15.4: `interp` surfaces for the `topo` dataset.

Delaunay triangulation; details are given in Ripley (1981, §4.3). The S implementation is the function `interp`; Akima's example is in datasets `akima.x`, `akima.y` and `akima.z`. The method is forced to interpolate the data, and has no flexibility at all to choose the smoothness of the surface. The arguments `ncp` and `extrap` control details of the method: see the on-line help for details. For Figure 15.4 we used

```
# R: library(akima) # replace interp by interp.old
par(mfrow = c(1, 2), pty=  "s")
topo.int <- interp(topo$x, topo$y, topo$z)
eqscplot(topo.int, xlab = "interp default", ylab = "", type = "n")
contour(topo.int, levels = seq(600, 1000, 25), add = T)
points(topo)
topo.mar <- list(x = seq(0, 6.5, 0.1), y = seq(0, 6.5, 0.1))
topo.int2 <- interp(topo$x, topo$y, topo$z, topo.mar$x, topo.mar$y,
                    ncp = 4, extrap = T)
eqscplot(topo.int2, xlab = "interp", ylab = "", type = "n")
contour(topo.int2, levels = seq(600, 1000, 25), add = T)
points(topo)
```

15.2   Kriging

Kriging is the name of a technique developed by Matheron in the early 1960s for mining applications, which has been independently discovered many times. Journel and Huijbregts (1978) give a comprehensive guide to its application in the mining industry. See also Chilès and Delfiner (1999). In its full form, *universal kriging*, it amounts to fitting a process of the form

$$Z(\boldsymbol{x}) = \boldsymbol{f}(\boldsymbol{x})^T \boldsymbol{\beta} + \epsilon(\boldsymbol{x})$$

by generalized least squares, predicting the value at $\boldsymbol{x}$ of both terms and taking their sum. Thus it differs from trend-surface prediction which predicts $\epsilon(\boldsymbol{x})$ by zero. In what is most commonly termed *kriging*, the trend surface is of degree zero, that is, a constant.

Our derivation of the predictions is given by Ripley (1981, pp. 48–50). Let $k(x) = [C(x, x_i)]$. The computational steps are as follows.

1. Form $K = [C(x_i, y_i)]$, with Choleski decomposition L.

2. Form F and Z.

3. Minimize $\|L^{-1}Z - L^{-1}F\beta\|^2$, reducing $L^{-1}F$ to R.

4. Form $W = Z - F\widehat{\beta}$, and y such that $L(L^T y) = W$.

5. Predict $Z(x)$ by $\widehat{Z}(x) = y^T k(x) + f(x)^T \widehat{\beta}$, with error variance given by $C(x, x) - \|e\|^2 + \|g\|^2$ where

$$Le = k(x), \qquad R^T g = f(x) - (L^{-1}F)^T e.$$

This recipe involves only linear algebra and so can be implemented in S, but our C version is about 10 times faster. For the topo data we have (Figure 15.5):

```
topo.ls <- surf.ls(2, topo)
trsurf <- trmat(topo.ls, 0, 6.5, 0, 6.5, 30)
eqscplot(trsurf, , xlab = "", ylab = "", type = "n")
contour(trsurf, levels = seq(600, 1000, 25), add = T)
points(topo); title("LS trend surface")

topo.gls <- surf.gls(2, expcov, topo, d = 0.7)
trsurf <- trmat(topo.gls, 0, 6.5, 0, 6.5, 30)
eqscplot(trsurf, , xlab = "", ylab = "", type = "n")
contour(trsurf, levels = seq(600, 1000, 25), add = T)
points(topo); title("GLS trend surface")

prsurf <- prmat(topo.gls, 0, 6.5, 0, 6.5, 50)
eqscplot(prsurf, , xlab = "", ylab = "", type = "n")
contour(prsurf, levels = seq(600, 1000, 25), add = T)
points(topo); title("Kriging prediction")
sesurf <- semat(topo.gls, 0, 6.5, 0, 6.5, 30)
eqscplot(sesurf, , xlab = "", ylab = "", type = "n")
contour(sesurf, levels = c(20, 25), add = T)
points(topo); title("Kriging s.e.")
```

Covariance estimation

To use either generalized least squares or kriging we have to know the covariance function C. We assume that

$$C(x, y) = c(d(x, y)) \tag{15.3}$$

where $d()$ is Euclidean distance. (An extension known as *geometric anisotropy* can be incorporated by rescaling the variables, as we did for the Mahalanobis

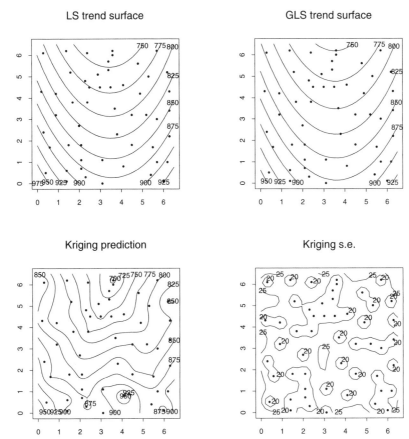

Figure 15.5: Trend surfaces by least squares and generalized least squares, and a kriged surface and standard error of prediction, for the topo dataset.

distance in Chapter 11.) We can compute a *correlogram* by dividing the distance into a number of bins and finding the covariance between pairs whose distance falls into that bin, then dividing by the overall variance.

Choosing the covariance is very much an iterative process, as we need the covariance of the residuals, and the fitting of the trend surface by generalized least squares depends on the assumed form of the covariance function. Furthermore, as we have residuals their covariance function is a biased estimator of c. In practice it is important to get the form right for small distances, for which the bias is least.

Although $c(0)$ must be one, there is no reason why $c(0+)$ should not be less than one. This is known in the kriging literature as a *nugget effect* since it could arise from a very short-range component of the process $Z(x)$. Another explanation is measurement error. In any case, if there is a nugget effect, the predicted surface will have spikes at the data points, and so effectively will not interpolate but smooth.

The kriging literature tends to work with the *variogram* rather than the covariance function. More properly termed the semi-variogram, this is defined by

$$V(\boldsymbol{x}, \boldsymbol{y}) = \frac{1}{2} E[Z(\boldsymbol{x}) - Z(\boldsymbol{y})]^2$$

and is related to C by

$$V(\boldsymbol{x}, \boldsymbol{y}) = \frac{1}{2}[C(\boldsymbol{x}, \boldsymbol{x}) + C(\boldsymbol{y}, \boldsymbol{y})] - C(\boldsymbol{x}, \boldsymbol{y}) = c(0) - c(d(\boldsymbol{x}, \boldsymbol{y}))$$

under our assumption (15.3). However, since different variance estimates will be used in different bins, the empirical versions will not be so exactly related. Much heat and little light emerges from discussions of their comparison.

There are a number of standard forms of covariance functions that are commonly used. A nugget effect can be added to each. The exponential covariance has

$$c(r) = \sigma^2 \exp -r/d$$

the so-called Gaussian covariance is

$$c(r) = \sigma^2 \exp -(r/d)^2$$

and the spherical covariance is in two dimensions

$$c(r) = \sigma^2 \left[1 - \frac{2}{\pi} \left(\frac{r}{d} \sqrt{1 - \frac{r^2}{d^2}} + \sin^{-1} \frac{r}{d} \right) \right]$$

and in three dimensions (but also valid as a covariance function in two)

$$c(r) = \sigma^2 \left[1 - \frac{3r}{2d} + \frac{r^3}{2d^3} \right]$$

for $r \leqslant d$ and zero for $r > d$. Note that this is genuinely local, since points at a greater distance than d from $\boldsymbol{x}$ are given zero weight at step 5 (although they do affect the trend surface).

We promised to relax the assumption of second-order stationarity slightly. As we only need to predict residuals, we only need a covariance to exist in the space of linear combinations $\sum a_i Z(\boldsymbol{x}_i)$ that are orthogonal to the trend surface. For degree 0, this corresponds to combinations with sum zero. It is possible that the variogram is finite, without the covariance existing, and there are extensions to more general trend surfaces given by Matheron (1973) and reproduced by Cressie (1991, §5.4). In particular, we can always add a constant to c without affecting the predictions (except perhaps numerically). Thus if the variogram v is specified, we work with covariance function $c = \text{const} - v$ for a suitably large constant. The main advantage is in allowing us to use certain functional forms that do not correspond to covariances, such as

$$v(d) = d^\alpha, 0 \leqslant \alpha < 2 \quad \text{or} \quad d^3 - \alpha d$$

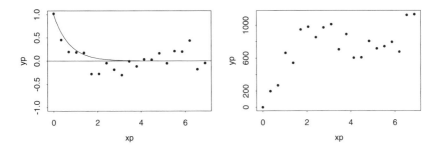

Figure 15.6: Correlogram (left) and variogram (right) for the residuals of `topo` dataset from a least-squares quadratic trend surface.

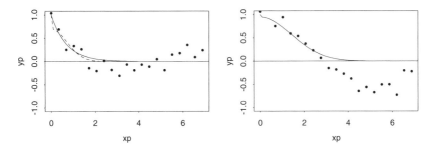

Figure 15.7: Correlograms for the `topo` dataset: (left) residuals from quadratic trend surface showing exponential covariance (solid) and Gaussian covariance (dashed); (right) raw data with fitted Gaussian covariance function.

The variogram $d^2 \log d$ corresponds to a thin-plate spline in $\mathbb{R}^2$ (see Wahba, 1990, and the review in Cressie, 1991, §3.4.5).

Our functions `correlogram` and `variogram` allow the empirical correlogram and variogram to be plotted and functions `expcov`, `gaucov` and `sphercov` compute the exponential, Gaussian and spherical covariance functions (the latter in two and three dimensions) and can be used as arguments to `surf.gls`. For our running example we have

```
topo.kr <- surf.ls(2, topo)
correlogram(topo.kr, 25)
d <- seq(0, 7, 0.1)
lines(d, expcov(d, 0.7))
variogram(topo.kr, 25)
```

See Figure 15.6. We then consider fits by generalized least squares.

```
## left panel of Figure 15.7
topo.kr <- surf.gls(2, expcov, topo, d=0.7)
correlogram(topo.kr, 25)
lines(d, expcov(d, 0.7))
lines(d, gaucov(d, 1.0, 0.3), lty = 3) # try nugget effect
```

```
## right panel
topo.kr <- surf.ls(0, topo)
correlogram(topo.kr, 25)
lines(d, gaucov(d, 2, 0.05))

## top row of Figure 15.8
topo.kr <- surf.gls(2, gaucov, topo, d = 1, alph = 0.3)
prsurf <- prmat(topo.kr, 0, 6.5, 0, 6.5, 50)
eqscplot(prsurf, , xlab = "fit", ylab = "", type = "n")
contour(prsurf, levels = seq(600, 1000, 25), add = T)
points(topo)
sesurf <- semat(topo.kr, 0, 6.5, 0, 6.5, 25)
eqscplot(sesurf, , xlab = "standard error", ylab = "", type = "n")
contour(sesurf, levels = c(15, 20, 25), add = T)
points(topo)

## bottom row of Figure 15.8
topo.kr <- surf.gls(0, gaucov, topo, d = 2, alph = 0.05,
                    nx = 10000)
prsurf <- prmat(topo.kr, 0, 6.5, 0, 6.5, 50)
eqscplot(prsurf, , xlab = "fit", ylab = "", type = "n")
contour(prsurf, levels = seq(600, 1000, 25), add = T)
points(topo)
sesurf <- semat(topo.kr, 0, 6.5, 0, 6.5, 25)
eqscplot(sesurf, , xlab = "standard error", ylab = "", type = "n")
contour(sesurf, levels = c(15, 20, 25), add = T)
points(topo)
```

We first fit a quadratic surface by least squares, then try one plausible covariance function (Figure 15.7). Re-fitting by generalized least squares suggests this function and another with a nugget effect, and we predict the surface from both. The first was shown in Figure 15.5, the second in Figure 15.8. We also consider not using a trend surface but a longer-range covariance function, also shown in Figure 15.8. (The small nugget effect is to ensure numerical stability as without it the matrix K is very ill-conditioned; the correlations at short distances are very near one. We increased nx for a more accurate lookup table of covariances.)

15.3 Point Process Analysis

A spatial point pattern is a collection of n points within a region $D \subset \mathbb{R}^2$. The number of points is thought of as random, and the points are considered to be generated by a stationary isotropic point process in $\mathbb{R}^2$. (This means that there is no preferred origin or orientation of the pattern.) For such patterns probably the most useful summaries of the process are the first and second moments of the counts $N(A)$ of the numbers of points within a set $A \subset D$. The first moment can be specified by a single number, the *intensity* λ giving the expected number of points per unit area, obviously estimated by n/a where a denotes the area of D.

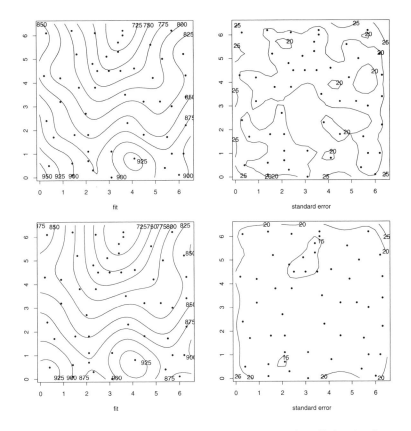

Figure 15.8: Two more kriged surfaces and standard errors of prediction for the topo dataset. The top row uses a quadratic trend surface and a nugget effect. The bottom row is without a trend surface.

The second moment can be specified by Ripley's K function. For example, $\lambda K(t)$ is the expected number of points within distance t of a point of the pattern. The benchmark of complete randomness is the Poisson process, for which $K(t) = \pi t^2$, the area of the search region for the points. Values larger than this indicate clustering on that distance scale, and smaller values indicate regularity. This suggests working with $L(t) = \sqrt{K(t)/\pi}$, which will be linear for a Poisson process.

We only have a single pattern from which to estimate K or L. The definition in the previous paragraph suggests an estimator of $\lambda K(t)$; average over all points of the pattern the number seen within distance t of that point. This would be valid but for the fact that some of the points will be outside D and so invisible. There are a number of edge-corrections available, but that of Ripley (1976) is both simple to compute and rather efficient. This considers a circle centred on the point x and passing through another point y. If the circle lies entirely within D, the point is counted once. If a proportion $p(x, y)$ of the circle lies within D, the point is counted as $1/p$ points. (We may want to put a limit on small p, to reduce

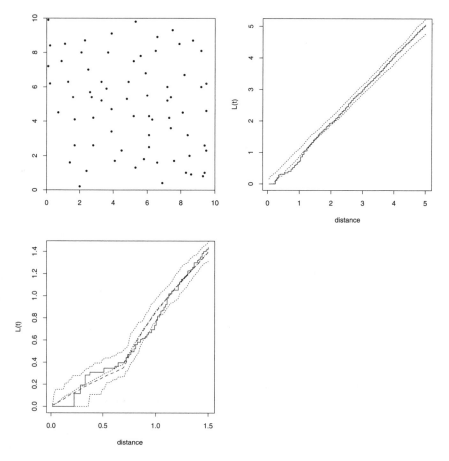

Figure 15.9: The Swedish pines dataset from Ripley (1981), with two plots of $L(t)$. That at the upper right shows the envelope of 100 binomial simulations, that at the lower left the average and the envelope (dotted) of 100 simulations of a Strauss process with $c = 0.2$ and $R = 0.7$. Also shown (dashed) is the average for $c = 0.15$. All units are in metres.

the variance at the expense of some bias.) This gives an estimator $\lambda \widehat{K}(t)$ which is unbiased for t up to the circumradius of D (so that it is possible to observe two points $2t$ apart). Since we do not know λ, we estimate it by $\hat{\lambda} = n/a$. Finally

$$\widehat{K}(t) = \frac{a}{n^2} \sum_{\boldsymbol{x} \in D, d(\boldsymbol{y}, \boldsymbol{x}) \leqslant t} \frac{1}{p(\boldsymbol{x}, \boldsymbol{y})}$$

and obviously we estimate $L(t)$ by $\sqrt{\widehat{K}(t)/\pi}$. We find that on square-root scale the variance of the estimator varies little with t.

Our example is the Swedish pines data from Ripley (1981, §8.6). This records 72 trees within a 10-metre square. Figure 15.9 shows that $\widehat{L}$ is not straight, and comparison with simulations from a binomial process (a Poisson process condi-

tioned on $N(D) = n$, so n independently uniformly distributed points within
D) shows that the lack of straightness is significant. The upper two panels of
Figure 15.9 were produced by the following code:

```
library(spatial)
pines <- ppinit("pines.dat")
par(mfrow = c(2, 2), pty = "s")
plot(pines, xlim = c(0, 10), ylim = c(0, 10),
     xlab = "", ylab = "", xaxs = "i", yaxs = "i")
plot(Kfn(pines,5), type = "s", xlab = "distance", ylab = "L(t)")
lims <- Kenvl(5, 100, Psim(72))
lines(lims$x, lims$l, lty = 2)
lines(lims$x, lims$u, lty = 2)
```

The function ppinit reads the data from the file and also the coordinates of
a rectangular domain D. The latter can be reset, or set up for simulations, by the
function ppregion. (It *must* be set for each session.) The function Kfn returns
an estimate of $L(t)$ and other useful information for plotting, for distances up to
its second argument fs (for full-scale).

The functions Kaver and Kenvl return the average and, for Kenvl, also
the extremes of K-functions (on L scale) for a series of simulations. The function Psim(n) simulates the binomial process on n points within the domain D,
which has already been set.

Alternative processes

We need to consider alternative point processes to the Poisson. One of the most
useful for regular point patterns is the so-called Strauss process, which is simulated by Strauss(n, c, r). This has a density of n points proportional to

$$c^{\text{number of } R\text{-close pairs}}$$

and so has $K(t) < \pi t^2$ for $t \leqslant R$ (and up to about $2R$). For $c = 0$ we
have a 'hard-core' process that never generates pairs closer than R and so can
be envisaged as laying down the centres of non-overlapping discs of diameter
$r = R$.

Figure 15.9 also shows the average and envelope of the L-plots for a Strauss
process fitted to the pines data by Ripley (1981). There the parameters were
chosen by trial-and-error based on a knowledge of how the L-plot changed with
(c, R). Ripley (1988) considers the estimation of c for known R by the pseudo-
likelihood. This is done by our function pplik and returns an estimate of about
$c = 0.15$ ('about' since it uses numerical integration). As Figure 15.9 shows, the
difference between $c = 0.2$ and $c = 0.15$ is small. We used the following code:

```
ppregion(pines)
plot(Kfn(pines, 1.5), type = "s",
     xlab = "distance", ylab = "L(t)")
lims <- Kenvl(1.5, 100, Strauss(72, 0.2, 0.7))
lines(lims$x, lims$a, lty = 2)
```

```
lines(lims$x, lims$l, lty = 2)
lines(lims$x, lims$u, lty = 2)
pplik(pines, 0.7)
lines(Kaver(1.5, 100, Strauss(72, 0.15, 0.7)), lty = 3)
```

The theory is given by Ripley (1988, p. 67). For a point $\xi \in D$ let $t(\xi)$ denote the number of points of the pattern within distance t of ξ. Then the pseudo-likelihood estimator solves

$$\frac{\int_D t(\xi)c^{t(\xi)}\,d\xi}{\int_D c^t(\xi)\,d\xi} = \frac{\#(R\text{--close pairs})}{n} = \frac{n\widehat{K}(R)}{a}$$

and the left-hand side is an increasing function of c. The function pplik uses the S-PLUS function uniroot to find a solution in the range $(0, 1]$.

Other processes for which simulation functions are provided are the binomial process (Psim(n)) and Matérn's sequential spatial inhibition process (SSI(n, r)), which sequentially lays down centres of discs of radius r that do not overlap existing discs.

Chapter 16

Optimization

Statisticians[1] often under-estimate the usefulness of general optimization methods in maximizing likelihoods and in other model-fitting problems. Not only are the general-purpose methods available in the S environments quick to use, they also often outperform the specialized methods that are available. A lot of the software we have illustrated in earlier chapters is based on the functions described in this. Code that seemed slow when the first edition was being prepared in 1993 now seems almost instant.

Many of the functions we describe in this chapter can be used as black boxes, but it *is* helpful to know some of the theory behind the methods. Although optimization theory and practice had an explosion in the 1960s and 1970s, there has been progress since and classic accounts such as Gill *et al.* (1981) and Fletcher (1987) are now somewhat outdated. The account we find authoritative is Nocedal and Wright (1999), and for a gentler introduction we recommend Nash (1990) or Monahan (2001).

Specialist methods for non-linear least-squares problems were discussed in Chapter 8.

16.1   Univariate Functions

The functions `optimize` and `uniroot` work with continuous functions of a single variable; `optimize` can find a (local) minimum (the default) or a (local) maximum, whereas `uniroot` finds a zero. Both search within an interval specified either by the argument `integer` or arguments `lower` and `upper`. Both are based on safeguarded polynomial interpolation (Brent, 1973; Nash, 1990).

Function `optimize` is used for cross-validatory bandwidth estimators, and `uniroot` is used to find the Sheather-Jones bandwidth estimator, the pseudo-likelihood estimator for a Strauss process in `pplik` and in several of the classical test functions.

---

[1]And not just statisticians; for example, it took the neural network community some years to appreciate this.

16.2   Special-Purpose Optimization Functions

Apart from the non-linear least-squares functions, there are no special-purpose optimization functions in the base S environments, but a few in contributed packages. Optimization texts usually start with linear, linearly-constrained problems, and these can often be solved efficiently by the simplex algorithm. There is an implementation of this in the `simplex` function in library section `boot` (by Angelo Canty).

Quadratic programming keeps the linear inequality constraints but optimizes a quadratic function. Quadratic programming is covered by library section `quadprog` (by Berwin Turlach).

There is a module S+NUOPT for S-PLUS interfacing to a fairly general simplex and interior-point optimization package.

16.3   General Optimization

Each of the S environments has a few functions for general optimization. This seems to be something of a historical accident, and in each case one function is normally preferred.

S-PLUS offers `ms`, `nlmin` and `nlminb`; of these, `nlminb` is preferred for new work.

R offers `nlm` and `optim`; the latter has several methods, including Nelder–Mead, the BFGS quasi-Newton method, conjugate gradients, box-constrained limited-memory BFGS and simulated annealing. There is a version of `optim` for S-PLUS in our `MASS` library section.

Practical issues with general optimization often have less to do with the optimizer used than with how carefully the problem is set up. In general it is worth supplying a function to calculate derivatives if you can, although it may be quicker in a once-off problem to let the software calculate numerical derivatives. It is worth ensuring that the problem is reasonably well scaled, so a unit step in any parameter have a comparable change in size to the objective, preferably about a unit change at the optimum. Functions `nlminb` and `optim` have parameters to allow the user to supply scaling hints.

It is normally not worth the effort to supply a function to calculate the Hessian matrix of second derivatives. However, sometimes it can be calculated automatically. We saw on page 215 the use of `deriv` to calculate first derivatives symbolically. This has a counterpart[2] `deriv3` to calculate Hessians. Functions `deriv` and `deriv3` are in principle extensible, and this has been done by us to handle our example.

An example: fitting a mixture model

The waiting times between eruptions in the data `geyser` are strongly bimodal. Figure 16.1 shows a histogram with a density estimate superimposed (with bandwidth chosen by the Sheather–Jones method described in Chapter 5).

---

[2]For S-PLUS written by David Smith and supplied in `MASS` .

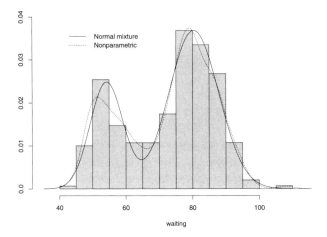

Figure 16.1: Histogram for the waiting times between successive eruptions for the "Old Faithful" geyser, with non-parametric and parametric estimated densities superimposed.

```
attach(geyser)
truehist(waiting, xlim = c(35, 115), ymax = 0.04, h = 5)
wait.dns <- density(waiting, n = 512, width = "SJ")
lines(wait.dns, lty = 2)
```

From inspection of this figure a mixture of two normal distributions would seem to be a reasonable descriptive model for the marginal distribution of waiting times. We now consider how to fit this by maximum likelihood. The observations are not independent since successive waiting times are strongly negatively correlated. In this section we propose to ignore this, both for simplicity and because ignoring this aspect is not likely to misrepresent seriously the information in the sample on the marginal distribution in question.

Useful references for mixture models include Everitt and Hand (1981); Titterington, Smith and Makov (1985) and McLachlan and Peel (2000). Everitt and Hand describe the EM algorithm for fitting a mixture model, which is simple,[3] but we consider here a more direct function minimization method that can be faster and more reliable (Redner and Walker, 1984; Ingrassia, 1992).

If y_i, $i = 1, 2, \ldots, n$ is a sample waiting time, the log-likelihood function for a mixture of two normal components is

$$L(\pi, \mu_1, \sigma_1, \mu_2, \sigma_2) = \sum_{i=1}^{n} \log\left[\frac{\pi}{\sigma_1}\phi\left(\frac{y_i - \mu_1}{\sigma_1}\right) + \frac{1 - \pi}{\sigma_2}\phi\left(\frac{y_i - \mu_2}{\sigma_2}\right)\right]$$

We estimate the parameters by minimizing $-L$.

It is helpful in this example to use both first- and second-derivative information and we use `deriv3` to produce an objective function. Both `deriv` and `deriv3` allow calls to `pnorm` and `dnorm` with only one argument. Thus calls

---

[3]It is implemented in `me`, page 320, and `mda`, page 341.

such as `pnorm(x,u,s)` must be written as `pnorm((x-u)/s)`. We can generate a function that calculates the summands of $-L$ and also returns first- and
second-derivative information for each summand by:

```
lmix2 <- deriv3(
    ~ -log(p*dnorm((x-u1)/s1)/s1 + (1-p)*dnorm((x-u2)/s2)/s2),
    c("p", "u1", "s1", "u2", "s2"),
    function(x, p, u1, s1, u2, s2) NULL)
```

which under S-PLUS took 14 seconds on the PC, much longer than any of the fits
below. (It was instant under R.)

Initial values for the parameters could be obtained by the method of moments
described in Everitt and Hand (1981, pp. 31ff), but for well-separated components
as we have here, we can choose initial values by reference to the plot. For the
initial value of π we take the proportion of the sample below the density low
point at 70.

```
> (p0 <- c(p = mean(waiting < 70), u1 = 50, s1 = 5, u2 = 80,
        s2 = 5))
     p u1 s1 u2 s2
0.36120 50  5 80  5
```

Using `nlminb` in S-PLUS

The most general minimization routine in S-PLUS is `nlminb`, which can find a
local minimum of a twice-differentiable function within a hypercube in parameter
space. Either the gradient or gradient plus Hessian can be supplied; if no gradient
is supplied it is approximated by finite differences. The underlying algorithm is
a quasi-Newton optimizer, or a Newton optimizer if the Hessian is supplied. We
can fit our mixture density (using zero, one and two derivatives) and enforcing the
constraints.

```
mix.obj <- function(p, x) {
    e <- p[1] * dnorm((x - p[2])/p[3])/p[3] +
         (1 - p[1]) * dnorm((x - p[4])/p[5])/p[5]
    -sum(log(e)) }
mix.nl0 <- nlminb(p0, mix.obj,
    scale = c(10, rep(1, 4)), lower = c(0, -Inf, 0, -Inf, 0),
    upper = c(1, rep(Inf, 4)), x = waiting)

lmix2a <- deriv(
    ~ -log(p*dnorm((x-u1)/s1)/s1 + (1-p)*dnorm((x-u2)/s2)/s2),
    c("p", "u1", "s1", "u2", "s2"),
    function(x, p, u1, s1, u2, s2) NULL)
mix.gr <- function(p, x) {
    u1 <- p[2]; s1 <- p[3]; u2 <- p[4]; s2 <- p[5]; p <- p[1]
    colSums(attr(lmix2a(x, p, u1, s1, u2, s2), "gradient")) }
mix.nl1 <- nlminb(p0, mix.obj, mix.gr,
    scale = c(10, rep(1, 4)), lower = c(0, -Inf, 0, -Inf, 0),
    upper = c(1, rep(Inf, 4)), x = waiting)
```

```
mix.grh <- function(p, x) {
    e <- lmix2(x, p[1], p[2], p[3], p[4], p[5])
    g <- colSums(attr(e, "gradient"))
    H <- colSums(attr(e, "hessian"), 2)
    list(gradient = g, hessian = H[row(H) <= col(H)]) }
mix.nl2 <- nlminb(p0, mix.obj, mix.grh, hessian = T,
    scale = c(10, rep(1, 4)), lower = c(0, -Inf, 0, -Inf, 0),
    upper = c(1, rep(Inf, 4)), x = waiting)
mix.nl2[c("parameter", "objective")]
$parameters:
       p      u1     s1     u2     s2
 0.30759 54.203 4.952 80.36 7.5076
$objective:
[1] 1157.5
```

We use a `scale` parameter to set the step length on the first parameter much smaller than the others, which can speed convergence. Generally it is helpful to have the scale set so that the range of uncertainty in (scale $\times$ parameter) is about one.

It is also possible to supply a separate function to calculate the Hessian; however it is supplied, it is a vector giving the lower triangle of the Hessian in row-first order (unlike S matrices).

Function `nlminb` only computes the Hessian at the solution if a means to compute Hessians is supplied (and even then it returns a scaled version of the Hessian). Thus it provides little help in using the observed information to provide approximate standard errors for the parameter estimates in a maximum likelihood estimation problem. Our function `vcov.nlminb` in MASS smooths over these niggles, and uses a finite-difference approximation to the Hessian if it is not available.

```
> sqrt(diag(vcov.nlminb(mix.nl0)))
[1] 0.030746 0.676833 0.539178 0.619138 0.509087
> sqrt(diag(vcov.nlminb(mix.nl1)))
[1] 0.030438 0.683067 0.518231 0.633388 0.507096
> sqrt(diag(vcov.nlminb(mix.nl2)))
[1] 0.030438 0.683066 0.518231 0.633387 0.507095
```

The (small) differences reflect a difference in philosophy: when no derivatives are available we seek a quadratic approximation to the negative log-likelihood over the scale of random variation in the parameter estimate rather than at the MLE, since the likelihood might not be twice-differentiable there.

Note that the theory used here is unreliable if the parameter estimate is close to or on the boundary; `vcov.nlminb` issues a warning.

We can now add the parametric density estimate to our original histogram plot.

```
dmix2 <- function(x, p, u1, s1, u2, s2)
            p * dnorm(x, u1, s1) + (1-p) * dnorm(x, u2, s2)
attach(as.list(mix.nl2$parameter))
```

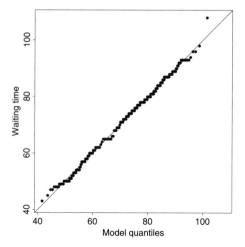

Figure 16.2: Sorted waiting times against normal mixture model quantiles for the 'Old Faithful' eruptions data.

```
wait.fdns <- list(x = wait.dns$x,
                  y = dmix2(wait.dns$x, p, u1, s1, u2, s2))
lines(wait.fdns)
par(usr = c(0, 1, 0, 1))
legend(0.1, 0.9, c("Normal mixture", "Nonparametric"),
       lty = c(1, 2), bty = "n")
```

The computations for Figure 16.1 are now complete.

The parametric and nonparametric density estimates are in fair agreement on the right component, but there is a suggestion of disparity on the left. We can informally check the adequacy of the parametric model by a Q-Q plot. First we solve for the quantiles using a reduced-step Newton method:

```
pmix2 <- deriv(~ p*pnorm((x-u1)/s1) + (1-p)*pnorm((x-u2)/s2),
               "x", function(x, p, u1, s1, u2, s2) {})
pr0 <- (seq(along = waiting) - 0.5)/length(waiting)
x0 <- x1 <- as.vector(sort(waiting)) ; del <- 1; i <- 0
while((i <- 1 + i) < 10 && abs(del) > 0.0005) {
  pr <- pmix2(x0, p, u1, s1, u2, s2)
  del <- (pr - pr0)/attr(pr, "gradient")
  x0 <- x0 - 0.5*del
  cat(format(del <- max(abs(del))), "\n")
}
detach()
par(pty = "s")
plot(x0, x1, xlim = range(x0, x1), ylim = range(x0, x1),
     xlab = "Model quantiles", ylab = "Waiting time")
abline(0, 1)
par(pty = "m")
```

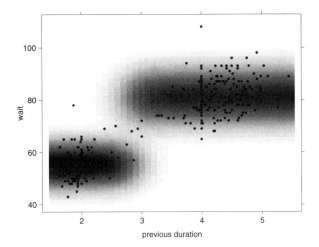

Figure 16.3: Waiting times for the next eruption against the previous duration, with the density of the conditional distribution predicted by a mixture model.

The plot is shown in Figure 16.2 and confirms the adequacy of the mixture model except perhaps for one outlier in the right tail. We saw in Figure 5.11 (on page 131) that the distribution of waiting time depends on the previous duration, and we can easily enhance our model to allow for this with the proportions depending logistically on the previous duration. We start at the fitted values of the simpler model.

```
lmix2r <- deriv(
    ~ -log((exp(a + b*y)*dnorm((x - u1)/s1)/s1 +
        dnorm((x - u2)/s2)/s2) / (1 + exp(a + b*y)) ),
    c("a", "b", "u1", "s1", "u2", "s2"),
    function(x, y, a, b, u1, s1, u2, s2) NULL)
p1 <- mix.nl2$parameters; tmp <- as.vector(p1[1])
p2 <- c(a = log(tmp/(1-tmp)), b = 0, p1[-1])

mix1.obj <- function(p, x, y) {
    q <- exp(p[1] + p[2]*y)
    q <- q/(1 + q)
    e <- q * dnorm((x - p[3])/p[4])/p[4] +
        (1 - q) * dnorm((x - p[5])/p[6])/p[6]
    -sum(log(e)) }
mix1.gr <- function(p, x, y) {
    a <- p[1]; b <- p[2]; u1 <- p[3]; s1 <- p[4];
    u2 <- p[5]; s2 <- p[6]
    colSums(attr(lmix2r(x, y, a, b, u1, s1, u2, s2), "gradient")) }
mix1.nl1 <- nlminb(p2, mix1.obj, mix1.gr,
    lower = c(-Inf, -Inf, -Inf, 0, -Inf, 0),
    upper = rep(Inf, 6), x = waiting[-1], y = duration[-299])
mix1.nl1[c("parameter", "objective")]
$parameters:
```

```
   a        b      u1      s1      u2      s2
16.14 -5.7363 55.138 5.6631 81.091 6.8376
$objective:
[1] 985.32
```

There is clearly a very significant improvement in the fit, a change in log-likelihood of 172 for one extra parameter. We can examine the fit by plotting the predictive densities for the next waiting time by (see Figure 16.3)

```
grid <- expand.grid(x = seq(1.5, 5.5, 0.1), y = seq(40, 110, 0.5))
grid$z <- exp(-lmix2r(grid$y, grid$x, 16.14, -5.74, 55.14,
                 5.663, 81.09, 6.838))
levelplot(z ~ x*y, grid, colorkey = F, at = seq(0, 0.075, 0.001),
          panel= function(...) {
            panel.levelplot(...)
            points(duration[-299], waiting[-1])
          }, xlab = "previous duration", ylab = "wait",
       col.regions = rev(trellis.par.get("regions")$col))
```

Using ms in S-PLUS

The ms function is a general function for minimizing quantities that can be written as sums of one or more terms. The call to ms is very similar to that for nls (see page 213), but the interpretation of the formula is slightly different. It has no response, and the quantity on the right-hand side specifies the entire sum to be minimized. The resulting object is of class ms, for which few method functions are available; the four main ones are point, coef, summary and profile. The fitted object has additional components, two of independent interest being

obj.ms$value the minimum value of the sum, and

obj.ms$pieces the vector of summands at the minimum.

The most common statistical use of ms is to minimize negative log-likelihood functions, thus finding maximum likelihood estimates. The fitting algorithm can make use of first- and second-derivative information, the first- and second-derivative arrays of the summands rather than of a model function. In the case of a negative log-likelihood function, these are arrays whose "row sums" are the negative score vector and observed information matrix, respectively. The first-derivative information is supplied as a gradient attribute of the formula, as described for nls in Section 8.2. Second-derivative information may be supplied via a hessian attribute. (The function deriv3 generates a model function with both gradient and hessian attributes to its return value.)

The convergence criterion for ms is that *one of* the relative changes in the parameter estimates is small, the relative change in the achieved value is small *or* that the optimized function value is close to zero. The latter is appropriate only for deviance-like quantities, and can be eliminated by setting the argument

```
control = ms.control(f.tolerance = -1)
```

We can trace the iterative process, using a trace function that produces just one line of output per iteration:

```
tr.ms <- function(info, theta, grad, scale, flags, fit.pars) {
    cat(round(info[3], 3), ":", signif(theta), "\n")
    invisible() }
```

Note that the trace function must have the same arguments as the standard trace function, `trace.ms`.

We can now fit the mixture model:

```
> wait.mix2 <- ms(~ lmix2(waiting, p, u1, s1, u2, s2),
                   start = p0, data = geyser, trace = tr.ms)
1233.991 : 0.361204 50 5 80 5
 . . . .
1157.542 : 0.307594 54.2027 4.952 80.3603 7.50763
```

The fitted model object has a component, `hessian`, giving the sum of the second derivative array over the observations and since the function we minimized is the negative of the log-likelihood, this component is the observed information matrix. Its inverse is calculated (but incorrectly labelled[4]) by `summary.ms`.

```
vmat <- summary(wait.mix2)$Information
rbind(est = coef(wait.mix2), se = sqrt(diag(vmat)))
            p       u1       s1       u2       s2
est 0.307594 54.20265 4.95200 80.36031 7.50764
 se 0.030438  0.68307 0.51823  0.63339 0.50709
```

Using `nlmin` in S-PLUS

The function `nlmin` finds an unconstrained local minimum of a function. For the mixture problem we can use

```
mix.f <- function(p) {
    e <- p[1] * dnorm((waiting - p[2])/p[3])/p[3] +
         (1 - p[1]) * dnorm((waiting - p[4])/p[5])/p[5]
    -sum(log(e)) }
nlmin(mix.f, p0, print.level = 1, max.iter = 25)
```

The argument `print.level = 1` provides some tracing of the iterations (here 16, one more than the default limit). There is no way to supply derivatives.

Using `optim` in R or S-PLUS

The robust Nelder–Mead method does well enough without derivatives and ignoring the constraints. To use the BFGS quasi-Newton method we need to supply some scaling information (`parscale` being a divisor). Function `optim` will always compute or estimate the Hessian at the optimum if argument `hessian` is true.

---

[4] And differently from the help page, which has `information`.

```
mix.obj <- function(p, x)
{
    e <- p[1] * dnorm((x - p[2])/p[3])/p[3] +
        (1 - p[1]) * dnorm((x - p[4])/p[5])/p[5]
    if(any(e <= 0)) Inf else -sum(log(e))
}
optim(p0, mix.obj, x = waiting)$par # Nelder-Mead
        p       u1       s1       u2       s2
 0.30735 54.19643  4.94834 80.36018  7.51105
optim(p0, mix.obj, x = waiting, method = "BFGS",
      control = list(parscale=c(0.1, rep(1, 4))))$par
        p       u1       s1       u2       s2
 0.30768 54.20466  4.95454 80.36302  7.50566


#same result with analytical derivatives.
optim(p0, mix.obj, mix.gr, x = waiting, method = "BFGS",
      control = list(parscale=c(0.1, rep(1, 4))))$par

mix.nl0 <- optim(p0, mix.obj, mix.gr, method = "L-BFGS-B",
                 hessian = T,
                 lower = c(0, -Inf, 0, -Inf, 0),
                 upper = c(1, rep(Inf, 4)), x = waiting)
rbind(est = mix.nl0$par, se = sqrt(diag(solve(mix.nl0$hessian))))
            p       u1       s1       u2       s2
est 0.307593 54.20268  4.95182 80.36035  7.5077
se  0.030438  0.68305  0.51819  0.63338  0.5071
```

The code to produce Figure 16.2 works unchanged in R.

A similar approach solves the extended problem.

```
mix1.obj <- function(p, x, y)
{
    q <- exp(p[1] + p[2]*y); q <- q/(1 + q)
    e <- q * dnorm((x - p[3])/p[4])/p[4] +
        (1 - q) * dnorm((x - p[5])/p[6])/p[6]
    if(any(e <= 0)) Inf else -sum(log(e))
}
p1 <- mix.nl0$par; tmp <- as.vector(p1[1])
p2 <- c(a = log(tmp/(1-tmp)), b = 0, p1[-1])
mix.nl1 <- optim(p2, mix1.obj, method = "L-BFGS-B",
                 lower = c(-Inf, -Inf, -Inf, 0, -Inf, 0),
                 upper = rep(Inf, 6), hessian = T,
                 x = waiting[-1], y = duration[-299])
rbind(est = mix.nl1$par, se = sqrt(diag(solve(mix.nl1$hessian))))
            a        b       u1       s1       u2       s2
est 16.1410 -5.7367 55.13756  5.66327 81.09077  6.83758
se   3.5738  1.2958  0.61221  0.48478  0.49925  0.35334
```

This converged in 26 evaluations of the log-likelihood.

Fitting GLM models

The Fisher scoring algorithm is not the only possible way to fit GLMs. The purpose of this section is to point out how easy it is to fit specific GLMs by direct maximization of the likelihood, and the code can easily be modified for other fit criteria.

Logistic regression

We consider maximum likelihood fitting of a binomial logistic regression. For other link functions just replace plogis and dlogis, for example by pnorm and dnorm for a probit regression.

```
logitreg <- function(x, y, wt = rep(1, length(y)),
                intercept = T, start = rep(0, p), ...)
{
    fmin <- function(beta, X, y, w) {
        p <- plogis(X %*% beta)
        -sum(2 * w * ifelse(y, log(p), log(1-p)))
    }
    gmin <- function(beta, X, y, w) {
        eta <- X %*% beta; p <- plogis(eta)
        -2 * (w *dlogis(eta) * ifelse(y, 1/p, -1/(1-p))) %*% X
    }
    if(is.null(dim(x))) dim(x) <- c(length(x), 1)
    dn <- dimnames(x)[[2]]
    if(!length(dn)) dn <- paste("Var", 1:ncol(x), sep="")
    p <- ncol(x) + intercept
    if(intercept) {x <- cbind(1, x); dn <- c("(Intercept)", dn)}
    if(is.factor(y)) y <- (unclass(y) != 1)
    fit <- nlminb(start, fmin, gmin, X = x, y = y, w = wt, ...)
    # R: fit <- optim(start, fmin, gmin, X = x, y = y, w = wt,
                    method = "BFGS", ...)
    names(fit$par) <- dn
    cat("\nCoefficients:\n"); print(fit$par)
    # R: use fit$value and fit$convergence
    cat("\nResidual Deviance:", format(fit$objective), "\n")
    cat("\nConvergence message:", fit$message, "\n")
    invisible(fit)
}

options(contrasts = c("contr.treatment", "contr.poly"))
X <- model.matrix(low ~ ., data = bwt)[, -1]
logitreg(X, bwt$low)
```

This can easily be modified to allow constraints on the parameters (just use arguments lower and/or upper in the call to logitreg) the handling of missing values by multiple imputation (Ripley, 1996), the possibility of errors in the recorded x or y values (Copas, 1988) and shrinkage and approximate Bayesian estimators, none of which can readily be handled by the glm algorithm.

It is always a good idea to check the convergence message, something `nlminb` makes particularly hard to do by not specifying all possible messages. In `optim` one can just check `fit$convergence == 0`.

Constrained Poisson regressions

Workers on the backcalculation of AIDS have used a Poisson regression model with identity link and non-negative parameters (for example, Rosenberg and Gail, 1991). The history of AIDS is divided into J time intervals $[T_{j-1}, T_j)$, and the data are counts Y_j of cases in each time interval. Then for a Poisson incidence process of rate ν, the counts are independent Poisson random variables with mean

$$EY_j = \mu_j = \int_0^{T_j} \left[F(T_j - s) - F(T_{j-1} - s) \right] \nu(s) \, ds$$

where F is the cumulative distribution of the incubation time. If we model ν by a step function

$$\nu(s) = \sum_i I(t_{i-1} \leqslant s < t_i) \beta_i$$

then $EY_j = \sum x_{ji} \beta_i$ for constants x_{ji}, which are considered known.

This is a Poisson regression with identity link, no intercept and constraint $\beta_i \geqslant 0$. The deviance is $2 \sum_j \mu_j - y_j - y_j \log(\mu_j / y_j)$.

```
AIDSfit <- function(y, z, start = rep(mean(y), ncol(z)), ...)
{
    deviance <- function(beta, y, z) {
        mu <- z %*% beta; 2 * sum(mu - y - y*log(mu/y)) }
    grad <- function(beta, y, z) {
        mu <- z %*% beta; 2 * t(1 - y/mu) %*% z }
    nlminb(start, deviance, grad, lower = 0, y = y, z = z, ...)
}
```

As an example, we consider the history of AIDS in Belgium using the numbers of cases for 1981 to 1993 (as reported by WHO in March 1994). We use two-year intervals for the step function and do the integration crudely using a Weibull model for the incubation times.

```
Y <- scan(n = 13)
12 14 33 50 67 74 123 141 165 204 253 246 240

library(nnet) # for class.ind
s <- seq(0, 13.999, 0.01); tint <- 1:14
X <- expand.grid(s, tint)
Z <- matrix(pweibull(pmax(X[,2] - X[,1],0), 2.5, 10), length(s))
Z <- Z[,2:14] - Z[,1:13]
Z <- t(Z) %*% class.ind(factor(floor(s/2))) * 0.01
round(AIDSfit(Y, Z)$par)
515    0 140 882    0    0    0
```

This has infections only in 1981–2 and 1985–8. We do not pursue the medical implications!

Appendix A

Implementation-Specific Details

This appendix contains details on how to set up an S environment, and other details that differ between the available S implementations.

A.1 Using S-PLUS under Unix / Linux

We use $ to denote the UNIX shell prompt, and assume that the commands to invoke S-PLUS is the default, Splus. Some installations may prefer Splus6.

To configure the command-line editor to use the cursor keys, set the environment variable S_CLEDITOR by

```
$ S_CLEDITOR=emacs; export S_CLEDITOR    ## sh or ksh or bash
$ setenv S_CLEDITOR emacs                ## csh and tcsh
```

Getting started

S-PLUS makes use of the file system and for each project we strongly recommend that you have a separate working directory to hold the files for that project.

The suggested procedure for the first occasion on which you use S-PLUS is:

1. Create a separate directory, say, SwS, for this project, which we suppose is 'Statistics with S', and make it your working directory.

   ```
   $ mkdir SwS
   $ cd SwS
   ```

   Copy any data files you need to use with S-PLUS to this directory.

2. Within the project directory run

   ```
   $ Splus CHAPTER
   ```

   which will create the subdirectories that S-PLUS uses under a directory .Data. (The .Data subdirectory is for use only by S-PLUS and hence has a 'dot name' which hides it from casual inspection.)

3. Start the S-PLUS system by

   ```
   $ Splus -e
   ```

An alternative mentioned on page 3 is to start the system with `Splus -g` which brings up the Java-based GUI interface.

4. At this point S commands may be issued (see later). The prompt is > unless the command is incomplete, when it is +. To use our software library issue

```
> library(MASS, first = T)
```

5. To quit the S program the command is

```
> q()
```

If you do not initialize the directory, the directory ~/MySwork is used, after initialization as an S-PLUS chapter if necessary. To keep projects separate, we strongly recommend that you *do* create a working directory.

For subsequent sessions the procedure is simpler: make SwS the working directory and start the program as before:

```
$ cd SwS
$ Splus
```

issue S commands, terminating with the command

```
> q()
```

On the other hand, to start a new project start at step 1.

Getting help

There are two ways to access the help system. One is from the command line as described on page 5. Function `help` will put up a program (`slynx`) in the terminal window running S-PLUS to view the HTML help file. If you prefer, a separate help window (which can be left up) can be obtained by

```
> help(var, window = T)
```

Using `help.start()` will bring up[1] a multi-panelled window listing the help topics available. This help system is shut down with `help.off()` and *not* by quitting the help window. If `help` or ? is used when this help system is running, the requests are sent to its window. This system is also used by the GUI console started by `Splus -g`.

Graphics hardcopy

There are several ways to produce a hardcopy of a plot on a suitable printer.

The function `dev.print` will copy the current plot to a printer device (default `postscript`) and allow its size, orientation, pointsize of text and so on to be set.

The `motif` device has a `Print` item on its `Graph` menu that will send a full-page copy of the current plot to the default printer. A little more control (the

---

[1]This uses Java and needs a machine with at least 128 Mb of memory to work comfortably. It may only work on a local X server.

orientation and the print command) is available from the `Printing` item on the
`Options` or `Properties` menu.

The function `printgraph` will copy the current plot to a **PostScript** or
LASERJET printer, and allows the size and orientation to be selected, as well
as paper size and printer resolution where appropriate.

It is normally possible to open an appropriate printer device and repeat the
plot commands, although this does preclude interacting with the plot on-screen.
This is sometimes necessary, as copying plots does not necessarily scale text fonts
and lines in exactly the same way.

Hardcopy to a file

It is very useful to be able to copy a plot to a file for inclusion in a paper or book
(as here). Since each of the hardcopy methods allows the printer output to be
re-directed to a file, there are many possibilities.

The simplest way is to edit the print command in the `Printing` item of a
`motif` device to be

```
cat > plot_file_name <
```

or to make a Bourne-shell command file `rmv` by

```
$ cat > rmv
mv $2 $1
^D
$ chmod +x rmv
```

place this in your path and use `rmv plot_file_name`. (This avoids leaving
around temporary files with names like `ps.out.0001.ps`.) Then click on Print
to produce the plot.

PostScript users will probably want Encapsulated PostScript (EPSF) format
files. These are produced automatically by the procedure in the last paragraph,
and also by setting both `onefile = F` and `print.it = F` as arguments to the
`postscript` device. Note that these are *not* EPSI files and do not include a preview image. It would be unusual to want an EPSF file in landscape format; select
'Portrait' on the `Printing` menu item, or use `horizontal = F` as argument to
the `postscript` device. The default pointsize (14) is set for a full-page landscape plot, and 10 or 11 are often more appropriate for a portrait plot. Set this in
the call to `postscript` or use `ps.options` (*before* the `motif` device is opened,
or use `ps.options.send` to alter the print settings of the current device).

Hardcopy from Trellis plots

The basic paradigm of a Trellis plot is to produce an object that the device 'prints',
that is, plots. Thus the simplest way to produce a hardcopy of a Trellis plot is to
switch to a printer device, 'print' the object again, and switch back. For example,

```
trellis.device()
p1 <- histogram(geyser$waiting)
p1      # plots it on screen
```

```
trellis.device("postscript", file = "hist.eps",
    onefile = F, print.it = F)
p1      # print the plot
dev.off()
```

However, it can be difficult to obtain precisely the same layout in this way (since this depends on the aspect ratio and size parameters), and it is impossible to interact with such a graph (for example, by using `identify`). Fortunately, the methods for hardcopy described on page 448 can still be used. It is important to set the options for the `postscript` device to match the colour schemes in use. For example, on UNIX with hardcopy via the `postscript` device we can use

```
ps.options(colors = colorps.trellis[, -1])
```

before the Trellis device is started. Then the `rmv` method and `dev.print` will use the Trellis (printer) colour scheme and produce colour PostScript output. Conversely, if greylevel PostScript output is required (for example, for figures in a book or article) we can use (for a `motif` device)

```
ps.options(colors = bwps.trellis, pointsize = 11)
trellis.settings.motif.bw <- trellis.settings.bwps
xcm.trellis.motif.bw <- xcm.trellis.motif.grey
trellis.device(color = F)
```

using `xcm.*` objects in our library MASS. This sets up a color screen device to mimic the 'black and white' (actually greylevel) PostScript device.

A.2 Using S-PLUS under Windows

There are versions of S-PLUS called 'Standard Edition' that lack the command-line interface and do not have the capabilities needed for use with this book.

Getting started

1. Create a new folder, say, SWS, for this project.

2. Copy any data files you need to use with S-PLUS to the folder SWS.

3. Select the folder representing the Start menu; precisely how is version-specific. Open the S-PLUS folder under Programs.

4. Create a duplicate copy of the S-PLUS for Windows icon, for example, using Copy from the Edit menu. Change the name of this icon to reflect the project for which it will be used.

5. Right-click on the new icon, and select Properties from the pop-up menu.

6. On the page labelled Shortcut, add at the end of the Target field S_PROJ= followed by the complete path to your directory. If that path contains spaces, enclose it in double quotes, as in

```
S_PROJ="c:\my work\S-Plus project"
```

Figure A.1: The dialog box to load the MASS library. Note that the check box is ticked.

If you have other files for the project, set the Start in field to their folder. If this is the same place as .Data, you can set S_PROJ=. on the target rather than repeating the path there.

7. Select the project's S-PLUS icon from the Start menu tree. You will be asked if .Data and .Prefs should be created. Click on OK. When the program has initialized, click on the Commands Window button with icon ▦ on the upper toolbar. If you always want a Commands Window at startup, select this from the menus via Options | General Settings... | Startup. (This setting is saved for the project on exiting the program.)

At this point S commands may be issued. The prompt is > unless the command is incomplete, when it is +. The usual Windows command-line recall and editing are available. To use our software library issue

```
library(MASS, first = T)
```

You can also do this from the Load Library... item on the File menu, which brings up a dialog box like Figure A.1.

You can exit the program from the File menu, and entering the command q() in the Commands window will also exit the program.

For subsequent sessions the procedure is much simpler; just launch S-PLUS by double-clicking on the project's icon. On the other hand, to start a new project start at step 1.

You can arrange for S-PLUS to allow the user to choose the project to work in when it is launched, by checking the option Prompt for project folder on the Startup tab in the dialog box from the General Settings... item on the Options menu. If this is checked, a dialog box will allow the user to choose the chapter (containing the .Data and .Prefs files).

graphsheet devices

Graphs can be redrawn by pressing function key F9. Hitting function key F2 with a graphsheet in focus zooms it to full screen. Click a mouse button or hit a key to revert.

Graphsheets can have multiple pages. The default is to use these pages for all the plots drawn within an S expression without any input, including code submitted from a script window. This is often helpful, but can be changed from the Page

Creation drop-down box on the Options... tab of the graphsheet properties dialog box (brought up by right-clicking on the background of a `graphsheet` device).

A `graphsheet` is a multi-purpose device, controlled by its `format` argument. For the default `format = ""` it provides a screen device. With `format = "printer"`, output is set to the current printer. With `format = "clipboard"`, a metafile is sent to the clipboard, and formats[2] `"BMP"`, `"GIF"`, `"JPG"` and `"TIF"` produce bitmaps. Formats `"WMF"` and `"EPS"` produce output on file as a Windows metafile and in PostScript, respectively. Copies to all of these formats (and many more) can also be obtained by using the Export Graph... item on the File menu which appears when a `graphsheet` has focus.

Graphical output is not drawn immediately on a `graphsheet` but delayed until the current S expression has finished or some input is required, for example from the keyboard or by `locator` or `identify` (see page 80). One way to overcome this is to add a call that asks for input at suitable points in your code, and a call to `guiLocator` with an argument of zero is usually the easiest, as this does nothing except flush the graphics queue.

Graphics hardcopy

There are several ways to produce a hardcopy of a (command-line) plot on a suitable printer or plotter. The simplest is to use the Print item on their File menu or the printer icon on the toolbar. This prints the window with focus, so bring the desired graphics window to the top first.

The function `dev.print` will copy the current plot to a printer device (default `win.printer`) and allow its size, orientation, pointsize of text and so on to be set.

It is normally possible to open an appropriate printer device (`postscript` or `win.printer`) and repeat the plot commands, although this does preclude interacting with the plot on-screen. This is sometimes necessary, as copying plots does not necessarily scale text fonts and lines in exactly the same way.

Hardcopy to a file

It is very useful to be able to copy a plot to a file for inclusion in a paper or book (as here). Since each of the hardcopy methods allows the printer output to be re-directed to a file, there are many possibilities.

The Print option on the File menu can select printing to a file. The graphics window can then be printed in the usual way. However, it is preferable to use the Export Graph... item on the File menu, which can save in a wide variety of graphics formats including Windows metafile and Encapsulated PostScript (with a preview image if desired).

On Windows the metafile format may be the most useful for plot files, as it is easily incorporated into other Windows applications while retaining full resolution. This is automatically used if the graphics device window is copied to the clipboard, and may also be generated by `graphsheet` or Export Graph....

PostScript users will probably want Encapsulated PostScript (EPSF) format files. These are produced by setting both `onefile = F` and `print.it = F` as

---

[2]There are others for less common formats.

arguments to the `postscript` device (which produces better output than using Export Graph...). Note that these do not include a preview image. It would be unusual to want an EPSF file in landscape format; select 'Portrait' on the `Printing` menu item, or use `horizontal = F` as argument to the `postscript` device. The default pointsize (14) is set for a full-page landscape plot, and 10 or 11 are often more appropriate for a portrait plot. Set this in the call to `postscript` or use `ps.options`.

A.3 Using R under Unix / Linux

Getting started

There is no need to prepare a directory for use with R, but it is desirable to store R sessions in separate directories.

1. Create a separate directory, say SwR, for this project, which we suppose is 'Statistics with R', and make it your working directory.

   ```
   $ mkdir SwR
   $ cd SwR
   ```

   Copy any data files you need to use with R to this directory.

2. Start the system with

   ```
   $ R
   ```

3. At this point S commands may be issued.[3] The default prompt is > unless the command is incomplete, when it is +. To use our software package issue

   ```
   library(MASS)
   ```

   (For users of S-PLUS who are used to adding `first = T`: packages are by default placed first.)

4. To quit the program the command is

   ```
   >  q()
   $
   ```

   You will be asked if you wish to save the workspace image. If you accept (type y) and command-line editing is operational, the command history will be saved in the file .Rhistory and (silently) reloaded at the beginning of the next session.

---

[3]Command-line editing should be available. You will probably find `man readline` describes it.

Getting help

There is both a text-based help facility as described on page 5, and an HTML-based help system started by

```
> help.start()
```

This puts up a home page in a browser, normally Netscape or Mozilla (which is started if not already running). If this help system is running, help requests are sent to the browser rather than to a pager in the terminal window. There is a Java-based search engine accessed from the home page.

Another way to search for information is the function `help.search`.

Graphics hardcopy

There are several ways to produce a hardcopy of a (command-line) plot on a suitable printer or plotter.

The function `dev.print` will copy the current plot to a printer device (default `postscript`) and allow its size, orientation, pointsize of text and so on to be set. In R, function `dev.print` copies to a printer whereas, `dev.copy2eps` make an encapulated PostScript file.

Bitmap graphics files can be produced by copying to a `png` or `jpeg` device,[4] or using `dev2bitmap` which makes a PostScript copy and post-processes it *via* `ghostscript`.

It is normally possible to open an appropriate printer device and repeat the plot commands, although this does preclude interacting with the plot on-screen. This is sometimes necessary, as copying plots does not necessarily scale text fonts and lines in exactly the same way.

A.4   Using R under Windows

A port of R to all modern versions of Windows is available. To use it, launch `bin\Rgui.exe` in one of the usual Windows ways. Perhaps the easiest is to create a shortcut to the executable, and set the Start in field to be the working directory you require, then double-click the shortcut. This will bring up its own console within which commands can be issued.

The appearance of the GUI is highly customizable; see the help for `Rconsole` for details. We prefer to use the `--sdi` option, for example.

In addition to text-based help, HTML-based help (invoked by `help.start`) and standard Windows compiled HTML help may be available if they were installed; see `?help` or the `README` for details.

---

[4]If R was built with support for these and an X11 server is available.

Graphics hardcopy

There are several ways to produce a hardcopy of a plot on a suitable printer or plotter. The simplest is to use the menu on the graphics window, which allows copying to a wide range of formats, including PostScript, Windows Metafile, PDF and several bitmap formats.

The function `dev.print` will copy the current plot to a printer device (default `postscript`, but printing directly only if the commands needed have been set; see `?postscript`) and allow its size, orientation, pointsize of text and so on to be set. For users with a non-PostScript printer the command `dev.print(win.print)` is the simplest way to print a copy of the current plot from the command-line.

Function `dev.copy2eps` copies to an encapulated PostScript file; there is also `savePlot` which will copy the plot to one of a range of formats.

It is normally possible to open an appropriate printer device and repeat the plot commands, although this does preclude interacting with the plot on-screen. This is sometimes necessary, as copying plots does not necessarily scale text fonts and lines in exactly the same way.

A.5   For Emacs Users

For users of GNU Emacs and Xemacs there is the independently developed ESS package available from

```
http://software.biostat.washington.edu/statsoft/ess/
```

which provides a comprehensive working environment for S programming. In particular it provides an editing environment tailored to S files with syntax highlighting. One can also interact with an S-PLUS or R process, submitting functions and extracting from the transcript.

ESS works almost equally well on Windows and on UNIX/Linux.

Appendix B

The S-PLUS GUI

The GUI under S-PLUS for Windows is highly configurable, but its default state is shown in Figure B.1. The top toolbar is constant, but the second toolbar and the menu items depend on the type of subwindow that has focus.

The object explorer and the commands window were selected by the two buttons on the top toolbar that are depressed. (What is opened when S-PLUS is launched is set during installation and can be set for each project from the Options menu.) To find out what the buttons mean, hover the mouse pointer over them and read the description in the bottom bar of the main S-PLUS window, or hold for longer and see the tooltip (if enabled).

Explanations of how to use a GUI are lengthy and appear in the S-PLUS guides and in the help system. We highlight a few points that we find make working with the GUI easier.

Subwindows

Instances of these types of subwindow can be launched from a button on the main toolbar or from the New or Open button. Some of the less obvious buttons are shown in Figure B.3.

The commands window

Commands typed in the commands window are executed immediately. Previous commands can be recalled by using the up and down arrow keys, and edited before submission (by pressing the return key). The commands history button (immediately to the right of the command window button) brings up a dialog with a list of the last few commands, which can be selected and re-submitted. When the commands window has focus, the second toolbar has just one button (with icon representing a pair of axes and a linear plot) that selects editable graphics. This is not recommended for routine use, as it may make the graphics very slow, and plots can be made editable later (see under 'object explorers'). It is also possible to launch a graphsheet with or without editable graphics from the command line by

```
graphsheet(object.mode = "object-oriented")
graphsheet(object.mode = "fast")
```

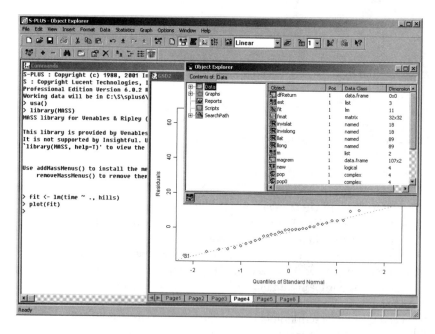

Figure B.1: A snapshot of the GUI interface showing three subwindows, from front to back an object explorer, a multi-tabbed graphsheet and a commands window. What is displayed as the lower toolbar depends on which subwindow is on top.

```
Script1 - program                                                    _□x

rlm.formula <-
function(formula, data = NULL, weights, ..., subset, na.action = na.fail,
    method = c("M", "MM", "model.frame"), wt.method = c("case", "inv.var"),
    model = TRUE, x.ret = TRUE, y.ret = FALSE, contrasts = NULL)
{
    mf <- match.call(expand.dots = FALSE)
    mf$method <- mf$model <- mf$x <- mf$y <- mf$contrasts <- mf$... <-
        NULL
    mf[[1]] <- as.name("model.frame")
    mf <- eval(mf, sys.parent())
    method <- match.arg(method)
    wt.method <- match.arg(wt.method)
    if(method == "model.frame")
        return(mf)
    mt <- attr(mf, "terms")
    y <- model.extract(mf, "response")
    x <- model.matrix(mt, mf, contrasts)
    xvars <- as.character(attr(mt, "variables"))
    if((yvar <- attr(mt, "response")) > 0)
        xvars <- xvars[ - yvar]
    xlev <- if(length(xvars) > 0) (
        xlev <- lapply(mf[xvars], levels)
        xlev[!sapply(xlev, is.null)]
    )
    weights <- model.extract(mf, weights)
```

Figure B.2: A script subwindow. The definition of `rlm.default` was inserted by highlighting the name (as shown), right-clicking and selecting Expand Inplace.

Figure B.3: Some buttons from the main toolbar. From left to right these give a data window, object explorer, history window, commands window, commands history window, 2D and 3D palettes and conditioning plots.

Script windows

You can use a script window rather than a commands window to enter S commands, and this may be most convenient for S programming. A new script window can be opened from the **New** file button or menu item, and presents a two-part subwindow as shown in Figure B.2. S commands can be typed into the top window and edited there. Pressing return submits the current line. Groups of commands can be selected (in the usual ways in **Windows**), and submitted by pressing the function key F10 or by the leftmost button on the second line (marked to represent a 'play' key). If text output is produced, this will (normally) appear in the bottom part of the subwindow. This output pane is cleared at each submission.

The input part of a script window is associated with a file, conventionally with extension `.ssc` and double-clicking on `.ssc` files in **Windows Explorer** ought to open them in a script window in **S-PLUS**, launching a new **S-PLUS** if none is running.

A script window can be launched from the command line to edit a function by the function `Edit` (note the difference from `edit`: the behaviour is closer to that of `fix`).

It is the help features that mark a scripts window as different from a commands window. Select a function name by double-clicking on it. Then help on that function is available by pressing the function key F1, and the right-click menu has items **Show Dialog...** and **Expand Inplace** to pop up a dialog box for the arguments of the function and to paste in the function body. There is a variety of convenient shortcuts for programmers: auto-indent, auto-insertion of right brace (`}`) and highlighting of matching left parentheses, brackets, braces and quotes (to `)` `]` `}` `"` and `'`).

Scripts can be saved as text files with extension `.ssc`; use the **Save** file menu item or button when the script window has focus. They can then be loaded into **S-PLUS** by opening them in **Explorer**.

More than one script window can be open at once. To avoid cluttering the screen, script windows can be hidden (and unhidden) from the **Windows** file menu. The **Hide** item hides the window that has focus, whereas the **Unhide...** provides a list of windows from which to select.

The **ESS** package mentioned on page 455 can also be used with **NTemacs** on **Windows**, and provides a programming editor that some may prefer (including both of us).

Report windows

A useful alternative to the output pane of a script window is a *report window* which records all the output directed to it and can be saved to a file. The contents of the report window can be edited, so mistakes can be removed and the contents annotated before saving.

Where output is sent is selected by the dialog box brought up by the Text Output Routing... item on the Options menu. This allows separate selections for normal output and warnings/errors; the most useful options are Default and Report.

Object explorers

There can be one or more object explorers on screen. They provide a two-panel view (see Figure B.1) that will be familiar from many Windows programs. Object views can be expanded down to component level. The right-click menu is context-sensitive; for example, for a linear model fit (of class lm) it has Summary, Plot, Predict and Coefficients items. Double-clicking on a data frame or vector will open it for editing or viewing in a spreadsheet-like *data window*.

If a graphsheet is selected and expanded it will first show its pages (if there are more than one) and then the plotted objects. If the object is labelled CompositeObject then the right-click menu will include the item Convert to Objects which will convert that plot to editable form.

Object explorers are highly customizable, both in the amount of detail in the right pane and in the databases and classes of objects to be shown. Right-clicking on the background of the left and right panes or on folders or using the Format menu will lead to dialog boxes to customize the format.

The ordering of items in the right pane can be puzzling; click on the heading of a column to sort on that column (as in Windows Explorer).

Data windows

A data window provides a spreadsheet-like view (see Figure 2.1 on page 22) of a data frame (or vector or matrix). The scrollbars scroll the table, but the headings remain visible. A region can be selected by dragging (and extended by shift-clicking); including the headings in the selection includes the whole row or column as appropriate.

Entries can be edited and rows and columns inserted or deleted in the usual spreadsheet styles. Toolbar buttons are provided for most of these operations, and for sorting by the selected column. Double-clicking in the top row of a column brings up a format dialog for that column; double-clicking in the top left cell brings up a format dialog for the window that allows the type font and size to be altered.

Appendix C

Datasets, Software and Libraries

The software and datasets used in this book are supplied with S-PLUS for Windows and also available on the World Wide Web. Point your browser at

    http://www.stats.ox.ac.uk/pub/MASS4/

The on-line instructions tell you how to install the software. In case of difficulty in accessing the software, please email MASS@stats.ox.ac.uk.

The on-line complements are available at this site as well as exercises (with selected answers) and printable versions of the on-line help for our software.

Note that this book assumes that you have access to an S environment. S-PLUS is a commercial product; please see http://www.insightful.com for details of its distribution channels. R is freely available from http://www.r-project.org and mirrors.

C.1   Our Software

Our software is packaged as four library sections (in S-PLUS's notation) or as a bundle VR of four packages for R. These are

MASS This contains all the datasets and a number of S functions, as well as a number of other datasets that we have used in learning or teaching.

nnet Software for feed-forward neural networks with a single hidden layer and for multinomial log-linear models.

spatial Software for spatial smoothing and the analysis of spatial point patterns. This directory contains a number of datasets of point patterns, described in the text file PP.files.

class Functions for nonparametric classification, by k-nearest neighbours and learning vector quantization, as well as Kohonen's SOM.

These are supplied with S-PLUS for Windows, and should be part of an R installation. Sources for use with S-PLUS for UNIX / Linux and updates for other systems can be found at the Web site listed above.

461

Caveat

These datasets and software are provided in good faith, but none of the authors, publishers or distributors warrant their accuracy nor can be held responsible for the consequences of their use.

We have tested the software as widely as we are able but it is inevitable that system dependencies will arise. We are unlikely to be in a position to assist with such problems.

The licences for the distribution and use of the software and datasets are given in the on-line distributions.

C.2   Using Libraries

A library section in S-PLUS is a convenient way to package S objects for a common purpose, and to allow these to extend the system.

The structure of a library section is the same as that of a working directory, but the `library` function makes sections much more convenient to use. Conventionally library sections are stored in a standard place, the subdirectory `library` of the main S-PLUS directory. Which library sections are available can be found by the `library` command with no argument; further information (the contents of the README file) on any section is given by

```
library(help = section_name)
```

and the library section itself is made available by

```
library(section_name)
```

This has two actions. It attaches the `.Data` subdirectory of the section at the end of the search path (having checked that it has not already been attached), and executes the function `.First.lib` if one exists within that `.Data` subdirectory.

Sometimes it is necessary to have functions in a library section that will replace standard system functions (for example, to correct bugs or to extend their functionality). This can be done by attaching the library section as the second dictionary on the search path with

```
library(section_name, first = T)
```

Of course, attaching other dictionaries with `attach` or other libraries with `first = T` will push previously attached libraries down the search path.

R calls library sections *packages* and by default attaches them first; otherwise they are functionally equivalent to S-PLUS library sections.

Private libraries in S-PLUS

So far we have only considered system-wide library sections installed under the main S-PLUS directory, which usually requires privileged access to the operating system. It is also possible to use a private library, by giving `library` the argument `lib.loc` or by assigning the object `lib.loc` in the current session

dictionary (frame 0). This should be a vector of directory names that are searched
in order for library sections before the system-wide library. For example, on one
of our systems we get

```
> assign(where = 0, "lib.loc", "/users/ripley/S/library")
> library()
Library "/users/ripley/S/library"
The following sections are available in the library:

SECTION          BRIEF DESCRIPTION

MASS             main library
nnet             neural nets
spatial          spatial statistics
class            classification

Library "/packages/splus/library"
The following sections are available in the library:

SECTION          BRIEF DESCRIPTION

chron            Functions to handle dates and times.
     ....
```

Because `lib.loc` is local to the session, it must be assigned for each session.
The `.First` function is often a convenient place to do so; Windows users can
put the `assign` line in a file `S.init` in the `local` folder of the main S-PLUS
directory.

Private libraries in R

In R terminology packages are installed in libraries. The easiest way to make use
of private libraries is to list their locations in the environment variable R_LIBS in
the same format as used by PATH (so separated by colons on UNIX, by semicolons
on Windows). We find it convenient to set this in the file `.Renviron` in the user's
home directory.

Sources of libraries

Many S users have generously collected their functions and datasets into library
sections and made them publicly available. An archive of sources for library sections is maintained at Carnegie-Mellon University. The World Wide Web address
is

```
http://lib.stat.cmu.edu/S/
```

colloquially known as `statlib`. There are several mirrors around the world.

Several of these sections have been mentioned in the text; where S-PLUS
library sections are not available from `statlib` their source is given at first mention.

The convention is to distribute library sections as 'shar' archives; these are text files that when used as scripts for the Bourne shell `sh` unpack to give all the files needed for a library section. Check files such as `Install` for installation instructions; these usually involve editing the `Makefile` and typing `make`. Beware that many of the instructions predate current versions of S-PLUS. (We provide a set of links to patched versions or updated instructions at our Web site.)

We have made available several of these library sections prepackaged for Windows users in `.zip` archives; check the WWW address

> `http://www.stats.ox.ac.uk/pub/MASS4/Winlibs`

Be aware that sections packaged for S-PLUS 4.x/2000 are incompatible with S-PLUS 6.x.

The equivalent R packages are available from CRAN

> `http://cran.r-project.org`
> `http://cran.us.r-project.org`
> `http://lib.stat.cmu.edu/R/CRAN/`

and other mirrors, normally as source code and pre-compiled for Windows.

References

Numbers in brackets [ ] are page references to citations.

Abbey, S. (1988) Robust measures and the estimator limit. *Geostandards Newsletter* **12**, 241–248. [124]

Aitchison, J. (1986) *The Statistical Analysis of Compositional Data*. London: Chapman & Hall. [77]

Aitchison, J. and Dunsmore, I. R. (1975) *Statistical Prediction Analysis*. Cambridge: Cambridge University Press. [339]

Aitkin, M. (1978) The analysis of unbalanced cross classifications (with discussion). *Journal of the Royal Statistical Society A* **141**, 195–223. [169, 170, 176]

Aitkin, M., Anderson, D., Francis, B. and Hinde, J. (1989) *Statistical Modelling in GLIM*. Oxford: Oxford University Press. [208]

Akaike, H. (1974) A new look at statistical model identification. *IEEE Transactions on Automatic Control* **AU–19**, 716–722. [174]

Akima, H. (1978) A method of bivariate interpolation and smooth surface fitting for irregularly distributed data points. *ACM Transactions on Mathematical Software* **4**, 148–159. [424]

Analytical Methods Committee (1987) Recommendations for the conduct and interpretation of co-operative trials. *The Analyst* **112**, 679–686. [279]

Analytical Methods Committee (1989a) Robust statistics — how not to reject outliers. Part 1. Basic concepts. *The Analyst* **114**, 1693–1697. [114, 124]

Analytical Methods Committee (1989b) Robust statistics — how not to reject outliers. Part 2. Inter-laboratory trials. *The Analyst* **114**, 1699–1702. [124, 281]

Andersen, P. K., Borgan, Ø., Gill, R. D. and Keiding, N. (1993) *Statistical Models Based on Counting Processes*. New York: Springer-Verlag. [353, 356]

Anderson, E. (1935) The irises of the Gaspe peninsula. *Bulletin of the American Iris Society* **59**, 2–5. [301]

Anderson, E. and ten others (1999) *LAPACK User's Guide*. Third Edition. Philadelphia: SIAM. [64]

Anderson, O. D. (1976) *Time Series Analysis and Forecasting. The Box-Jenkins Approach*. London: Butterworths. [406]

Atkinson, A. C. (1985) *Plots, Transformations and Regression*. Oxford: Oxford University Press. [86, 151]

Atkinson, A. C. (1986) Comment: Aspects of diagnostic regression analysis. *Statistical Science* **1**, 397–402. [152]

Atkinson, A. C. (1988) Transformations unmasked. *Technometrics* **30**, 311–318. [152, 153]

Azzalini, A. and Bowman, A. W. (1990) A look at some data on the Old Faithful geyser. *Applied Statistics* **39**, 357–365. [113]

Bartholomew, D. J. and Knott, M. (1999) *Latent Variable Analysis and Factor Analysis*. Second Edition. London: Arnold. [323]

Bates, D. M. and Chambers, J. M. (1992) Nonlinear models. Chapter 10 of Chambers and Hastie (1992). [211]

Bates, D. M. and Watts, D. G. (1988) *Nonlinear Regression Analysis and Its Applications*. New York: John Wiley and Sons. [211]

Becker, R. A. (1994) A brief history of S. In *Computational Statistics: Papers Collected on the Occasion of the 25th Conference on Statistical Computing at Schloss Reisenburg*, eds P. Dirschedl and R. Osterman, pp. 81–110. Heidelberg: Physica-Verlag. [1]

Becker, R. A., Chambers, J. M. and Wilks, A. R. (1988) *The NEW S Language*. New York: Chapman & Hall. (Formerly Monterey: Wadsworth and Brooks/Cole.). [2]

Becker, R. A., Cleveland, W. S. and Shyu, M.-J. (1996) The visual design and control of Trellis display. *Journal of Computational and Graphical Statistics* **5**, 123–155. [69]

Bie, O., Borgan, Ø. and Liestøl, K. (1987) Confidence intervals and confidence bands for the cumulative hazard rate function and their small sample properties. *Scandinavian Journal of Statistics* **14**, 221–233. [357]

Bishop, C. M. (1995) *Neural Networks for Pattern Recognition*. Oxford: Clarendon Press. [243]

Bishop, Y. M. M., Fienberg, S. E. and Holland, P. W. (1975) *Discrete Multivariate Analysis*. Cambridge, MA: MIT Press. [199, 325]

Bloomfield, P. (2000) *Fourier Analysis of Time Series: An Introduction*. Second Edition. New York: John Wiley and Sons. [387, 392, 394, 396]

Bollerslev, T. (1986) Generalized autoregressive conditional heteroscedasticity. *Journal of Econometrics* **51**, 307–327. [416]

Borgan, Ø. and Liestøl, K. (1990) A note on confidence intervals and bands for the survival function based on transformations. *Scandinavian Journal of Statistics* **17**, 35–41. [357]

Boser, B. E., Guyon, I. M. and Vapnik, V. N. (1992) A training algorithm for optimal margin classifiers. In *Proceeedings of the 5th Annual ACM Workshop on Computational Learning Theory.*, ed. D. Haussler, pp. 144–152. ACM Press. [344]

Bowman, A. and Azzalini, A. (1997) *Applied Smoothing Techniques for Data Analysis: The Kernel Approach with S-Plus Illustrations*. Oxford: Oxford University Press. [126, 211, 249]

Box, G. E. P. and Cox, D. R. (1964) An analysis of transformations (with discussion). *Journal of the Royal Statistical Society B* **26**, 211–252. [170, 171]

Box, G. E. P., Hunter, W. G. and Hunter, J. S. (1978) *Statistics for Experimenters*. New York: John Wiley and Sons. [116, 117, 168, 169]

Breiman, L. and Friedman, J. H. (1985) Estimating optimal transformations for multiple regression and correlations (with discussion). *Journal of the American Statistical Association* **80**, 580–619. [237]

Breiman, L., Friedman, J. H., Olshen, R. A. and Stone, C. J. (1984) *Classification and Regression Trees*. New York: Chapman & Hall / CRC Press. (Formerly Monterey: Wadsworth and Brooks/Cole.). [251, 257]

Brent, R. (1973) *Algorithms for Minimization Without Derivatives*. Englewood Cliffs, NJ: Prentice-Hall. [435]

Breslow, N. E. and Clayton, D. G. (1993) Approximate inference in generalized linear mixed models. *Journal of the American Statistical Association* **88**, 9–25. [294, 297, 298]

Brockwell, P. J. and Davis, R. A. (1991) *Time Series: Theory and Methods*. Second Edition. New York: Springer-Verlag. [387]

Brockwell, P. J. and Davis, R. A. (1996) *Introduction to Time Series and Forecasting*. New York: Springer-Verlag. [387]

Bruckner, L. A. (1978) On Chernoff faces. In *Graphical Representation of Multivariate Data*, ed. P. C. C. Wang, pp. 93–121. New York: Academic Press. [314]

Bryan, J. G. (1951) The generalized discriminant function: mathematical foundation and computational routine. *Harvard Educational Review* **21**, 90–95. [333]

Campbell, J. Y., Lo, A. W. and MacKinlay, A. C. (1997) *The Econometrics of Financial Markets*. Princeton, New Jersey: Princeton University Press. [414]

Campbell, N. A. and Mahon, R. J. (1974) A multivariate study of variation in two species of rock crab of genus *Leptograpsus*. *Australian Journal of Zoology* **22**, 417–425. [302]

Cao, R., Cuevas, A. and González-Manteiga, W. (1994) A comparative study of several smoothing methods in density estimation. *Computational Statistics and Data Analysis* **17**, 153–176. [129]

Chambers, J. M. (1998) *Programming with Data. A Guide to the S Language*. New York: Springer-Verlag. [2, 53]

Chambers, J. M. and Hastie, T. J. eds (1992) *Statistical Models in S*. New York: Chapman & Hall. (Formerly Monterey: Wadsworth and Brooks/Cole.). [2, 188, 197, 466, 467]

Chernoff, H. (1973) The use of faces to represent points in k-dimensional space graphically. *Journal of the American Statistical Association* **68**, 361–368. [314]

Chilès, J.-P. and Delfiner, P. (1999) *Geostatistics. Modeling Under Spatial Uncertainty*. New York: Wiley. [425]

Ciampi, A., Chang, C.-H., Hogg, S. and McKinney, S. (1987) Recursive partitioning: A versatile method for exploratory data analysis in biostatistics. In *Biostatistics*, eds I. B. McNeil and G. J. Umphrey, pp. 23–50. New York: Reidel. [256]

Clark, L. A. and Pregibon, D. (1992) Tree-based models. Chapter 9 of Chambers and Hastie (1992). [251, 255]

Clayton, D. G. (1996) Generalized linear mixed models. In *Markov Chain Monte Carlo in Practice*, Chapter 16, pp. 275–301. London: Chapman & Hall. [296]

Cleveland, R. B., Cleveland, W. S., McRae, J. E. and Terpenning, I. (1990) STL: A seasonal-trend decomposition procedure based on loess (with discussion). *Journal of Official Statistics* **6**, 3–73. [403]

Cleveland, W. S. (1993) *Visualizing Data*. Summit, NJ: Hobart Press. [69, 89, 178]

Cleveland, W. S., Grosse, E. and Shyu, W. M. (1992) Local regression models. Chapter 8 of Chambers and Hastie (1992). [423]

Collett, D. (1991) *Modelling Binary Data*. London: Chapman & Hall. [189, 190, 208]

Collett, D. (1994) *Modelling Survival Data in Medical Research*. London: Chapman & Hall. [353]

Comon, P. (1994) Independent component analysis — a new concept? *Signal Processing* **36**, 287–314. [313]

Copas, J. B. (1988) Binary regression models for contaminated data (with discussion). *Journal of the Royal Statistical Society series B* **50**, 225–266. [445]

Cortes, C. and Vapnik, V. (1995) Support-vector networks. *Machine Learning* **20**, 273–297. [344]

Cox, D. R. (1972) Regression models and life-tables (with discussion). *Journal of the Royal Statistical Society B* **34**, 187–220. [365]

Cox, D. R., Hinkley, D. V. and Barndorff-Nielsen, O. E. eds (1996) *Time Series Models. In econometric, finance and other fields*. London: Chapman & Hall. [474, 478]

Cox, D. R. and Oakes, D. (1984) *Analysis of Survival Data*. London: Chapman & Hall. [353, 354]

Cox, D. R. and Snell, E. J. (1981) *Applied Statistics. Principles and Examples*. London: Chapman & Hall. [199]

Cox, D. R. and Snell, E. J. (1989) *The Analysis of Binary Data*. Second Edition. London: Chapman & Hall. [194, 208]

Cox, T. F. and Cox, M. A. A. (2001) *Multidimensional Scaling*. Second Edition. Chapman & Hall / CRC. [306, 308]

Cressie, N. A. C. (1991) *Statistics for Spatial Data*. New York: John Wiley and Sons. [419, 428, 429]

Cristianini, N. and Shawe-Taylor, J. (2000) *An Introduction to Support Vector Machines and other kernel-based learning methods*. Cambridge: Cambridge University Press. [344]

Cromwell, J. B., Labys, W. C. and Terraza, M. (1994) *Univariate Tests for Time Series Models*. Thousand Oaks, CA: Sage. [416]

Cybenko, G. (1989) Approximation by superpositions of a sigmoidal function. *Mathematics of Controls, Signals, and Systems* **2**, 303–314. [245]

Daniel, C. and Wood, F. S. (1980) *Fitting Equations to Data*. Second Edition. New York: John Wiley and Sons. [272]

Darroch, J. N. and Ratcliff, D. (1972) Generalized iterative scaling for log-linear models. *Annals of Mathematical Statistics* **43**, 1470–1480. [185, 203]

Davidian, M. and Giltinan, D. M. (1995) *Nonlinear Models for Repeated Measurement Data*. London: Chapman & Hall. [272]

Davies, P. L. (1993) Aspects of robust linear regression. *Annals of Statistics* **21**, 1843–1899. [159]

Davison, A. C. and Hinkley, D. V. (1997) *Bootstrap Methods and Their Application*. Cambridge: Cambridge University Press. [133, 134, 137, 138, 164]

Davison, A. C. and Snell, E. J. (1991) Residuals and diagnostics. Chapter 4 of Hinkley *et al.* (1991). [189]

Dawid, A. P. (1982) The well-calibrated Bayesian (with discussion). *Journal of the American Statistical Association* **77**, 605–613. [349]

Dawid, A. P. (1986) Probability forecasting. In *Encyclopedia of Statistical Sciences*, eds S. Kotz, N. L. Johnson and C. B. Read, volume 7, pp. 210–218. New York: John Wiley and Sons. [349]

Deming, W. E. and Stephan, F. F. (1940) On a least-squares adjustment of a sampled frequency table when the expected marginal totals are known. *Annals of Mathematical Statistics* **11**, 427–444. [185]

Devijver, P. A. and Kittler, J. V. (1982) *Pattern Recognition: A Statistical Approach.* Englewood Cliffs, NJ: Prentice-Hall. [341]

Diaconis, P. and Shahshahani, M. (1984) On non-linear functions of linear combinations. *SIAM Journal of Scientific and Statistical Computing* **5**, 175–191. [239]

Diggle, P. J. (1983) *Statistical Analysis of Spatial Point Patterns.* London: Academic Press. [419]

Diggle, P. J. (1990) *Time Series: A Biostatistical Introduction.* Oxford: Oxford University Press. [387, 388]

Diggle, P. J., Liang, K.-Y. and Zeger, S. L. (1994) *Analysis of Longitudinal Data.* Oxford: Clarendon Press. [278, 293, 294, 295, 299]

Dixon, W. J. (1960) Simplified estimation for censored normal samples. *Annals of Mathematical Statistics* **31**, 385–391. [122]

Duda, R. O. and Hart, P. E. (1973) *Pattern Classification and Scene Analysis.* New York: John Wiley and Sons. [199]

Duda, R. O., Hart, P. E. and Stork, D. G. (2001) *Pattern Classification.* Second Edition. New York: John Wiley and Sons. [331]

Edwards, D. (2000) *Introduction to Graphical Modelling.* Second Edition. New York: Springer. [199]

Efron, B. (1982) *The Jackknife, the Bootstrap, and Other Resampling Plans.* Philadelphia: Society for Industrial and Applied Mathematics. [134]

Efron, B. and Tibshirani, R. (1993) *An Introduction to the Bootstrap.* New York: Chapman & Hall. [133]

Eilers, P. H. and Marx, B. D. (1996) Flexible smoothing with B-splines and penalties. *Statistical Science* **11**, 89–121. [384]

Ein-Dor, P. and Feldmesser, J. (1987) Attributes of the performance of central processing units: A relative performance prediction model. *Communications of the ACM* **30**, 308–317. [177]

Emerson, J. W. (1998) Mosaic displays in S-PLUS: a general implementation and a case study. *Statistical Computing and Graphics Newsletter* **9**(1), 17–23. [325]

Engle, R. F. (1982) Autoregressive conditional heteroscedasticity with estimates of the variance of the United Kingdom inflation. *Econometrica* **50**, 987–1007. [414]

Evans, M. and Swartz, T. (2000) *Approximating Integrals via Monte Carlo and Deterministic Methods.* Oxford: Oxford University Press. [296, 298]

Everitt, B. S. and Hand, D. J. (1981) *Finite Mixture Distributions.* London: Chapman & Hall. [437, 438]

Fan, J. and Gijbels, I. (1996) *Local Polynomial Modelling and its Applications*. London: Chapman & Hall. [132]

Feigl, P. and Zelen, M. (1965) Estimation of exponential survival probabilities with concomitant information. *Biometrics* **21**, 826–838. [354]

Fernandez, C. and Steel, M. F. J. (1999) Multivariate Student t-regression models: Pitfalls and inference. *Biometrika* **86**, 153–167. [110]

Finney, D. J. (1971) *Probit Analysis*. Third Edition. Cambridge: Cambridge University Press. [208]

Firth, D. (1991) Generalized linear models. Chapter 3 of Hinkley *et al.* (1991). [183, 185, 187]

Fisher, R. A. (1925) Theory of statistical estimation. *Proceedings of the Cambridge Philosophical Society* **22**, 700–725. [186]

Fisher, R. A. (1936) The use of multiple measurements in taxonomic problems. *Annals of Eugenics (London)* **7**, 179–188. [301, 332]

Fisher, R. A. (1940) The precision of discriminant functions. *Annals of Eugenics (London)* **10**, 422–429. [325]

Fleming, T. R. and Harrington, D. P. (1981) A class of hypothesis tests for one and two sample censored survival data. *Communications in Statistics* **A10**, 763–794. [358]

Fleming, T. R. and Harrington, D. P. (1991) *Counting Processes and Survival Analysis*. New York: John Wiley and Sons. [353, 356, 366, 370]

Fletcher, R. (1987) *Practical Methods of Optimization*. Second Edition. Chichester: John Wiley and Sons. [435]

Flury, B. and Riedwyl, H. (1981) Graphical representation of multivariate data by means of asymmetrical faces. *Journal of the American Statistical Association* **76**, 757–765. [314]

Freedman, D. and Diaconis, P. (1981) On the histogram as a density estimator: L_2 theory. *Zeitschrift für Wahrscheinlichkeitstheorie und verwandte Gebiete* **57**, 453–476. [112]

Friedman, J. H. (1984) SMART user's guide. Technical Report 1, Laboratory for Computational Statistics, Department of Statistics, Stanford University. [239]

Friedman, J. H. (1987) Exploratory projection pursuit. *Journal of the American Statistical Association* **82**, 249–266. [302]

Friedman, J. H. (1991) Multivariate adaptive regression splines (with discussion). *Annals of Statistics* **19**, 1–141. [235]

Friedman, J. H. and Silverman, B. W. (1989) Flexible parsimonious smoothing and additive modeling (with discussion). *Technometrics* **31**, 3–39. [234]

Friedman, J. H. and Stuetzle, W. (1981) Projection pursuit regression. *Journal of the American Statistical Association* **76**, 817–823. [239]

Friendly, M. (1994) Mosaic displays for multi-way contingency tables. *Journal of the American Statistical Association* **89**, 190–200. [325]

Friendly, M. (2000) *Visualizing Categorical Data*. Cary, NC: SAS Institute. [325]

Fuller, W. A. (1987) *Measurement Error Models*. New York: John Wiley and Sons. [278]

Funahashi, K. (1989) On the approximate realization of continuous mappings by neural networks. *Neural Networks* **2**, 183–192. [245]

Gabriel, K. R. (1971) The biplot graphical display of matrices with application to principal component analysis. *Biometrika* **58**, 453–467. [311, 312]

Gallant, A. R. (1987) *Nonlinear Statistical Models.* New York: John Wiley and Sons. [211]

Gamerman, D. (1997a) *Markov Chain Monte Carlo: stochastic simulation for Bayesian inference.* London: Chapman & Hall. [296]

Gamerman, D. (1997b) Sampling from the posterior distribution in generalized linear mixed models. *Statistics and Computing* **7**, 57–68. [296]

Gehan, E. A. (1965) A generalized Wilcoxon test for comparing arbitrarily singly-censored samples. *Biometrika* **52**, 203–223. [354]

Geisser, S. (1993) *Predictive Inference: An Introduction.* New York: Chapman & Hall. [339]

Gentle, J. E. (1998) *Numerical Linear Algebra for Applications in Statistics.* New York: Springer-Verlag. [62]

Gill, P. E., Murray, W. and Wright, M. H. (1981) *Practical Optimization.* London: Academic Press. [435]

Goldstein, H. (1995) *Multilevel Statistical Models.* Second Edition. London: Edward Arnold. [271]

Golub, G. H. and Van Loan, C. F. (1989) *Matrix Computations.* Second Edition. Baltimore: Johns Hopkins University Press. [62, 63, 422]

Goodman, L. A. (1978) *Analyzing Qualitative/Categorical Data: Log-Linear Models and Latent-Structure Analysis.* Cambridge, MA: Abt Books. [199]

de Gooijer, J. G., Abraham, B., Gould, A. and Robinson, L. (1985) Methods for determining the order of an autoregressive-moving average process: A survey. *International Statistical Review* **53**, 301–329. [398]

Gordon, A. D. (1999) *Classification.* Second Edition. London: Chapman & Hall / CRC. [315, 316]

Gower, J. C. and Hand, D. J. (1996) *Biplots.* London: Chapman & Hall. [313, 326, 329, 330]

Grambsch, P. and Therneau, T. M. (1994) Proportional hazards tests and diagnostics based on weighted residuals. *Biometrika* **81**, 515–526. [371]

Green, P. J. and Silverman, B. W. (1994) *Nonparametric Regression and Generalized Linear Models. A Roughness Penalty Approach.* London: Chapman & Hall. [228, 230]

Greenacre, M. (1992) Correspondence analysis in medical research. *Statistical Methods in Medical Research* **1**, 97–117. [328, 329]

Haberman, S. J. (1978) *Analysis of Qualitative Data. Volume 1: Introductory Topics.* New York: Academic Press. [199]

Haberman, S. J. (1979) *Analysis of Qualitative Data. Volume 2: New Developments.* New York: Academic Press. [199]

Hampel, F. R., Ronchetti, E. M., Rousseeuw, P. J. and Stahel, W. A. (1986) *Robust Statistics. The Approach Based on Influence Functions.* New York: John Wiley and Sons. [120, 142]

Hand, D., Mannila, H. and Smyth, P. (2001) *Principles of Data Mining.* Cambridge, MA: The MIT Press. [301, 331]

Hand, D. J., Daly, F., McConway, K., Lunn, D. and Ostrowski, E. eds (1994) *A Handbook of Small Data Sets*. London: Chapman & Hall. [139, 272]

Harrington, D. P. and Fleming, T. R. (1982) A class of rank test procedures for censored survival data. *Biometrika* **69**, 553–566. [358]

Hartigan, J. A. (1975) *Clustering Algorithms*. New York: John Wiley and Sons. [318]

Hartigan, J. A. (1982) Classification. In *Encyclopedia of Statistical Sciences*, eds S. Kotz, N. L. Johnson and C. B. Read, volume 2, pp. 1–10. New York: John Wiley and Sons. [331]

Hartigan, J. A. and Kleiner, B. (1981) Mosaics for contingency tables. In *Computer Science and Statistics: Proceedings of the 13th Symposium on the Interface*, ed. W. F. Eddy, pp. 268–273. New York: Springer-Verlag. [325]

Hartigan, J. A. and Kleiner, B. (1984) A mosaic of television ratings. *American Statistician* **38**, 32–35. [325]

Hartigan, J. A. and Wong, M. A. (1979) A K-means clustering algorithm. *Applied Statistics* **28**, 100–108. [318]

Hastie, T. J. and Tibshirani, R. J. (1990) *Generalized Additive Models*. London: Chapman & Hall. [211, 230, 234]

Hastie, T. J. and Tibshirani, R. J. (1996) Discriminant analysis by Gaussian mixtures. *Journal of the Royal Statistical Society Series B* **58**, 158–176. [341]

Hastie, T. J., Tibshirani, R. J. and Friedman, J. (2001) *The Elements of Statistical Learning. Data Mining Inference and Prediction*. New York: Springer-Verlag. [211, 229, 313, 331, 344, 345]

Hauck, Jr., W. W. and Donner, A. (1977) Wald's test as applied to hypotheses in logit analysis. *Journal of the American Statistical Association* **72**, 851–853. [197]

Heiberger, R. M. (1989) *Computation for the Analysis of Designed Experiments*. New York: John Wiley and Sons. [282]

Henrichon, Jr., E. G. and Fu, K.-S. (1969) A nonparametric partitioning procedure for pattern classification. *IEEE Transactions on Computers* **18**, 614–624. [251]

Hertz, J., Krogh, A. and Palmer, R. G. (1991) *Introduction to the Theory of Neural Computation*. Redwood City, CA: Addison-Wesley. [243]

Hettmansperger, T. P. and Sheather, S. J. (1992) A cautionary note on the method of least median squares. *American Statistician* **46**, 79–83. [159]

Hinkley, D. V., Reid, N. and Snell, E. J. eds (1991) *Statistical Theory and Modelling. In Honour of Sir David Cox, FRS*. London: Chapman & Hall. [468, 470]

Hoaglin, D. C., Mosteller, F. and Tukey, J. W. eds (1983) *Understanding Robust and Exploratory Data Analysis*. New York: John Wiley and Sons. [115, 473]

Hornik, K., Stinchcombe, M. and White, H. (1989) Multilayer feedforward networks are universal approximators. *Neural Networks* **2**, 359–366. [245]

Hosmer, D. W. and Lemeshow, S. (1999) *Applied Survival Analysis. Regression Modeling of Time to Event Data*. New York: John Wiley and Sons. [353]

Hosmer, Jr., D. W. and Lemeshow, S. (1989) *Applied Logistic Regression*. New York: John Wiley and Sons. [194]

Hsu, J. C. (1996) *Multiple Comparison: Theory and Methods*. London: Chapman & Hall. [181]

Huber, P. J. (1981) *Robust Statistics*. New York: John Wiley and Sons. [120]

Huber, P. J. (1985) Projection pursuit (with discussion). *Annals of Statistics* **13**, 435–525. [302]

Huet, S., Bouvier, A., Gruet, M.-A. and Jolivet, E. (1996) *Statistical Tools for Nonlinear Regression. A Practical Guide with S-PLUS Examples*. New York: Springer-Verlag. [211]

Hyndman, R. J. and Fan, Y. (1996) Sample quantiles in statistical packages. *American Statistician* **50**, 361–365. [S function from http://www-personal.buseco.monash.edu.au/~hyndman/Rlibrary/misc/quantile.htm]. [112]

Hyvärinen, A., Karhunen, J. and Oja, E. (2001) *Independent Component Analysis*. New York: John Wiley and Sons. [313]

Hyvärinen, A. and Oja, E. (2000) Independent component analysis. algorithms and applications. *Neural Networks* **13**, 411–430. [313]

Iglewicz, B. (1983) Robust scale estimators and confidence intervals for location. In Hoaglin *et al.* (1983), pp. 405–431. [121]

Ingrassia, S. (1992) A comparison between the simulated annealing and the EM algorithms in normal mixture decompositions. *Statistics and Computing* **2**, 203–211. [437]

Inselberg, A. (1984) The plane with parallel coordinates. *The Visual Computer* **1**, 69–91. [314]

Jackson, J. E. (1991) *A User's Guide to Principal Components*. New York: John Wiley and Sons. [305]

Jardine, N. and Sibson, R. (1971) *Mathematical Taxonomy*. London: John Wiley and Sons. [306, 316]

John, P. W. M. (1971) *Statistical Design and Analysis of Experiments*. New York: Macmillan. [282]

Jolliffe, I. T. (1986) *Principal Component Analysis*. New York: Springer-Verlag. [305, 312]

Jones, M. C., Marron, J. S. and Sheather, S. J. (1996) A brief survey of bandwidth selection for density estimation. *Journal of the American Statistical Association* **91**, 401–407. [129]

Jones, M. C. and Sibson, R. (1987) What is projection pursuit? (with discussion). *Journal of the Royal Statistical Society A* **150**, 1–36. [302]

Journel, A. G. and Huijbregts, C. J. (1978) *Mining Geostatistics*. London: Academic Press. [425]

Kalbfleisch, J. D. and Prentice, R. L. (1980) *The Statistical Analysis of Failure Time Data*. New York: John Wiley and Sons. [353, 354, 355, 364, 371, 377]

Kaluzny, S. and Vega, S. C. (1997) *S+SPATIALSTATS*. New York: Springer-Verlag. [419]

Kaufman, L. and Rousseeuw, P. J. (1990) *Finding Groups in Data. An Introduction to Cluster Analysis*. New York: John Wiley and Sons. [306, 315, 316]

Kent, J. T., Tyler, D. E. and Vardi, Y. (1994) A curious likelihood identity for the multivariate t-distribution. *Communications in Statistics—Simulation and Computation* **23**, 441–453. [337]

Klein, J. P. (1991) Small sample moments of some estimators of the variance of the Kaplan-Meier and Nelson-Aalen estimators. *Scandinavian Journal of Statistics* **18**, 333–340. [356]

Klein, J. P. and Moeschberger, M. L. (1997) *Survival Analysis. Techniques for Censored and Truncated Data*. New York: Springer-Verlag. [353]

Knerr, S., Personnaz, L. and Dreyfus, G. (1990) Single-layer learning revisited: a stepwise procedure for building and training a neural network. In *Neuro-computing: Algorithms, Architectures and Applications*, eds F. Fogelman Soulié and J. Hérault. Berlin: Springer-Verlag. [346]

Knuth, D. E. (1968) *The Art of Computer Programming, Volume 1: Fundamental Algorithms*. Reading, MA: Addison-Wesley. [48]

Kohonen, T. (1990) The self-organizing map. *Proceedings IEEE* **78**, 1464–1480. [341]

Kohonen, T. (1995) *Self-Organizing Maps*. Berlin: Springer-Verlag. [310, 341]

Kooperberg, C., Bose, S. and Stone, C. J. (1997) Polychotomous regression. *Journal of the American Statistical Association* **92**, 117–127. [235]

Kooperberg, C. and Stone, C. J. (1992) Logspline density estimation for censored data. *Journal of Computational and Graphical Statistics* **1**, 301–328. [132]

Krause, A. and Olson, M. (2000) *The Basics of S and S-PLUS*. Second Edition. New York: Springer-Verlag. [3]

Krzanowski, W. J. (1988) *Principles of Multivariate Analysis. A User's Perspective*. Oxford: Oxford University Press. [301, 331]

Laird, N. M. (1996) Longitudinal panel data: an overview of current methodology. Chapter 4, pp. 143–175 of Cox *et al.* (1996). [293, 299]

Lam, L. (2001) *An Introduction to S-PLUS for Windows*. Amsterdam: CANdienstein. [3]

Lange, K. L., Little, R. J. A. and Taylor, J. M. G. (1989) Robust statistical modeling using the t distribution. *Journal of the American Statistical Association* **84**, 881–896. [110]

Lauritzen, S. L. (1996) *Graphical Models*. Oxford: Clarendon Press. [199]

Lawless, J. F. (1987) Negative binomial and mixed Poisson regression. *Canadian Journal of Statistics* **15**, 209–225. [206]

Lawley, D. N. and Maxwell, A. E. (1971) *Factor Analysis as a Statistical Method*. Second Edition. London: Butterworths. [322]

Lee, T. W. (1998) *Independent Component Analysis: Theory and Applications*. Dordrecht: Kluwer Academic Publishers. [313]

Liang, K.-Y. and Zeger, S. L. (1986) Longitudinal data analysis using generalized linear models. *Biometrika* **73**, 13–22. [299]

Linder, A., Chakravarti, I. M. and Vuagnat, P. (1964) Fitting asymptotic regression curves with different asymptotes. In *Contributions to Statistics. Presented to Professor P. C. Mahalanobis on the Occasion of his 70th Birthday*, ed. C. R. Rao, pp. 221–228. Oxford: Pergamon Press. [218, 219]

Lindstrom, M. J. and Bates, D. M. (1990) Nonlinear mixed effects models for repeated measures data. *Biometrics* **46**, 673–687. [287]

Ljung, G. M. and Box, G. E. P. (1978) On a measure of lack of fit in time series models. *Biometrika* **65**, 553–564. [401]

Loader, C. (1999) *Local Regression and Likelihood*. New York: Springer-Verlag. [132]

Ludbrook, J. (1994) Repeated measurements and multiple comparisons in cardiovascular research. *Cardiovascular Research* **28**, 303–311. [288]

Mackenzie, T. and Abrahamowicz, M. (1996) B-splines without divided differences. *Student* **1**, 223–230. [228]

Macnaughton-Smith, P., Williams, W. T., Dale, M. B. and Mockett, L. G. (1964) Dissimilarity analysis: A new technique of hierarchical sub-division. *Nature* **202**, 1034–1035. [317]

MacQueen, J. (1967) Some methods for classification and analysis of multivariate observations. In *Proceedings of the Fifth Berkeley Symposium on Mathematical Statistics and Probability*, eds L. M. Le Cam and J. Neyman, volume 1, pp. 281–297. Berkeley, CA: University of California Press. [318]

Madsen, M. (1976) Statistical analysis of multiple contingency tables. Two examples. *Scandinavian Journal of Statistics* **3**, 97–106. [199]

Mallows, C. L. (1973) Some comments on C_p. *Technometrics* **15**, 661–675. [174]

Marazzi, A. (1993) *Algorithms, Routines and S Functions for Robust Statistics*. Pacific Grove, CA: Wadsworth and Brooks/Cole. [161]

Mardia, K. V., Kent, J. T. and Bibby, J. M. (1979) *Multivariate Analysis*. London: Academic Press. [301, 331]

Matheron, G. (1973) The intrinsic random functions and their applications. *Advances in Applied Probability* **5**, 439–468. [428]

McCullagh, P. (1980) Regression models for ordinal data (with discussion). *Journal of the Royal Statistical Society Series B* **42**, 109–142. [204]

McCullagh, P. and Nelder, J. A. (1989) *Generalized Linear Models*. Second Edition. London: Chapman & Hall. [183, 185, 186, 204, 208, 209, 210, 275, 299]

McCulloch, W. S. and Pitts, W. (1943) A logical calculus of ideas immanent in neural activity. *Bulletin of Mathematical Biophysics* **5**, 115–133. [244]

McLachlan, G. J. (1992) *Discriminant Analysis and Statistical Pattern Recognition*. New York: John Wiley and Sons. [301, 331]

McLachlan, G. J. and Peel, D. (2000) *Finite Mixture Models*. New York: John Wiley and Sons. [437]

Michie, D. (1989) Problems of computer-aided concept formation. In *Applications of Expert Systems 2*, ed. J. R. Quinlan, pp. 310–333. Glasgow: Turing Institute Press / Addison-Wesley. [253]

Miller, R. G. (1981) *Simultaneous Statistical Inference*. New York: Springer-Verlag. [181]

Monahan, J. F. (2001) *Numerical Methods of Statistics*. Cambridge: Cambridge University Press. [62, 296, 435]

Moran, M. A. and Murphy, B. J. (1979) A closer look at two alternative methods of statistical discrimination. *Applied Statistics* **28**, 223–232. [339]

Morgan, J. N. and Messenger, R. C. (1973) THAID: *a Sequential Search Program for the Analysis of Nominal Scale Dependent Variables*. Survey Research Center, Institute for Social Research, University of Michigan. [251]

Morgan, J. N. and Sonquist, J. A. (1963) Problems in the analysis of survey data, and a proposal. *Journal of the American Statistical Association* **58**, 415–434. [251]

Mosteller, F. and Tukey, J. W. (1977) *Data Analysis and Regression.* Reading, MA: Addison-Wesley. [115, 317]

Murrell, P. and Ihaka, R. (2000) An approach to providing mathematical annotation in plots. *Journal of Computational and Graphical Statistics* **9**, 582–599. [82]

Murtagh, F. and Hernández-Pajares, M. (1995) The Kohonen self-organizing map method: An assessment. *Journal of Classification* **12**, 165–190. [311]

Murthy, S. K. (1998) Automatic construction of decision trees from data: a multi-disciplinary survey. *Data Mining and Knowledge Discovery* **2**, 345–389. [251]

Nagelkerke, N. J. D. (1991) A note on a general definition of the coefficient of determination. *Biometrika* **78**, 691–692. [368]

Nash, J. C. (1990) *Compact Numerical Methods for Computers. Linear Algebra and Function Minimization.* Second Edition. Bristol: Adam Hilger. [422, 435]

Nelson, D. B. (1991) Conditional heteroscedasticity in asset pricing: a new approach. *Econometrica* **59**, 347–370. [416]

Nelson, W. D. and Hahn, G. J. (1972) Linear estimation of a regression relationship from censored data. Part 1 — simple methods and their application (with discussion). *Technometrics* **14**, 247–276. [355, 364]

Nocedal, J. and Wright, S. J. (1999) *Numerical Optimization.* New York: Springer-Verlag. [435]

Nolan, D. and Speed, T. (2000) *Stat Labs. Mathematical Statistics Through Applications.* New York: Springer-Verlag. [vi]

Park, B.-U. and Turlach, B. A. (1992) Practical performance of several data-driven bandwidth selectors (with discussion). *Computational Statistics* **7**, 251–285. [129]

Peto, R. (1972) Contribution to the discussion of the paper by D. R. Cox. *Journal of the Royal Statistical Society B* **34**, 205–207. [366]

Peto, R. and Peto, J. (1972) Asymptotically efficient rank invariant test procedures (with discussion). *Journal of the Royal Statistical Society A* **135**, 185–206. [359]

Pinheiro, J. C. and Bates, D. M. (2000) *Mixed-Effects Models in S and S-PLUS.* New York: Springer-Verlag. [271]

Plackett, R. L. (1974) *The Analysis of Categorical Data.* London: Griffin. [199]

Prater, N. H. (1956) Estimate gasoline yields from crudes. *Petroleum Refiner* **35**, 236–238. [272]

Priestley, M. B. (1981) *Spectral Analysis and Time Series.* London: Academic Press. [387, 392]

Quinlan, J. R. (1979) Discovering rules by induction from large collections of examples. In *Expert Systems in the Microelectronic Age*, ed. D. Michie. Edinburgh: Edinburgh University Press. [251]

Quinlan, J. R. (1983) Learning efficient classification procedures and their application to chess end-games. In *Machine Learning*, eds R. S. Michalski, J. G. Carbonell and T. M. Mitchell, pp. 463–482. Palo Alto: Tioga. [251]

Quinlan, J. R. (1986) Induction of decision trees. *Machine Learning* **1**, 81–106. [251]

Quinlan, J. R. (1993) *C4.5: Programs for Machine Learning.* San Mateo, CA: Morgan Kaufmann. [251]

Ramsey, F. L. and Schafer, D. W. (1997) *The Statistical Sleuth. A Course in Methods of Data Analysis*. Belmont, CA: Duxbury Press. [vi, 295]

Ramsey, F. L. and Schafer, D. W. (2002) *The Statistical Sleuth. A Course in Methods of Data Analysis*. Second Edition. Belmont, CA: Duxbury Press. [vi]

Rao, C. R. (1948) The utilization of multiple measurements in problems of biological classification. *Journal of the Royal Statistical Society series B* **10**, 159–203. [333]

Rao, C. R. (1971a) Estimation of variance and covariance components—MINQUE theory. *Journal of Multivariate Analysis* **1**, 257–275. [281]

Rao, C. R. (1971b) Minimum variance quadratic unbiased estimation of variance components. *Journal of Multivariate Analysis* **1**, 445–456. [281]

Rao, C. R. (1973) *Linear Statistical Inference and Its Applications*. Second Edition. New York: John Wiley and Sons. [170]

Rao, C. R. and Kleffe, J. (1988) *Estimation of Variance Components and Applications*. Amsterdam: North-Holland. [281]

Redner, R. A. and Walker, H. F. (1984) Mixture densities, maximum likelihood and the EM algorithm. *SIAM Review* **26**, 195–239. [437]

Reynolds, P. S. (1994) Time-series analyses of beaver body temperatures. In *Case Studies in Biometry*, eds N. Lange, L. Ryan, L. Billard, D. Brillinger, L. Conquest and J. Greenhouse, Chapter 11. New York: John Wiley and Sons. [411]

Ripley, B. D. (1976) The second-order analysis of stationary point processes. *Journal of Applied Probability* **13**, 255–266. [431]

Ripley, B. D. (1981) *Spatial Statistics*. New York: John Wiley and Sons. [86, 419, 424, 425, 426, 432, 433]

Ripley, B. D. (1988) *Statistical Inference for Spatial Processes*. Cambridge: Cambridge University Press. [419, 433, 434]

Ripley, B. D. (1993) Statistical aspects of neural networks. In *Networks and Chaos — Statistical and Probabilistic Aspects*, eds O. E. Barndorff-Nielsen, J. L. Jensen and W. S. Kendall, pp. 40–123. London: Chapman & Hall. [243, 245]

Ripley, B. D. (1994) Neural networks and flexible regression and discrimination. In *Statistics and Images 2*, ed. K. V. Mardia, volume 2 of *Advances in Applied Statistics*, pp. 39–57. Abingdon: Carfax. [245]

Ripley, B. D. (1996) *Pattern Recognition and Neural Networks*. Cambridge: Cambridge University Press. [199, 211, 243, 251, 257, 258, 301, 302, 308, 331, 337, 338, 339, 341, 347, 445]

Ripley, B. D. (1997) Classification (update). In *Encyclopedia of Statistical Sciences*, eds S. Kotz, C. B. Read and D. L. Banks, volume Update 1. New York: John Wiley and Sons. [331]

Robert, C. P. and Casella, G. (1999) *Monte Carlo Statistical Methods*. New York: Springer-Verlag. [296]

Roberts, S. and Tarassenko, L. (1995) Automated sleep EEG analysis using an RBF network. In *Neural Network Applications*, ed. A. F. Murray, pp. 305–322. Dordrecht: Kluwer Academic Publishers. [310]

Robinson, G. K. (1991) That BLUP is a good thing: The estimation of random effects (with discussion). *Statistical Science* **6**, 15–51. [276]

Roeder, K. (1990) Density estimation with confidence sets exemplified by superclusters and voids in galaxies. *Journal of the American Statistical Association* **85**, 617–624. [129]

Rosenberg, P. S. and Gail, M. H. (1991) Backcalculation of flexible linear models of the human immunodeficiency virus infection curve. *Applied Statistics* **40**, 269–282. [446]

Ross, G. J. S. (1990) *Nonlinear Estimation.* New York: Springer-Verlag. [211]

Rousseeuw, P. J. (1984) Least median of squares regression. *Journal of the American Statistical Association* **79**, 871–881. [336]

Rousseeuw, P. J. and Leroy, A. M. (1987) *Robust Regression and Outlier Detection.* New York: John Wiley and Sons. [120, 156, 157, 336]

Sammon, J. W. (1969) A non-linear mapping for data structure analysis. *IEEE Transactions on Computers* **C-18**, 401–409. [307]

Santner, T. J. and Duffy, D. E. (1989) *The Statistical Analysis of Discrete Data.* New York: Springer-Verlag. [198]

Schall, R. (1991) Estimation in generalized linear models with random effects. *Biometrika* **78**, 719–727. [297]

Scheffé, H. (1959) *The Analysis of Variance.* New York: John Wiley and Sons. [282]

Schoenfeld, D. (1982) Partial residuals for the proportional hazards model. *Biometrika* **69**, 239–241. [371]

Schwarz, G. (1978) Estimating the dimension of a model. *Annals of Statistics* **6**, 461–464. [276]

Scott, D. W. (1979) On optimal and data-based histograms. *Biometrika* **66**, 605–610. [112]

Scott, D. W. (1992) *Multivariate Density Estimation. Theory, Practice, and Visualization.* New York: John Wiley and Sons. [112, 126, 128, 130]

Searle, S. R., Casella, R. and McCulloch, C. E. (1992) *Variance Components.* New York: John Wiley and Sons. [271]

Seber, G. A. F. and Wild, C. J. (1989) *Nonlinear Regression.* New York: John Wiley and Sons. [211]

Sethi, I. K. and Sarvarayudu, G. P. R. (1982) Hierarchical classifier design using mutual information. *IEEE Transactions on Pattern Analysis and Machine Intelligence* **4**, 441–445. [251]

Sheather, S. J. and Jones, M. C. (1991) A reliable data-based bandwidth selection method for kernel density estimation. *Journal of the Royal Statistical Society B* **53**, 683–690. [129]

Shephard, N. (1996) Statistical aspects of ARCH and stochastic volatility. Chapter 1, pp. 1–67 of Cox *et al.* (1996). [414, 416]

Shumway, R. H. and Stoffer, D. S. (2000) *Time Series Analysis and its Applications.* New York: Springer-Verlag. [387]

Silverman, B. W. (1986) *Density Estimation for Statistics and Data Analysis.* London: Chapman & Hall. [126, 127, 130]

Simonoff, J. S. (1996) *Smoothing Methods in Statistics.* New York: Springer-Verlag. [126, 132, 211]

Smith, G. A. and Stanley, G. (1983) Clocking g: relating intelligence and measures of timed performance. *Intelligence* **7**, 353–368. [323]

Snijders, T. A. B. and Bosker, R. J. (1999) *Multilevel Analysis. An Introduction to Basic and Advanced Multilevel Modelling.* London: Sage. [271, 277]

Solomon, P. J. (1984) Effect of misspecification of regression models in the analysis of survival data. *Biometrika* **71**, 291–298. [368]

Spence, R. (2001) *Information Visualization.* Harlow: Addison-Wesley. [69]

Stace, C. (1991) *New Flora of the British Isles.* Cambridge: Cambridge University Press. [252]

Staudte, R. G. and Sheather, S. J. (1990) *Robust Estimation and Testing.* New York: John Wiley and Sons. [120, 122, 134, 152]

Stevens, W. L. (1948) Statistical analysis of a non-orthogonal tri-factorial experiment. *Biometrika* **35**, 346–367. [185]

Taylor, S. J. (1986) *Modelling Financial Time Series.* Chichester: John Wiley and Sons. [416]

Thall, P. F. and Vail, S. C. (1990) Some covariance models for longitudinal count data with overdispersion. *Biometrics* **46**, 657–671. [294]

Therneau, T. M. and Atkinson, E. J. (1997) An introduction to recursive partitioning using the RPART routines. Technical report, Mayo Foundation. [251, 258]

Therneau, T. M. and Grambsch, P. M. (2000) *Modeling Survival Data. Extending the Cox Model.* New York: Springer-Verlag. [353]

Thisted, R. A. (1988) *Elements of Statistical Computing. Numerical Computation.* New York: Chapman & Hall. [62]

Thomson, D. J. (1990) Time series analysis of Holocene climate data. *Philosophical Transactions of the Royal Society A* **330**, 601–616. [402]

Tibshirani, R. (1988) Estimating transformations for regression via additivity and variance stabilization. *Journal of the American Statistical Association* **83**, 394–405. [237]

Titterington, D. M., Smith, A. F. M. and Makov, U. E. (1985) *Statistical Analysis of Finite Mixture Distributions.* Chichester: John Wiley and Sons. [437]

Tsiatis, A. A. (1981) A large sample study of Cox's regression model. *Annals of Statistics* **9**, 93–108. [356]

Tufte, E. R. (1983) *The Visual Display of Quantitative Information.* Cheshire, CT: Graphics Press. [69]

Tufte, E. R. (1990) *Envisioning Information.* Cheshire, CT: Graphics Press. [69]

Tufte, E. R. (1997) *Visual Explanations.* Cheshire, CT: Graphics Press. [69]

Tukey, J. W. (1960) A survey of sampling from contaminated distributions. In *Contributions to Probability and Statistics*, eds I. Olkin, S. Ghurye, W. Hoeffding, W. Madow and H. Mann, pp. 448–485. Stanford: Stanford University Press. [121]

Upton, G. J. G. and Fingleton, B. J. (1985) *Spatial Data Analysis by Example.* Volume 1. Chichester: John Wiley and Sons. [419]

Vapnik, V. N. (1995) *The Nature of Statistical Learning Theory.* New York: Springer-Verlag. [344]

Vapnik, V. N. (1998) *Statistical Learning Theory*. New York: John Wiley and Sons. [344]

Velleman, P. F. and Hoaglin, D. C. (1981) *Applications, Basics, and Computing of Exploratory Data Analysis*. Boston: Duxbury. [115]

Venables, W. N. and Ripley, B. D. (2000) *S Programming*. New York: Springer-Verlag. [2, 12, 20, 29, 41, 56, 60, 67, 68]

Vinod, H. (1969) Integer programming and the theory of grouping. *Journal of the American Statistical Association* **64**, 506–517. [319]

Vonesh, E. F. and Chinchilli, V. M. (1997) *Linear and Nonlinear Models for the Analysis of Repeated Measurements*. New York: Marcel Dekker. [272]

Wahba, G. (1990) *Spline Models for Observational Data*. Philadelphia: SIAM. [232, 429]

Wahba, G., Wang, Y., Gu, C., Klein, R. and Klein, B. (1995) Smoothing spline ANOVA for exponential families, with application to the Wisconsin epidemiological study of diabetic retinopathy. *Annals of Statistics* **23**, 1865–1895. [232]

Wand, M. P. and Jones, M. C. (1995) *Kernel Smoothing*. London: Chapman & Hall. [126, 129, 130, 231]

Webb, A. (1999) *Statistical Pattern Recognition*. London: Arnold. [331]

Wegman, E. J. (1990) Hyperdimensional data analysis using parallel coordinates. *Journal of the American Statistical Association* **85**, 664–675. [314]

Whittaker, J. (1990) *Graphical Models in Applied Multivariate Statistics*. Chichester: John Wiley and Sons. [199]

Wilkinson, G. N. and Rogers, C. E. (1973) Symbolic description of factorial models for analysis of variance. *Applied Statistics* **22**, 392–399. [4, 139]

Wilkinson, L. (1999) *The Grammar of Graphics*. New York: Springer-Verlag. [69, 314]

Williams, E. J. (1959) *Regression Analysis*. New York: John Wiley and Sons. [222]

Wilson, S. R. (1982) Sound and exploratory data analysis. In *COMPSTAT 1982, Proceedings in Computational Statistics*, eds H. Caussinus, P. Ettinger and R. Tamassone, pp. 447–450. Vienna: Physica-Verlag. [301]

Wolfinger, R. and O'Connell, M. (1993) Generalized linear mixed models: a pseudo-likelihood approach. *Journal of Statistical Computation and Simulation* **48**, 233–243. [298]

Yandell, B. S. (1997) *Practical Data Analysis for Designed Experiments*. London: Chapman & Hall. [181]

Yates, F. (1935) Complex experiments. *Journal of the Royal Statistical Society (Supplement)* **2**, 181–247. [282]

Yates, F. (1937) *The Design and Analysis of Factorial Experiments*. Technical communication No. 35. Harpenden, England: Imperial Bureau of Soil Science. [282]

Yohai, V., Stahel, W. A. and Zamar, R. H. (1991) A procedure for robust estimation and inference in linear regression. In *Directions in Robust Statistics and Diagnostics, Part II*, eds W. A. Stahel and S. W. Weisberg. New York: Springer-Verlag. [161]

Zeger, S. L., Liang, K.-Y. and Albert, P. S. (1988) Models for longitudinal data. a generalized estimating equations approach. *Biometrics* **44**, 1049–1060. [300]

Index

Entries in `this font` are names of S objects. Page numbers in **bold** are to the most comprehensive treatment of the topic.